Confronting Emerging Zoonoses

Akio Yamada • Laura H. Kahn • Bruce Kaplan
Thomas P. Monath • Jack Woodall • Lisa Conti
Editors

Confronting Emerging Zoonoses

The One Health Paradigm

Springer

Editors
Akio Yamada
Department of Veterinary Medical Science
Graduate School of Agricultural and Life Sciences
The University of Tokyo
Tokyo, Japan

Bruce Kaplan
Sarasota, FL, USA

Jack Woodall
Rio de Janeiro, Brazil

Laura H. Kahn
Woodrow Wilson School of Public and International Affairs
Princeton University
Princeton, NJ, USA

Thomas P. Monath
Hookipa BioTech AG
Townsend, MA, USA

Lisa Conti
Florida Department of Agriculture and Consumer Services
Tallahassee, FL, USA

ISBN 978-4-431-55119-5 ISBN 978-4-431-55120-1 (eBook)
DOI 10.1007/978-4-431-55120-1
Springer Tokyo Heidelberg New York Dordrecht London

Library of Congress Control Number: 2014950637

Printed on acid-free paper

Springer is part of Springer Science+Business Media (www.springer.com)

Preface

According to the United Nations, the global population is estimated to reach 8.1 billion in 2025 and increase to 9.6 billion in 2050. In developing countries, the population is projected to increase to 8.2 billion in 2050. Sustainably supporting these unprecedented populations will require integrating a "One Health" approach in medicine, public health, and agriculture.

In a recent presentation by Jimmy Smith, the Director General of the International Livestock Research Institute, it was projected that total livestock production would increase 92 % by 2050. To feed these enormous numbers of humans and livestock, 61 million hectares of additional cropland would be needed worldwide. Direct and indirect interactions between different species of animals, including humans, are anticipated to increase, leading to the emergence of more zoonotic pathogens. These novel pathogens might have the capacity for interhuman spread and even pandemic potential among humans. To mitigate the burden of emerging infectious zoonotic diseases, efficacious monitoring of diseases of wild and domesticated animals would be required. This interspecies disease surveillance would enable early disease detection at the interface of humans, livestock, and wildlife and would promote rapid response capabilities.

Although emerging infectious diseases pose health threats to people regardless of their economic status, the poor in many developing countries suffer the brunt of the burden from "neglected zoonotic diseases." In these countries, the poor depend upon small livestock like goats and poultry for their livelihoods, but many of these animals harbor zoonotic diseases. Industrialized countries, in contrast, have much better biocontainment capabilities in their livestock production. Prevention is critical but is often hampered by the diversion of necessary resources for diseases with higher priorities in the human medical communities such as heart disease, cancer, and diabetes.

Furthermore, the provincial silo approach to zoonotic and neglected diseases makes matters more difficult because prevention and response measures are not implemented using a collaborative, interdisciplinary One Health approach. These diseases, which affect all species, require that disciplines with expertise in different areas work together. One Health is a concept that underpins the multidisciplinary or transdisciplinary approaches to zoonotic diseases. This is equally applicable to other health and health care categories that fall under the One Health Umbrella (http://www.onehealthinitiative.com/OneHealth2) such as comparative medicine/translational medicine.

Contributors to this book provide an overview of the current understandings of zoonotic and emerging infectious diseases using examples where the One Health approach was successfully applied. The book also highlights some of the challenges societies face in confronting several specific zoonotic diseases. A chapter is included on comparative medicine to demonstrate the broad scope of the One Health concept.

This book is dedicated to those studying zoonotic diseases and comparative medicine in both human and veterinary medicine, to those involved in the prevention and control of zoonotic infections, and to those in the general public interested in the visionary field of One Health.

Akio Yamada
Laura H. Khan
Bruce Kaplan
Thomas P. Monath
Jack Woodall
Lisa Conti

Contents

Part III The Successes and Challenges of Implementing One Health

Part I
The Importance of a One Health Approach to Emerging Zoonotic Diseases

Chapter 1
The Origin of Human Pathogens

Gabriel Trueba

Abstract Modern human infectious diseases are thought to have originated in domestic animals during the Neolithic period or afterwards. However, recent genetic, phylogeographic and molecular clock analyses of microbial genomes point to a much older Paleolithic origin (2.5 million to 10,000 years ago) and suggest that many of these pathogens coevolved with ancestral hominids in Africa. Another group of human pathogens seems to have derived recently from non-human hominids.

Keywords Evolution • Molecular clock • Phylogeny • Zoonosis

1.1 Introduction

Some scientists contend that many modern human infectious diseases arose during the Neolithic period or afterwards due to close contact with domestic animals and their pathogens (Diamond 1999; Pearce-Duvet 2006; Wolfe et al. 2007). There is indeed evidence of a recent origin of measles virus from bovine rinderpest virus (Furuse et al. 2010), and bubonic plague (*Yersinia pestis*) from *Y. pseudotuberculosis*, a zoonotic bacterium carried by rodents (Cui et al. 2013). As a consequence of this view, the origins of many human infectious diseases, such as tuberculosis, malaria, smallpox, and pertussis have focused on domestic animals and environments outside of Africa. However, the combination of genetics, molecular clock analysis and phylogeography provide evidence that some of these diseases arose much earlier in the Paleolithic period and probably when our hominid ancestors were still isolated in Africa (Forster 2004). Some of these findings are in agreement with paleontological and archeological discoveries.

G. Trueba (✉)
Instituto de Microbiología Universidad, San Francisco de Quito, Quito, Ecuador
e-mail: gtrueba@usfq.edu.ec

A. Yamada et al. (eds.), *Confronting Emerging Zoonoses*,
DOI 10.1007/978-4-431-55120-1_1

1.2 Older Origins

Tuberculosis was the typical example of a disease thought to have originated in the Neolithic period and transmitted from cattle to humans (Diamond 1999). This scenario implied that the human *Mycobacterium tuberculosis* bacterium derived from a bovine *M. bovis* ancestor.

Recent analyses of genomic deletions and nucleotide polymorphisms suggests that the ancestor of *M. tuberculosis* Complex (MTCP) may actually have originated from a group of smooth mycobacteria found in humans in East Africa (Gutierrez et al. 2005; Wirth et al. 2008). Molecular clock analysis of strains belonging to MTCP indicates that the common ancestor may have existed 2.6–2.8 million years ago (Gutierrez et al. 2005). This ancient date places the origin of the disease at a time when our hominid ancestors were living in Africa. Furthermore, there is evidence of a genetic bottleneck of *M. tuberculosis* occurring about 40,000 years ago, possibly the result of the human migration out of African to Eurasia (Gutierrez et al. 2005; Wirth et al. 2008). In congruence with the molecular data, tuberculosis lesions and mycobacterial DNA have been found in pre-Columbian Americans (Stone et al. 2009) who most likely did not have contact with domesticated cattle. Furthermore, rather than evolving from bovines as originally hypothesized, recent analyses of genome deletions and phylogeny suggest that *M. bovis* (and other animal adapted MTCP mycobacteria) derived from a mycobacteria more similar to the human adapted *M. tuberculosis* than to *M. bovis*. These studies clearly contradict the notion that the human *M. tuberculosis* derived from *M. bovis* (Smith et al. 2009; Garnier et al. 2003; Wirth et al. 2008). Nevertheless the evolutionary relationship between *M. tuberculosis* and *M bovis* seems to be complex; current evidence suggest that the most recent common ancestor of MTCP may have evolved into two groups: *M. tuberculosis* and a group of mycobacteria infecting different animal species; *M. bovis* may have emerged from the group of animal adapted mycobacteria (Smith et al. 2009).

The presence of *M. bovis* in badgers and its possible connections with livestock have been investigated using whole genome sequencing analysis and showed that the transmission of *M. bovis* between cattle and badgers is recent, however the direction of the transmission remains unresolved (Biek et al. 2012).

Another mycobacterial disease, leprosy was first recorded in humans around 600 B.C. in India (Stone et al. 2009). Based on historic documents, it was thought that this disease was later brought to Europe during Greek military campaigns (Stone et al. 2009). In support of this recent origin is an absence of leprosy in pre-Columbian Americans (Stone et al. 2009), and little genetic variation among isolates of *Mycobacteria leprae* (Gómez-Valero et al. 2007; Monot et al. 2005), the causative agent of this infectious disease. By contrast, phylogeography using single nucleotide polymorphism (SNP) analyses point to *M. leprae* originating in Africa during the Paleolithic (Monot et al. 2005; Monot et al. 2009). This ancient date suggests that the current presence of little genomic variation may be due to a recent bottleneck (Monot et al. 2009; Monot et al. 2009), possibly due to *M. leprae*'s low rate of infection (Smith 1904). This low infection rate could also explain the absence of leprosy in pre-Columbian Americans, even though their ancestors may have themselves been infected. Interestingly, a mycobacterial species (*M. lepromatosis*) has been discovered recently in Mexico and the Caribbean. However, this bacterium seems to have diverged around 10 million years ago from *M. leprae* and may have been brought to the American continent by the first human immigrants (Han et al. 2009). Molecular clock analysis suggests that the ancestor of *M. leprae* diverged from *M. tuberculosis* around 66 million years ago, prior to the origins of the genus *Homo*, 2.5 million years ago (Forster 2004). Analysis of non-synonymous nucleotide substitutions suggests that *M. leprae* underwent genomic decay between 10 to 20 million years ago (Gómez-Valero et al. 2007; Monot et al. 2009). See Sect. 1.4 below.

Other diseases whose origins have been subjected to major debates are human treponematoses, which include syphilis (caused by *Treponema pallidum* subsp. *pallidum*), bejel (caused by *T. pallidum* subsp. *endemicum*), yaws (caused by *T. pallidum* subsp. *pertenue*) and pinta (caused by *T. pallidum* subsp. *carateum*) (Scolnik et al. 2003). The genomes of the four subspecies display very few differences, suggesting a recent common *Treponema* ancestor. However *T. pallidum* seems to be a pathogen that has co-evolved with human ancestors; typical yaws-like lesions have been found in prehistoric human bones and ancestral hominids, indicating a Paleolithic origin of treponematosis (Rothschild and Rothschild 1996).

Most of the debate has focused on the origins and spread of syphilis. Recent phylogenetic and SNP analyses of treponemal genes suggest that a New World lineage of *T. pallidum* subsp. *pertenue* may be the origin of the subspecies *pallidum* (Harper et al. 2008). Also there are interesting parallelisms between human treponematosis and similar infections in primates (please see Sect. 1.4) The change from casual to venereal route of transmission in *Treponema pallidum* remains a puzzle. However, non-venereal *T. pallidum* subsp. *pertenue* has been found to cause chancre-like (syphilis-like) lesions in Guyanese indigenous people (Scolnik et al. 2003). *Neisseria gonorroheae*, another venereal pathogen, may have evolved from a linage of *Neisseria meningitides* (upper respiratory tract inhabitant) during the Neolithic (Saunders et al. 1999) and it may be related to the emergence of large villages. In this case there is no evidence of recent zoonotic origin of pathogenic of human *Neisseria* which may suggest co-evolution within hominids.

Bordetella pertussis, the etiologic agent of whooping cough, was thought to have originated recently from *Bordetella bronchiseptica* infecting domestic animals such as pigs and dogs (Diamond 1999; Pearce-Duvet 2006; Wolfe et al. 2007). Although analysis of DNA sequences of multiple loci (MLST) indicated that *B. pertussis* evolved from *B. bronchiseptica* (Diavatopoulos et al. 2005), recent molecular clock estimations suggest that the divergence time between *B. pertussis* and *B. bronchiseptica* associated with domestic animals was 1.1 to 5.6 million years ago (Diavatopoulos et al. 2005), before the origin of *Homo sapiens*, 0.2 million years ago (Forster 2004) and therefore also before the domestication of pigs and dogs. Genomic decay in *B. pertussis* may have been the result of evolution among ancestral hominids and adaptation to these hosts (Diavatopoulos et al. 2005; Bentley and Parkhill 2004). Additionally, human strains of *B. parapertussis* (a bacterium causing less severe whooping cough in humans) have diverged from animal *B. bronchiseptica* 0.7 to 3.5 million years ago and from a different clade than *B. parapertussis* isolated from domestic animals, however a recent report suggests that strains of *B. parapertussis* of human and ovine origins may be more closely related than previously described (Park et al. 2012)

Some *B. bronchisepica* infecting humans seem to belong to different lineages from those infecting domestic animals whereas other clades appear to be zoonotic (Diavatopoulos et al. 2005). Human pathogenic *Bordetella* may have originated from related bacteria infecting other animals during the Paleolithic period (Diavatopoulos et al. 2005).

Variola (Smallpox) virus was also thought to have originated in domestic animals (Diamond 1999; Pearce-Duvet 2006; Wolfe et al. 2007). However correlation of phylogenetic analysis and historical records indicate that variola virus belongs to a different lineage than cowpox virus (Li et al. 2007). Recent analysis suggests that this virus originated in an African rodent during the Paleolithic period (Li et al. 2007), however other authors have presented evidence of a more recent divergence (approximately 3,500 years ago) between the virus from the African rodent and the variola virus (Babkin and Babkina 2012).

1.3 Evidence of Co-evolution with African Primates

Mycobacterium leprae has been found in wild primates showing signs of leprosy (Hubbard et al. 1991; Clark-Curtiss and Walsh 1989). The significance of non-human primate leprosy is unknown because of lack of genetic information of the etiologic agents. At this point it is not possible to decipher if these primates, like armadillos (Truman et al. 2011), contracted *M. leprae* from humans or vice versa, or whether this bacterium belongs to a distinct but phylogenetically related linage (co-evolution within primates). While it is unknown whether this leprosy from non-human primates could be passed to humans, there is some evidence that leprosy from armadillos could be zoonotic in some regions (Truman et al. 2011).

Similarly, *T. pallidum* subsp. *pertenue* infection rates are high in both humans and primates in yaws-endemic areas of West Africa (Centurion-Lara et al. 2006). A simian yaws-like skin disease caused by a variant closely related to the human *T. pallidum* subsp. *pertenue*, which does not appear to be the result of recent cross infection from humans, has been described (Harper et al. 2008; Centurion-Lara et al. 2006), although inoculation with the simian strain can cause a yaws-like infection in humans suggesting that cross species transference is also possible (Harper et al. 2008). Additionally, *T. pallidum* subsp.*pertenue* has been reported to cause genital ulcerations in African primates (Knauf et al. 2011) and the tropism to genital epithelia of these non-human primate strains could be an example of parallel evolution. These data suggest that skin treponematosis may have evolved within African primates and ancestral hominids.

Other microorganisms which may have co-evolved with African hominids are herpesviruses (McGeoch et al. 2005), papillomaviruses (Gottschling et al. 2007), *Helicobacter pylori* (Linz et al. 2007), *Streptococcus pneumoniae* (Kilian et al. 2008), *Taenia solium*, *T. saginata* (Hoberg et al. 2001), and even human intestinal microbiota (Ochman et al. 2010).

1.4 Evolving to Human-Specific Pathogens

Pathogen crossing of host-species barriers is a common occurrence in natural environments (zoonosis and anthroponosis). However, acquiring traits enabling efficient transmission within a given host species is a more unusual event. Evolutionary adaptation of many bacterial pathogens to a specific host transmission may be accompanied by a trade off which reduces the competence to cross host species barriers and often involves genome decay (Bentley and Parkhill 2004).

Adaptation to transmission within a new host (i.e. human to human transmision) seems to occur more frequently in pathogens infecting phylogenetically related hosts (Davies and Pedersen 2008; Holmes 2008). Molecular similarities between cells from closely related animal species may facilitate this adaptation process (Holmes 2008). The recent evolution of the human pathogens such as hepatitis B virus (Simmonds and Midgley 2005; Vartanian et al. 2002; Tatematsu et al. 2009), HIV (Keele et al. 2006), HTLV (Wolfe et al. 2005), falciparum malaria (Liu et al. 2010) from African primates follows this pattern (Fig. 1.1). Close contact of modern humans with their genetically closer hominid species such as *Homo neanderthalensis* (Green et al. 2010) (or other archaic humans) may have also played a role in the introduction of some of these infectious diseases to modern humans (Fig. 1.1). For instance, the possible emergence of the typhoid fever agent *S. enterica* serovar Typhi (Kidgell et al. 2002) and *E. coli* (Pupo et al. 2000) roughly coincides with the time estimates of the interaction between *Homo sapiens* and *H. neanderthalensis* (Green et al. 2010).

Close contact with phylogenetically distant animals can also result in the evolution of new host-specific pathogens; animal pathogens becoming human pathogens (measles, bubonic plague, smallpox, and swine and avian influenza) and human pathogens becoming pathogens of other animal species (taeniasis, tuberculosis and infectious gastritis).

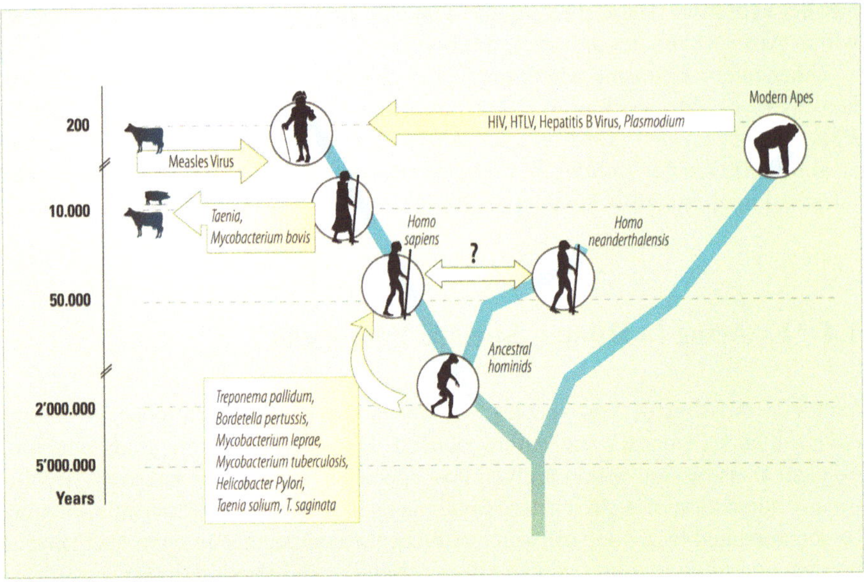

Fig. 1.1 Probable origins of human-specific infectious diseases. *Arrows* indicate suggested direction of the original transmission

1.5 Conclusions

Many problems exist in inferring phylogenetic relationships (Felsenstein 2003) and molecular clock estimations (Feng et al. 2008; Graur and Martin 2004; Sharp and Simmonds 2011; Bromham and Penny 2003), however it is possible to conclude that recent evidences (molecular clocks analyses, single nucleotide polymorphisms and phylogeography) suggest that many modern human infectious disease did not arise from domestic animals during the Neolithic period as previously thought. Rather, these diseases have a much older origin (Paleolithic period); some of them may have co-evolved with hominids whereas others may have originated in wild animals. In other cases, such as measles and possibly parapertussis, the evidence suggests a Neolithic origin from zoonotic agents present in domestic animals (Sharp and Simmonds 2011; Park et al. 2012). Despite the recent evidence that many human-infectious diseases have originated in primates, the study of infectious diseases of primates (especially non-human hominids) is still a neglected field of research. African primates, including human's closest relatives (chimpanzees and bonobos) also share the same habitats in many regions of central Africa. Most importantly, many people in this region consume ape meat and are exposed to blood and other fluids from these animals (Wolfe et al. 2005). African primates remain an untapped source of information required to complete the puzzle of the mechanisms of origin and evolution of many human pathogens.

As the genomic data from a wider population of microbial pathogens and commensals from humans and other animals are available, we will have a better understanding of the mechanisms that govern transmission and microbial adaptation to different animal species. The discovery of the factors involved in host species crossing and the evolution of human specific pathogens may help the identification of human activities, which can potentially promote the emergence of new infectious diseases.

There is a complex relationship between human activity, animal microbial diseases, human microbial diseases and the environment. A better understanding of the evolutionary processes implicated in the emergence of new diseases may require the collaboration of many disciplines, an idea promoted by the One Health concept.

Acknowledgements I would like to thank Micah Dunthorn Ph.D. who collaborated in the edition of this chapter and the support of Universidad San Francisco de Quito. My thanks also go to Ricardo Vásquez Bsc. identification useful here) for the figure.

References

Babkin IV, Babkina IN (2012) A retrospective study of the orthopoxvirus molecular evolution. Infect Genet Evol 12:1597–15604

Bentley SD, Parkhill J (2004) Comparative genomic structure of prokaryotes. Annu Rev Genet 38:771–791

Biek R, O'Hare A, Wright D et al (2012) Whole genome sequencing reveals local transmission patterns of *Mycobacterium bovis* in sympatric cattle and badger populations. PLoS Pathog 8 (11):e1003008. doi:10.1371/journal.ppat.1003008

Bromham L, Penny D (2003) The modern molecular clock. Nat Rev Genet 4:216–224

Centurion-Lara A, Molini BJ, Godornes C et al (2006) Molecular differentiation of *Treponema pallidum* subspecies. J Clin Microbiol 44:3377–3380

Clark-Curtiss JE, Walsh GP (1989) Conservation of genomic sequences among isolates of *Mycobacterium leprae*. J Bacteriol 171:4844–4851

Cui Y, Yu C, Yan Y et al (2013) Historical variations in mutation rate in an epidemic pathogen, *Yersinia pestis*. Proc Natl Acad Sci U S A 110:577–582

Davies TJT, Pedersen AB (2008) Phylogeny and geography predict pathogen community similarity in wild primates and humans. Phylog Proc R SocB 275:1695–1701

Diamond J (1999) Guns, germs and steel: the fates of human societies. W. W. Norton & Company, New York

Diavatopoulos DA, Cummings CA, Schouls LM et al (2005) *Bordetella pertussis*, the causative agent of whooping cough, evolved from a distinct, human-associated lineage of *B. bronchiseptica*. PLoS Pathog 1(4):e45

Felsenstein J (2003) Inferring phylogenies, 2nd edn. Sinauer Associates, Sutherland

Feng L, Reeves PR, Lan R et al (2008) A recalibrated molecular clock and independent origins for the cholera pandemic clones. PLoS One 3(12):e4053. doi:10.1371/journal.pone.0004053

Forster P (2004) Ice Ages and the mitochondrial DNA chronology of human dispersals: a review. Philos Trans R Soc Lond B Biol Sci 359:255–264

Furuse Y, Suzuki A, Oshitani H (2010) Origin of measles virus: divergence from rinderpest virus between the 11th and 12th centuries. Virol J 7:52. doi:10.1186/1743-422X-7-52

Garnier T, Eiglmeier K, Camus JC et al (2003) The complete genome sequence of *Mycobacterium bovis*. Proc Natl Acad Sci U S A 100:7877–7882
Gómez-Valero L, Rocha EPC, Latorre A, Silva F (2007) Reconstructing the ancestor of *Mycobacterium leprae*: the dynamics of gene loss and genome reduction. Genome Res 17:1178–1185
Gottschling M, Stamatakis A, Nindl I et al (2007) Multiple evolutionary mechanisms drive papillomavirus diversification. Mol Biol Evol 24:1242–1258
Graur D, Martin W (2004) Reading the entrails of chickens: molecular timescales of evolution and the illusion of precision. Trends Genet 20:80–86
Green RE, Krause J, Briggs AW et al (2010) A draft sequence of the Neandertal genome. Science 328:710–722
Gutierrez MC, Brisse S, Brosch R et al (2005) Ancient origin and gene mosaicism of the progenitor of *Mycobacterium tuberculosis*. PLoS Pathog 1(1):e5. doi:10.1371/journal.ppat.0010005
Han XY, Sizer KC, Thompson EJ et al (2009) Comparative sequence analysis of *Mycobacterium leprae* and the new leprosy-causing *Mycobacterium lepromatosis*. J Bacteriol 191:6067–6074
Harper KN, Ocampo PS, Steiner BM et al (2008) On the origin of the treponematoses: a phylogenetic approach. PLoS Negl Trop Dis 2(1):e148. doi:10.1371/journal.pntd.0000148
Hoberg EP, Alkire NL, de Queiroz A, Jones A (2001) Out of Africa: origins of the *Taenia* tapeworms in humans. Proc Biol Sci 268:781–787
Holmes EC (2008) Evolutionary history and phylogeography of human viruses. Annu Rev Microbiol 62:307–328
Hubbard GB, Lee R, Eichberg W et al (1991) Spontaneous leprosy in a chimpanzee (*Pan troglodytes*). Vet Pathol 28:546–548
Keele BF, Van Heuverswyn F, Li Y et al (2006) Chimpanzee reservoirs of pandemic and nonpandemic HIV-1. Science 313:523–526
Kidgell C, Reichard U, Wain J et al (2002) *Salmonella typhi*, the causative agent of typhoid fever, is approximately 50,000 years old. Infect Genet Evol 2:39–45
Kilian M, Poulsen K, Blomqvist T et al (2008) Evolution of *Streptococcus pneumoniae* and its close commensal relatives. PLoS One 3(7):e2683. doi:10.1371/journal.pone.0002683
Knauf S, Batamuzi EK, Mlengeya T et al (2011) *Treponema* infection associated with genital ulceration in wild baboons. Vet Path. doi:10.1177/0300985811402839
Li Y, Carroll DS, Gardner SN et al (2007) On the origin of smallpox: correlating variola phylogenics with historical smallpox records. Proc Natl Acad Sci U S A 104:15787–15792
Linz B, Balloux F, Moodley Y et al (2007) An African origin for the intimate association between humans and *Helicobacter pylori*. Nature 445:915–918
Liu W, Li Y, Learn GH et al (2010) Origin of the human malaria parasite *Plasmodium falciparum* in gorillas. Nature 467:420–425
McGeoch DJ, Gatherer D, Dolan A (2005) On phylogenetic relationships among major lineages of the Gammaherpesvirinae. J Gen Virol 86:307–316
Monot M, Honoré N, Garnier T et al (2005) On the origin of leprosy. Science 308:1040–1042
Monot M, Honoré N, Garnier T et al (2009) Comparative genomic and phylogeographic analysis of Mycobacterium leprae. Nat Genet 41:1282–1289
Ochman H, Worobey M, Kuo C-H, Ndjango J-BN, Peeters M et al (2010) Evolutionary relationships of wild hominids recapitulated by gut microbial communities. PLoS Biol 8(11): e1000546. doi:10.1371/journal.pbio.1000546
Park J, Zhang Y, Buboltz AM et al (2012) Comparative genomics of the classical *Bordetella* subspecies: the evolution and exchange of virulence-associated diversity amongst closely related pathogens. BMC Genomics 13:545. doi:10.1186/1471-2164-13-545
Pearce-Duvet JM (2006) The origin of human pathogens: evaluating the role of agriculture and domestic animals in the evolution of human disease. Biol Rev Camb Philos Soc 8:369–382

Pupo GM, Lan R, Reeves PR (2000) Multiple independent origins of *Shigella* clones of *Escherichia coli* and convergent evolution of many of their characteristics. Proc Natl Acad Sci U S A 97:10567–10572
Rothschild BM, Rothschild C (1996) Treponematoses - origins and 1.5 million years of transition. Hum Evol 11:225–232
Saunders NJ, Hood DW, Moxon ER (1999) Bacterial evolution: bacteria play pass the gene. Curr Biol 9:180–183
Scolnik D, Aronson L, Lovinsky R et al (2003) Efficacy of a targeted, oral penicillin–based yaws control program among children living in rural South America. Clin Infec Dis 36:1232–1238
Sharp PM, Simmonds P (2011) Evaluating the evidence for virus/host co-evolution. Curr Opin Virol 1:436–441
Simmonds P, Midgley S (2005) Recombination in the genesis and evolution of hepatitis B virus genotypes. J Virol 79:15467–15476
Smith T (1904) Leprosy. J Mass Assoc Boards Health 14:214–217
Smith NH, Hewinson RG, Kremer K, Brosch R, Gordon SV (2009) Myths and misconceptions: the origin and evolution of *Mycobacterium tuberculosis*. Nat Rev Microbiol 7:537–544
Stone AC, Wilbur AK, Buikstra JE, Roberts CA (2009) Tuberculosis and leprosy in perspective. Am J Phys Anthropol 49:66–94
Tatematsu K, Tanaka Y, Kurbanov F et al (2009) Genetic variant of hepatitis B virus divergent from known human and ape genotypes isolated from a Japanese patient and provisionally assigned to new genotype. J Virol 83:10538–10547
Truman RW, Singh P, Sharma R et al (2011) Probable zoonotic leprosy in the southern United States. N Engl J Med 364:1626–1633
Vartanian JP, Pineau P, Henry M et al (2002) Identification of a hepatitis B virus genome in wild chimpanzees (*Pan troglodytes schweinfurthi*) from East Africa indicates a wide geographical dispersion among Equatorial African primates. J Virol 76:11155–11158
Wirth T, Hildebrand F, Allix-Béguec C et al (2008) Origin, spread and demography of the Mycobacterium tuberculosis complex. PLoS Pathog 4(9):e1000160. doi:10.1371/journal.ppat.1000160
Wolfe ND, Heneine W, Carr JK et al (2005) Emergence of unique primate T-lymphotropic viruses among central African bushmeat hunters. Proc Natl Acad Sci U S A 102:7994–7999
Wolfe ND, Dunavan CP, Diamond J (2007) Origins of major human infectious diseases. Nature 17:279–283

Chapter 2
Drivers of Emerging Zoonotic Infectious Diseases

Peter W. Horby, Ngo Thi Hoa, Dirk U. Pfeiffer, and Heiman F.L. Wertheim

Abstract This chapter discusses drivers of emerging infectious diseases (EID) of humans that have an origin in other vertebrate animals (zoonoses). This is a broad topic, worthy of a book in its own right. This chapter will therefore provide only an overview of key concepts of drivers of the emergence of zoonotic diseases, and particularly infectious diseases with a major disease burden in humans. As the authors mainly work in Asia, the focus of this chapter is Asia, but many of the lessons learned in this region are likely to apply elsewhere.

More than 60 % of the world population live in Asia, a region with some of the fastest developing economies in the world. Yet, despite tremendous advances, infectious diseases still remain a major burden for the human population in Asia. Of the estimated 2.1 million deaths in children aged less than 5 years in Southeast Asia in 2010, 47 % are attributable to infectious causes (Liu et al., Lancet 379:2151–2161, 2012). As such, Asia is both vulnerable to imported EIDs and a global focus of major social and environmental change that may facilitate the emergence and dissemination of new pathogens. However, it would be too simplistic to present the extensive changes in Asia as inevitably increasing the risk of

---

This work is in part based on an article published in Emerging Infectious Diseases in June 2013.

P.W. Horby, MBBS, DTM&H, FFPH, FRCP, PhD (✉)
Epidemic Diseases Research Group, Centre for Tropical Medicine and Global Health, Nuffield Department of Medicine, University of Oxford, Old Road campus, Roosevelt Drive, Headington, Oxford, OX3 7FZ, UK
e-mail: peter.horby@ndm.ox.ac.uk

N.T. Hoa • H.F.L. Wertheim
Wellcome Trust Major Overseas Program, Oxford University Clinical Research Unit, National Hospital for Tropical Disease, 78 Giai Phong Street, Hanoi, Vietnam

Nuffield Department of Clinical Medicine, Centre for Tropical Medicine, University of Oxford, Oxford, UK
e-mail: heiman.wertheim@gmail.com

D.U. Pfeiffer
Veterinary Epidemiology, Economics and Public Health Group, Royal Veterinary College, London, UK

A. Yamada et al. (eds.), *Confronting Emerging Zoonoses*,
DOI 10.1007/978-4-431-55120-1_2

EIDs. Some aspects of socio-economic change might serve to reduce the overall risk of infectious disease emergence, but all ecosystem changes have the potential to provide new opportunities for microorganisms to spill-over into human populations.

Keywords Driver • Emerging infectious disease • One health • Zoonosis

2.1 The Concept of Emergence

Emerging infectious diseases are often perceived to be 'novel' pathogens of humans, but EIDs are more broadly defined as diseases that are increasing in their incidence, geographic or host species range, or their impact (due for instance to the acquisition of resistance to antimicrobial drugs or new virulence factors) (Fauci and Morens 2012). This encompasses both newly recognised pathogens and known pathogens that are 're-emerging' (Box 1). The development and introduction over time of new, more sensitive technologies for identifying microbial diversity, such as molecular methods to identify elements of the microbial genome, coupled with geographical heterogeneities in the availability of these technologies, make it difficult to accurately assess the rate at which new infectious diseases emerge. The difficulty is further compounded by changes over time in the ease of publishing articles in the bio-medical literature. Attempts have nevertheless been made, and 335 emerging infectious disease events were identified between 1940 and 2004, an average of five per year, with a peak in the 1980s associated with the emergence of HIV/AIDs and its associated opportunistic infections (Jones et al. 2008).

New human pathogens must come from somewhere, and unsurprisingly they most often arise from animals; with which we share genes, physiology, microorganisms, and environments. Around 60 % of current human infectious diseases are zoonotic, and those that exclusively infect humans were probably shared with other animal species at some time in the past (e.g. measles, mumps, dengue, pertussis, hepatitis B). In addition, around 60 % of recent emerging infectious diseases of humans have arisen from animals, with an estimated 72 % of these having their origin in wildlife species (Jones et al. 2008). The recognition of the shared susceptibility of humans and animals to many diseases has led to the concept of One Health. The infection of humans by the microorganisms of animals is a natural consequence of ecology and evolution (Karesh et al. 2012). Microbes are an abundant life form and ecological factors (specifically the environmental living conditions) critically define the niches that they inhabit. Microbes that normally colonise or infect animals are able to 'spill over' to humans when they possess the ability and are given the opportunity to exploit a similar niche (e.g. the gut or bloodstream) of humans. The same is true for microbes of wild animals that spill over to domesticated animals. Once provided this opportunity to infect a new host, a process of Darwinian competition will select those natural variants of the microorganism that are best able to survive and replicate in the new environment. In this way microorganisms adapt to humans, and may become so well adapted as to become exclusive pathogens of humans (Fig. 2.1) (Wolfe et al. 2007).

Box 1. Categories of Infectious Diseases (Fauci, NEJM, 2012)

1. *Established infectious diseases*: endemic diseases that have been around for a sufficient amount of time to allow for a relatively stable and predictable level of morbidity and mortality. Examples are common diarrheal pathogens, drug-susceptible malaria, tuberculosis, helminthic and other parasitic diseases;
2. *Newly emerging infectious diseases*: diseases that recently have been detected in the human host for the first time. Nipah virus, severe acute respiratory syndrome virus (SARS), human metapneumovirus (hMPV), and new influenza subtypes (swine H1N1, avian H5N1, or avian H7N9) are examples in this category. Often RNA viruses causing respiratory diseases are in this category.
3. *Re-emerging infectious diseases*: Diseases in this category can be subclassified as follows:

   - Diseases that appear either in new regions, as we have seen for West Nile virus (WNV) in the Americas.
   - Already known diseases that have become drug-resistant. Recent examples are drug resistant bacterial infections (penicillin resistant pneumococcal pneumonia, carbapenem resistant hospital acquired infections), drug resistant malaria, and oseltamivir resistant influenza.
   - Already known diseases that reappear after apparent control or elimination, or under unusual circumstances. Examples in category are the deliberate release of anthrax in the USA in 2001, or the reappearance of dengue on Florida, USA.

2.2 Shared Ecologies

The conditions or events (the ‘drivers’) that result in the successful cross–over of an animal microbe into humans are not well characterised, but emergence is often precipitated by changes to ecological or biological systems (Wilcox and Colwell 2005). Such changes include altered patterns of contact between wild and domestic animals (e.g. Nipah virus), of direct human and wild animal contact (e.g. HIV, Ebola), and changes in species abundance or diversity (e.g. Hantavirus; Lyme disease). Species diversity, including the diversity of insect vectors and pathogenic microorganisms, increases towards the equator (Guernier et al. 2004), and Jones et al. found a correlation between the emergence of zoonotic pathogens and the diversity of mammalian wildlife species (Jones et al. 2008). Whilst high animal host and pathogen species diversity may be associated with a high burden of infectious diseases and an increased risk of disease emergence, biodiversity loss may, perhaps counter-intuitively, be associated with increased disease transmission. Biodiversity

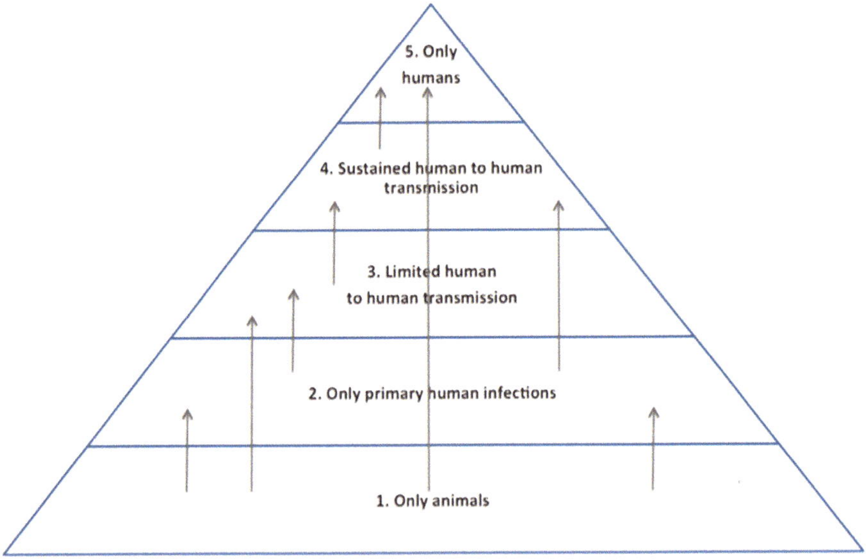

Fig. 2.1 Five stages of how exclusive animal pathogens can adapt to and infect humans (Wolfe et al. 2007). Stage 1: The pathogen is still exclusively infecting animals. Stage 2: The pathogen has been transmitted from animals to humans under natural conditions but not yet from human to human. Stage 3: There is limited transmission from animals to humans and between humans. These are often severe and lethal diseases due to for instance filoviruses (e.g. Ebola). Stage 4: Animal disease that have sustained transmission between humans (e.g. influenza). Stage 5: A microbe that exclusively infects humans (e.g. measles, syphilis)

loss directly disrupts the functioning and stability of ecosystems, producing effects that can extend well beyond the particular lost species (Hooper et al. 2012). Changes in species abundance and diversity may favour pathogen amplification and 'spill-over' through a variety of mechanisms, including reduced predation and competition resulting in increased abundance of competent hosts, and the loss of 'buffering species' leading to increased contact between amplifying host species and compatible pathogens (Keesing et al. 2010). Tropical regions with a rich pool of existing and potential pathogens that are increasingly connected, but also experiencing high rates of ecosystem disruption and biodiversity loss, may therefore be at a particularly high risk of disease emergence.

Human pathogens occasionally re-emerge as a result of dynamics that are beyond the control of humans. For example the Hantavirus outbreaks in the Southwestern U.S. in 1991–1992 and 1997–1998 have been attributed to changes in the abundance of infected rodents following periods of heavy snow and rainfall and vegetation growth leading to abundant production of rodent food. However, many ecological disturbances resulting in an EID seem to originate in the direct actions of humans. A wide range of anthropogenic factors have been linked to infectious disease emergence, including changes in land-use, travel, trade, and demographics. Notably, most of these associations are speculative and supported by little hard data because ecological and biological systems are highly complex and multi-layered (Woolhouse 2011). Demonstrating or predicting the impact of particular conditions or events on the functioning of a system is difficult, with further inference of the impact of any changes on the risk of pathogen emergence posing a formidable challenge.

2.3 Socioeconomic Development and Altered Ecosystems

Socioeconomic development is associated with large increases in demand for natural resources. The demand for water, wood, pulp, agricultural land, living space, roads, minerals and power has had an enormous impact on the landscapes of Asia. Deforestation occurred throughout the 1990s and the area of primary forest in Asia has continued to decline (FAO 2012). Deforestation, forest fragmentation, and afforestation are all alterations in habitat, which change species composition and the interaction between wild animals, domestic animals, insect vectors and humans, providing new opportunities for microbial transmission and potential emergence. There are well-documented examples of deforestation and forest encroachment resulting in increases in infectious diseases, such as yellow fever, Mayaro, and Chagas disease in the Americas (Saker et al. 2004). The clearing of forest and planting of large cacao plantations was linked to the emergence of Oropouche virus in Brazil. In Asia there are already very high pressures on productive land, and the peak in land-use change in Asia has probably passed. Many areas are now in an era of increasing intensification of land productivity. This intensification is driven largely by demographic pressures, which are predicted to result in a 70 % increase in food production by 2050, with decreased consumption of grains and increased demand for meats, fruits and vegetables (FAO 2011a, b).

The increased demand for food, and meat in particular, when combined with demands for natural resources from industry and domestic consumers, and river damming for hydroelectric power, is resulting in a large increase in stress on water resources (FAO 2011b). The consequences of intensified agricultural production include the depletion and degradation of river and groundwater, reduced soil quality, and biodiversity loss. A direct and predictable effect of reduced access to clean water for low-income families is an increase in the risk of water-washed diseases (diseases that increase when the availability of water for personal hygiene is limited e.g. diarrhoeal and respiratory infections, trachoma), and water-borne diseases such as typhoid and hepatitis E. However, unquantified risks arising from the intensification of agriculture are pollution of freshwater with pesticides and fertilizers, loss of biodiversity, and land abandonment by small-scale farmers. The potential consequences of these changes on the risk of emergence of zoonotic infections have not been assessed.

2.4 Wildlife Trade

Wild animals are an important source of food (bush meat) in some developing countries and bush meat has been implicated in the emergence of HIV, and the spill over of monkeypox, Nipah and Ebola virus (Brashares et al. 2011). Whilst the reservoir of the SARS coronavirus is thought to be bats, wild civet cats traded for food are thought to have acted as an intermediate host, transmitting the SARS virus to humans through live animal markets (Li et al. 2005). Wild animal products are popular in Asia as traditional medicines, tonics, delicacies, or as symbols of wealth. Although all ten countries in the Association of Southeast Asian Nations (ASEAN) are signatories to the Convention on International Trade in Endangered Species of Wild Fauna and Flora (CITES), Asia continues to host the largest illegal wildlife trade in the world (Rosen and Smith 2010). The smuggling of H5N1-infected birds of prey into Europe, the frequent smuggling of bush meat from Africa into the U.S., and the importation into the U.S. of pet rodents infected with monkeypox show that both the legal and illegal trade in wild animals and wild animal products is a potential conduit for the international spread of zoonotic pathogens (Bair-Brake et al. 2013; Van Borm et al. 2005)

2.5 Urbanization, Consumer Behavior, and Market Chains

Between 2011 and 2050 the world population is expected to increase by 2.3 billion (a 32 % increase), and the increase will be concentrated in urban areas of developing countries (UN-DESA 2012). Whilst mega-cities (cities with a population of at least ten million) receive a lot of attention, most urban dwellers live in small cities, with half of the global urban population in 2011 living in cities of less than 500,000 people (UN-DESA 2012). This can be perceived positively as cities generally offer better economic opportunities, better educational opportunities, better living conditions, better nutrition, better sanitation, and therefore better health than underdeveloped rural areas. At the same time it is likely to mean that in many low to middle income countries, health and veterinary infrastructure in rural areas will not improve, or may even deteriorate, adversely affecting the likelihood of the early detection of EID. The demand for food in urban centres will increase and result in livestock and their products being transported over large distances from a wider catchment area, and thereby increase risk of spread and amplification of EID. Urbanisation is one facet of changing human sociocultural systems, which also includes changing consumer demands and dietary habits (Janes et al. 2012). The consequence is a spatial concentration of people and animals; not necessarily co-located, but connected through increasingly complex networks of rural and peri-urban farms and markets, distributors, agricultural workers, and consumers.

2.6 Livestock Production Systems

2.6.1 Intensification of Production

Due to the increase in global human population and economic development, demand for livestock products has risen dramatically over the last 50 years, with the per capita consumption of meat in developing countries more than tripling since the early 1960s and egg consumption increasing fivefold (FAO 2011a, b). The increased demand for meat has been met by more intensive and geographically concentrated production of livestock, especially pigs and poultry (Steinfeld et al. 2006). Much of this has been through expansion of both the number of small-scale production units and large commercial farms. High-density monoculture of domestic animals is a form of low biodiversity that poses a particular threat for the spread of infectious diseases from farmed animals to humans. Where domesticated animals are a conduit of spread from wild animals to humans, high density livestock production may promote spread of zoonotic diseases. Genetic diversity within an individual host species is important since genetic diversity limits the potential for devastating epidemics (King and Lively 2012).

The Nipah virus outbreak in Malaysia and Singapore in 1998–1999 is a good example. Once Nipah virus crossed from wild bats to domestic swine, an explosive outbreak in high-density swine farms resulted in widespread exposure of humans and over 250 human cases of encephalitis (Pulliam et al. 2012). Other examples where intensified livestock production practices may have led to emergence of a zoonosis include:

- *Streptococcus suis* causes severe sepsis and meningitis in humans and is associated with areas of intensive pig production (see Fig. 2.2). Risk factors for human infection include swine slaughtering and the eating of undercooked pig products (Wertheim et al. 2009). Outbreaks in swine herds of porcine reproductive and respiratory syndrome virus also potentially increases the rate of invasive *S. suis* infection in swine, which in turn leads to an increased risk of *S. suis* infection in humans (Hoa et al. 2013).
- Highly pathogenic avian influenza A subtype H5N1 crossed-over from wild aquatic birds (the natural reservoir of influenza A viruses) to humans via massive amplification in domestic poultry.
- The human Q-fever epidemic in the Netherlands during 2007–2010 caused by the bacterium *Coxiella burnetii* is thought to have arisen when economic drivers led to an increased density of dairy goat farming, which resulted in amplification of *Coxiella burnetii* prevalence and consequent increased spill- over to humans (Roest et al. 2011; Georgiev et al. 2013).

The classical foodborne diseases such as *E. coli*, campylobacteriosis and salmonellosis associated with livestock products have been a significant problem in high-income countries for some time (CDC 2011; Painter et al. 2013). One of the key factors, in particular in the case of *Campylobacter*, has been the integration of

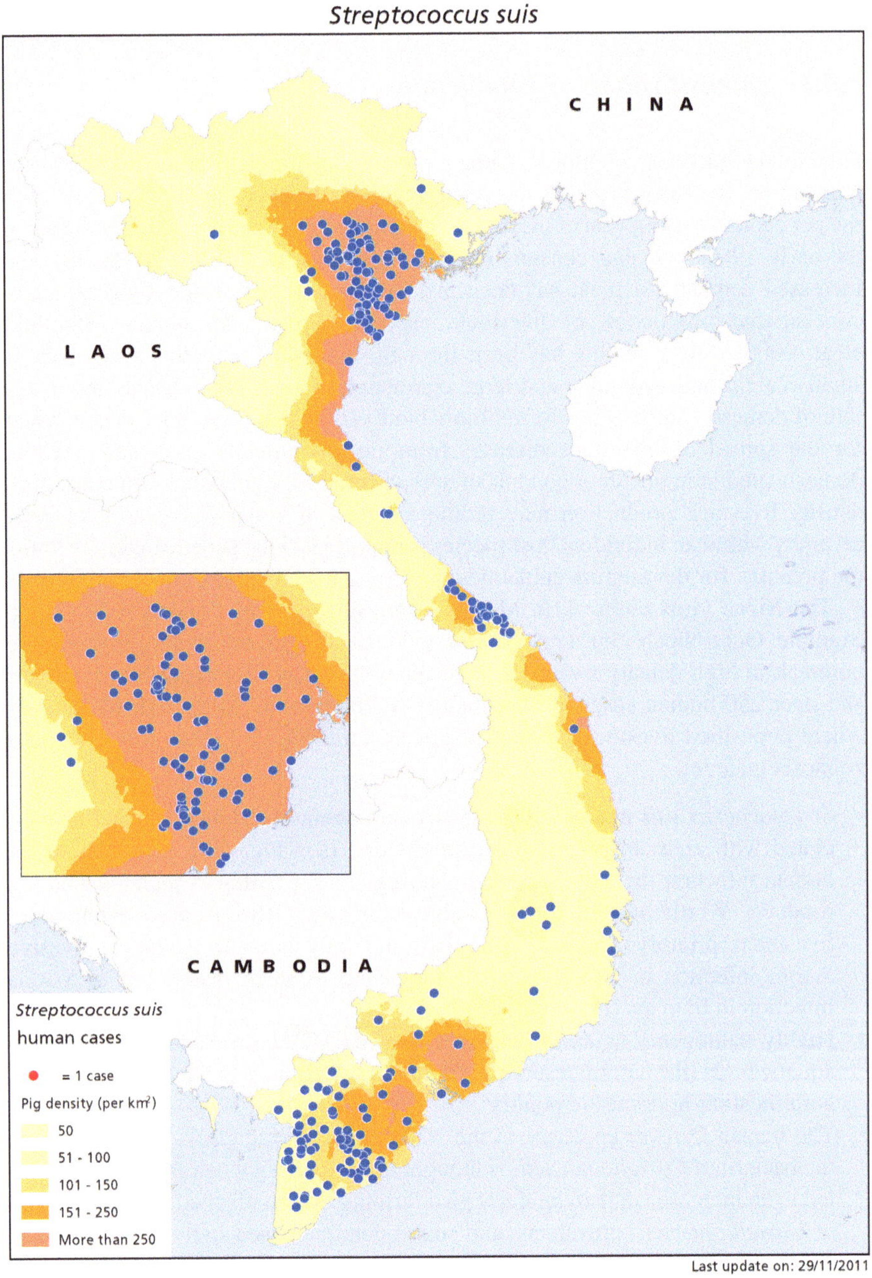

Fig. 2.2 Number of swine per km^2 and human *Streptococcus suis* infections in Vietnam from 2005–2011. Pig density data are based on Vietnam Government Statistics Office data 2006

highly intensive poultry production with poultry processing systems facilitating cross-contamination of meat (Moore et al. 2005; Nyachuba 2010). It is notable that some foodborne pathogens cause subclinical infections in their animal hosts (e.g., *Campylobacter* in poultry); therefore there is no direct incentive for farmers to control them. In addition, subsequent attribution of the source of foodborne disease in an affected human is notoriously difficult. It is likely that the global incidence of foodborne illness will rise as a result of increased industrial production of poultry in the emerging economies. The situation will be exacerbated by the emergence of antimicrobial resistant pathogens of foodborne pathogens (Sahin et al. 2012; Luangtongkum et al. 2009; Altekruse and Tollefson 2003).

Despite the examples above, the intensification of livestock farming often entails more effective separation of domestic and wild animals, improved standards of animal health and welfare, reduced movement and species mixing: all of which reduce the risk of EIDs. Reducing contact between domestic and wild animals, whether the wild animals are in their natural habitat or captive, is a key strategy of the Food and Agriculture Organization (FAO) to reduce risk to human health, and is part of the wider FAO strategy of enhancing 'biosecurity'. Improving biosecurity in farms is a major challenge since a large proportion of farming in Asia is based on small-scale backyard production, and there is often a mix of commercial and backyard farming in any one location. Achieving improvements in biosecurity without adversely affecting the livelihoods of small-scale farmers requires an approach to risk management that is adapted to their socio-economic context. The longer-term vision is to restructure the livestock production sector towards a more commercialised and controlled system, where controls benefit animal health, human health, and commercial profitability. But it has also been recognised that small-scale producers will continue to have a key role in providing food security in developing economies, and in fact their importance may increase due to their more efficient use of natural resources compared with large-scale industrial production (Sjauw-Koen-Fa 2012; Quan 2011).

2.6.2 Intensity and Complexity of Animal Markets

Food production has become a complex national, regional and global network of food value chains (Ercsey-Ravasz et al. 2012; Dickson-Hoyle and Reenberg 2009). Many countries have begun to more tightly regulate live animal trade in the wake of Bovine Spongiform Encephalopathy (BSE), SARS and avian influenza A/H5N1, but complex and poorly regulated food manufacturing and distribution chains still offer ample opportunities for disease outbreaks. In endemic regions, live bird markets are frequently contaminated with highly pathogenic avian influenza A subtype H5N1 (Davis et al. 2010; Samaan et al. 2011; Wan et al. 2011). Contaminated markets can become reservoirs and the source of infection for poultry and humans.

Measures to reduce the risk in live animal markets include the separation of species, not allowing animals to remain in the market more than 24 h, not allowing poultry to exit the market alive, improved cleaning and disinfection, and weekly rest-days, when all animals are removed and the market is thoroughly cleaned. The re-emergence of brucellosis, one of the world's most common zoonoses, is thought to be the result of higher risk of transmission through increased within- and between country trade of susceptible livestock and their products (Seleem et al. 2010). The development of denser regional road transport networks may be especially important since, compared to air travel, roads offer a more egalitarian form of connectivity that includes animals and goods as well as humans. This provides opportunities for pathogens to disperse beyond their traditional niche by increasing opportunities for informal trade in live animals and their products (Eisenberg et al. 2006).

Veterinary medical authorities often struggle to enforce compliance with regulations, and it is common practice for farmers and traders to adapt their production and trading behaviors to avoid adverse economic consequences of regulations aimed at controlling zoonotic diseases. As a result, informal trade networks may intensify and re-structure in unpredictable ways during disease outbreaks.

Increasing intensification of animal husbandry in Asia may be a trade-off between a lower risk of emergence events, as animals will be 'healthier' and better isolated, but should an EID event occur, an increased risk of massive amplification in large, naïve monocultures. Perhaps the greatest risk arises when a limited number of intensive livestock production units are surrounded by substantial backyard farming with little or no biosecurity, thereby linking extensive and weakly regulated value chains with global food systems.

2.6.3 Antibiotics and Other Practices

Livestock production practices can have major unintended and unanticipated impacts on the risk of zoonotic infections. Perhaps the most notable example is BSE. The feeding of ruminant-derived meat and bone meal to cattle, and possibly changes in practices for rendering animal tissue, led to an epidemic of BSE in cattle, that was followed by an epidemic of a fatal neurological disease (variant Creutzfeldt-Jacob Disease) in humans (Taylor and Woodgate 2003).

In contemporary livestock production, the use of antibiotics is a significant concern. Almost 40 % (23/59) of EIDs identified in Asia since 1940 represent the emergence of a new pattern of antimicrobial resistance (Jones et al. 2008). Bacteria in some areas show alarmingly high rates of resistance to anti-bacterial agents, and the recent emergence of the New-Delhi Metallo-beta-lactamase-1 (NDM-1) resistance gene, conferring resistance to a last resort antibiotic, and its rapid dissemination to other regions highlights the serious threat to human health of antimicrobial resistance (Khan and Nordmann 2012). Antibiotics are used extensively in the livestock and aquaculture sector to treat or prevent infections, or as growth promoters. Non-therapeutic use of antibiotics as growth promoters involves the prolonged administration of sub-therapeutic doses. This practise has a demonstrable effect on the emergence and prevalence of resistant microorganisms in food animals and their environment, as well as resulting in the excretion of antibiotics into the environment, where environmental bacteria may be subject to antibiotic selection pressures (Marshall and Levy 2011).

Synthetic antibiotics such as quinolones are quite stable in the environment over long periods of time. Heavy metal contamination of animal food may also play a role in selection of antimicrobial resistance. Whilst there remains some debate about the overall impact of these findings on human health, it is clear that the continued use of non-therapeutic antibiotics in an agriculture industry that is rapidly increasing in scale and intensity, has potential for becoming a very real threat through the inability to prevent/cure disease in production animals and the consequences for human food security as well as the transmission, for example, of resistant food-borne bacterial pathogens to humans.

Some recent examples concern transmission of methicillin resistant *Staphylococcus aureus* (MRSA) from pigs to humans and ESBL (extended spectrum beta-lactamase, an important resistance mechanism against most beta-lactam antibiotics) positive *E. coli* on chicken meat for human consumption (Verkade et al. 2012; Kluytmans et al. 2013). Swine-associated MRSA is now contributing significantly to invasive disease in patients in the Netherlands (Verkade et al. 2012).

Antibiotic production and consumption continues to increase, often in an uncontrolled way, accelerating the evolution of antibiotic resistance, which then spreads rapidly across the globe. Since the 1990s few new antibiotics have been developed and we are on the cusp of returning to an era of untreatable infections. The consequences of antibiotic use in animals therefore require better surveillance, research and regulation.

2.7 Conclusions

Infectious diseases continue to be an important cause of human and animal morbidity and mortality worldwide. Whilst important health advances (e.g. hygiene and vaccination) have been made, infectious diseases are dynamic and resilient, and continue to challenge local, national and global public health systems (Fauci and Morens 2012).

The recognition of the linkage between anthropogenic changes, animals, and disease emergence has resulted in repeated calls for a more holistic, interdisciplinary, and integrated approach to the study of infectious diseases. The One Health approach is one initiative aimed at realizing this aspiration. The task ahead is to understand how social, economic, and environmental changes are altering the landscape of infectious disease risks for both animals and humans, and how future emerging risks may be mitigated. A more analytical approach to these emergence events requires characterisation of the attributes of natural and artificial ecological systems before and after disturbance, and the functional relationship between changes in system attributes and pathogen emergence (Woolhouse 2011).

References

Altekruse SF, Tollefson LK (2003) Human campylobacteriosis: a challenge for the veterinary profession. J Am Vet Med Assoc 223:445–452

Bair-Brake H, Bell T, Higgins A et al (2013) Is that a rodent in your luggage? A mixed method approach to describe bushmeat importation into the United States. Zoonoses Public Health 61(2):97–104

Brashares JS, Golden CD, Weinbaum KZ, Barrett CB, Okello GV (2011) Economic and geographic drivers of wildlife consumption in rural Africa. Proc Natl Acad Sci U S A 108:13931–13936

Centers for Disease Control and Prevention (2011) Vital signs: incidence and trends of infection with pathogens transmitted commonly through food–foodborne diseases active surveillance network, 10 U.S. sites, 1996–2010. MMWR Morb Mortal Wkly Rep 60:749–755

Davis CT, Balish AL, O'Neill E et al (2010) Detection and characterization of clade 7 high pathogenicity avian influenza H5N1 viruses in chickens seized at ports of entry and live poultry markets in Vietnam. Avian Dis 54:307–312

Dickson-Hoyle S, Reenberg A (2009) The shrinking globe: globalisation of food systems and the changing geographies of livestock production. Danish J Geogr 109:105–112

Eisenberg JN, Cevallos W, Ponce K et al (2006) Environmental change and infectious disease: how new roads affect the transmission of diarrheal pathogens in rural Ecuador. Proc Natl Acad Sci U S A 103:19460–19465

Ercsey-Ravasz M, Toroczkai Z, Lakner Z, Baranyi J (2012) Complexity of the international agro-food trade network and its impact on food safety. PLoS One 7:e37810

FAO (2011) The state of food and agriculture 2010–2011. Rome: Food and Agriculture Organisation of the United Nations

FAO (2011) The state of the world's land and water resources for food and agriculture. Rome: Food and Agriculture Organization of The United Nations and Earthscan

FAO (2012) State of the world's forests, Rome. Rome: Food and Agriculture Organisation of the United Nations

Fauci AS, Morens DM (2012) The perpetual challenge of infectious diseases. N Engl J Med 366:454–461

Georgiev M, Afonso A, Neubauer H et al (2013) Q fever in humans and farm animals in four European countries, 1982 to 2010. Euro Surveill 18:pii: 20407

Guernier V, Hochberg ME, Guegan JF (2004) Ecology drives the worldwide distribution of human diseases. PLoS Biol 2:e141

Hoa NT, Chieu TT, Do Dung S et al (2013) Streptococcus suis and porcine reproductive and respiratory syndrome, Vietnam. Emerg Infect Dis 19:331–333

Hooper DU, Adair EC, Cardinale BJ et al (2012) A global synthesis reveals biodiversity loss as a major driver of ecosystem change. Nature 486:105–108

Janes CR, Corbett KK, Jones JH, Trostle J (2012) Emerging infectious diseases: the role of social sciences. Lancet 380:1884–1886

Jones KE, Patel NG, Levy MA et al (2008) Global trends in emerging infectious diseases. Nature 451:990–993

Karesh WB, Dobson A, Lloyd-Smith JO et al (2012) Ecology of zoonoses: natural and unnatural histories. Lancet 380:1936–1945

Keesing F, Belden LK, Daszak P et al (2010) Impacts of biodiversity on the emergence and transmission of infectious diseases. Nature 468:647–652
Khan AU, Nordmann P (2012) Spread of carbapenemase NDM-1 producers: the situation in India and what may be proposed. Scand J Infect Dis 44:531–535
King KC, Lively CM (2012) Does genetic diversity limit disease spread in natural host populations? Heredity 109:199–203
Kluytmans JA, Overdevest IT, Willemsen I et al (2013) Extended-spectrum beta-lactamase-producing Escherichia coli from retail chicken meat and humans: comparison of strains, plasmids, resistance genes, and virulence factors. Clin Infect Dis 56:478–487
Li W, Shi Z, Yu M et al (2005) Bats are natural reservoirs of SARS-like coronaviruses. Science 310:676–679
Liu L, Johnson HL, Cousens S et al (2012) Global, regional, and national causes of child mortality: an updated systematic analysis for 2010 with time trends since 2000. Lancet 379:2151–2161
Luangtongkum T, Jeon B, Han J, Plummer P, Logue CM, Zhang Q (2009) Antibiotic resistance in Campylobacter: emergence, transmission and persistence. Future Microbiol 4:189–200
Marshall BM, Levy SB (2011) Food animals and antimicrobials: impacts on human health. Clin Microbiol Rev 24:718–733
Moore JE, Corcoran D, Dooley JS et al (2005) Campylobacter. Vet Res 36:351–382
Nguyen BV, Nguyen HT, Olowokure B, Wertheim H (eds) (2014) Atlas Communicable Diseases Vietnam from 2000 to 2011. National Institute of Hygiene and Epidemiology, Hanoi
Nyachuba DG (2010) Foodborne illness: is it on the rise? Nutr Rev 68:257–269
Painter JA, Hoekstra RM, Ayers T et al (2013) Attribution of foodborne illnesses, hospitalizations, and deaths to food commodities by using outbreak data, United States, 1998–2008. Emerg Infect Dis 19:407–415
Pulliam JR, Epstein JH, Dushoff J et al (2012) Agricultural intensification, priming for persistence and the emergence of Nipah virus: a lethal bat-borne zoonosis. J R Soc Interface 9:89–101
Quan J (2011) A future for small-scale farming. London, UK
Roest HI, Tilburg JJ, van der Hoek W et al (2011) The Q fever epidemic in the Netherlands: history, onset, response and reflection. Epidemiol Infect 139:1–12
Rosen GE, Smith KF (2010) Summarizing the evidence on the international trade in illegal wildlife. Ecohealth 7:24–32
Sahin O, Fitzgerald C, Stroika S et al (2012) Molecular evidence for zoonotic transmission of an emergent, highly pathogenic Campylobacter jejuni clone in the United States. J Clin Microbiol 50:680–687
Saker L, Lee K, Cannito B, Gilmore A, Campbell-Lendrum D (2004) Globalization and infectious diseases; a review of the linkages. World Health Organization, Geneva
Samaan G, Gultom A, Indriani R, Lokuge K, Kelly PM (2011) Critical control points for avian influenza A H5N1 in live bird markets in low resource settings. Prev Vet Med 100:71–78
Seleem MN, Boyle SM, Sriranganathan N (2010) Brucellosis: a re-emerging zoonosis. Vet Microbiol 140:392–398
Sjauw-Koen-Fa AR (2012) Framework for an inclusive food strategy. Cooperatives - a key for Smallholder Inclusion in Valeu Chains Utrecht. Rabobank Nederland, The Netherlands
Steinfeld H, Wassenaar T, Jutzi S (2006) Livestock production systems in developing countries: status, drivers, trends. Rev Sci Tech 25:505–516
Taylor DM, Woodgate SL (2003) Rendering practices and inactivation of transmissible spongiform encephalopathy agents. Rev Sci Tech 22:297–310
UN-DESA (2012) World urbanization prospects: the 2011 revision
Van Borm S, Thomas I, Hanquet G et al (2005) Highly pathogenic H5N1 influenza virus in smuggled Thai eagles, Belgium. Emerg Infect Dis 11:702–705
Verkade E, Bergmans AM, Budding AE et al (2012) Recent emergence of Staphylococcus aureus clonal complex 398 in human blood cultures. PLoS One 7:e41855
Wan XF, Dong L, Lan Y et al (2011) Indications that live poultry markets are a major source of human H5N1 influenza virus infection in China. J Virol 85:13432–13438

Wertheim HFL, Nghia HDT, Taylor W, Schultsz C (2009) Streptococcus suis: an emerging human pathogen. Clin Infect Dis 48:6

Wilcox B, Colwell R (2005) Emerging and reemerging infectious diseases: biocomplexity as an interdisciplinary paradigm. Ecohealth 2(4):244–257

Wolfe ND, Dunavan CP, Diamond J (2007) Origins of major human infectious diseases. Nature 447:279–283

Woolhouse M (2011) How to make predictions about future infectious disease risks. Philos Trans R Soc Lond B Biol Sci 366:2045–2054

Chapter 3
Biodiversity and Emerging Zoonoses

Serge Morand, Katharine Owers, and Frédéric Bordes

Abstract This chapter examines the links between biodiversity and worldwide patterns of zoonotic disease outbreaks. We find that biodiversity appears to be a major contributing factor determining the diversity of human infectious diseases, and that zoonotic disease outbreaks should be considered with the indices of human development and biodiversity loss. Increasing population and wealth (i.e., GDP) both threaten biodiversity and favour epidemics, leading us to conduct an analysis showing the deleterious impacts of development (and globalization) on health. Our results support this premise in contrast to most studies showing generally negative effects of biodiversity reduction on the spread of infectious diseases. Biodiversity has a complex relationship with human infectious diseases, serving both as a source of pathogens and as a regulating factor, keeping the incidence down.

Keywords Biodiversity • Dilution effect • Infectious diseases • Outbreaks • Species richness • Zoonoses

3.1 Introduction

The number of emerging infectious diseases (EID) has increased during the last century (Wilcox and Gubler 2005; Wolfe et al. 2007; Jones et al. 2008), a change often attributed to growing human-induced environmental changes (Daszak et al. 2000; Chivian 2003; Chivian and Bernstein 2004; Patz et al. 2004; Wilcox and Colwell 2005; Lafferty 2009). Concurrently, the number of reported infectious disease outbreaks has also dramatically increased during the last 60 years (Morand and Waret-Szkuta 2012; Morand et al. 2013). The explanatory factors reported for these trends are generally associated with ongoing global changes

S. Morand (✉)
CNRS-CIRAD, Centre d'Infectiologie Christophe Mérieux du Laos, Vientiane, Lao PDR
e-mail: serge.morand@univ-montp2.fr

K. Owers • F. Bordes
Institut des Sciences de l'Evolution, CNRS-IRD-UM2, Université de Montpellier 2, Montpellier, France

A. Yamada et al. (eds.), *Confronting Emerging Zoonoses*,
DOI 10.1007/978-4-431-55120-1_3

such as climate change and variability, global trade and economic development, land use change and biodiversity loss, which have been occurring on an unprecedented scale (MEA 2005).

Increased biodiversity may correlate to more numerous human infectious diseases, as a statistical correlation has been observed between the number of birds and mammals (used as a proxy of biodiversity) and the number of human pathogens (Dunn et al. 2010). The density of human pathogens increases towards lower latitudes, and indeed, bird and mammalian species richness are highest in the tropical zones (Schipper et al. 2008). A similar trend has been observed for human pathogens (Guernier et al. 2004; Dunn et al. 2010) and rodent-borne diseases (Bordes et al. 2011).

However, biodiversity loss has been positively associated with human infectious disease emergence. This occurs through increased interactions between humans, their domestic animals and wildlife as a result of habitat fragmentation and changes in wildlife species richness and community composition. Several studies have shown that a reduction in biodiversity at the local level may lead to an increase in the prevalence and transmission rates of certain diseases, such as Lyme disease or West Nile fever (Keesing et al. 2010). The decrease in biodiversity seems to be associated with the loss of the ecosystem's capacity to buffer the spread of pathogens (Keesing et al. 2006).

Taken together these two patterns seem a priori opposite if not contradictory. From this perspective this chapter re-examines the links between biodiversity and patterns of infectious disease number, but also on zoonotic disease outbreaks, by considering the prediction that the number of zoonotic disease outbreaks should be associated with indices of human development and biodiversity loss. In fact, number of infectious diseases and zoonotic outbreaks are different things, even though most human infectious diseases are zoonotic. Such global trends should be discussed in relation to the ecological and epidemiological mechanisms that function at the local scale, which will be re-examined with a review on the effect of biodiversity reduction on the spread of infectious diseases. Finally, we discuss how biodiversity is related to emerging zoonotic diseases.

3.2 Re-examining the Link Between Biodiversity and the Number of Infectious Diseases

3.2.1 Data and First Analyses to Identify Highly Correlated Variables

Data on infectious diseases were obtained from GIDEON (Global Infectious Diseases and Epidemiology Network, www.gideononline.com), which contains information on the presence and occurrence of epidemics of human infectious diseases in each country. This source has regularly been used in comparative studies of pathogen diversity and epidemics (Guernier et al 2004; Dunn et al. 2010; Morand et al. 2013). We collected data on socioeconomic, demographic, biodiversity and environmental variables to use as explanatory factors for the disease data. Our variables included country area and population, *per capita* gross domestic product (GDP), *per capita* health expenditure, forested surface area and forest as a proportion of each country's total area from the World Bank (worldbank.org/ddp/home.do); latitude, area, and elevation from the Harvard Center for International Development (Gallup et al. 1999); values of mean temperature and precipitation from WorldClim (Hijmans et al. 2005); values for evapotranspiration, a proxy of biological productivity, from the FAO (www.fao.org); the number of languages spoken in a country, as a measure of cultural diversity and complexity, which was found to be linked with biodiversity at threat (Sutherland 2003), was obtained from the Ethnologue (www.ethnologue.com); bird and mammal species richness came from Bird Life International (http://birdlife.org/datazone/home) and from the International Union for Conservation of Nature (www.iucnredlist.org/initiatives/mammals); and species at threat from the International Union for Conservation of Nature. Maps of bird and mammal species richness, number of threatened species, number of languages, and number of infectious diseases show that each of these variables vary widely between countries and seem to correlate each other (Fig. 3.1, but see below).

Per capita healthcare expenditure and the number of diseases for which a survey has been conducted in a country (GIDEON) were used as measures of investigation effort. To avoid potential confounding effects of nation size, we converted all variables potentially linked to nation size, such as number of infectious diseases, bird and mammal richness, number of languages, number of species at threat to a ratio based on nation size (per square kilometre).

A Principal Component Analysis (PCA) allowed the identification of highly correlated variables such as latitude and temperature (Fig. 3.2a), biodiversity and number of languages, and GPD and health expenditure. These results were confirmed using cross correlation analyses (Fig. 3.2b, c). The closest associations were found between bird and mammal species richness, language diversity and number of infectious diseases. This is in agreement with studies showing that cultural biodiversity is linked with natural biodiversity (Collard and Foley 2002; Harmon and Maffi 2002; Maffi 2005) and with biodiversity under threat (Sutherland 2003). In contrast to the correlation of number of diseases with bio- and cultural diversity, the number of outbreaks was more closely associated to the number of species at threat (Fig. 3.2a).

This first result permits us to test several hypotheses using general linear models (GLMs) to explain bird and mammal species richness, number of species at threat, number of diseases, and number zoonotic of outbreaks as a function of potential explanatory variables selected from the above PCA.

3.2.2 Main Variables Linked to Biodiversity

We performed general linear modelling to explain the richness of birds and mammals with human population density, mean temperature, mean precipitation, evapotranspiration (i.e. the amount of water removed from a soil surface through

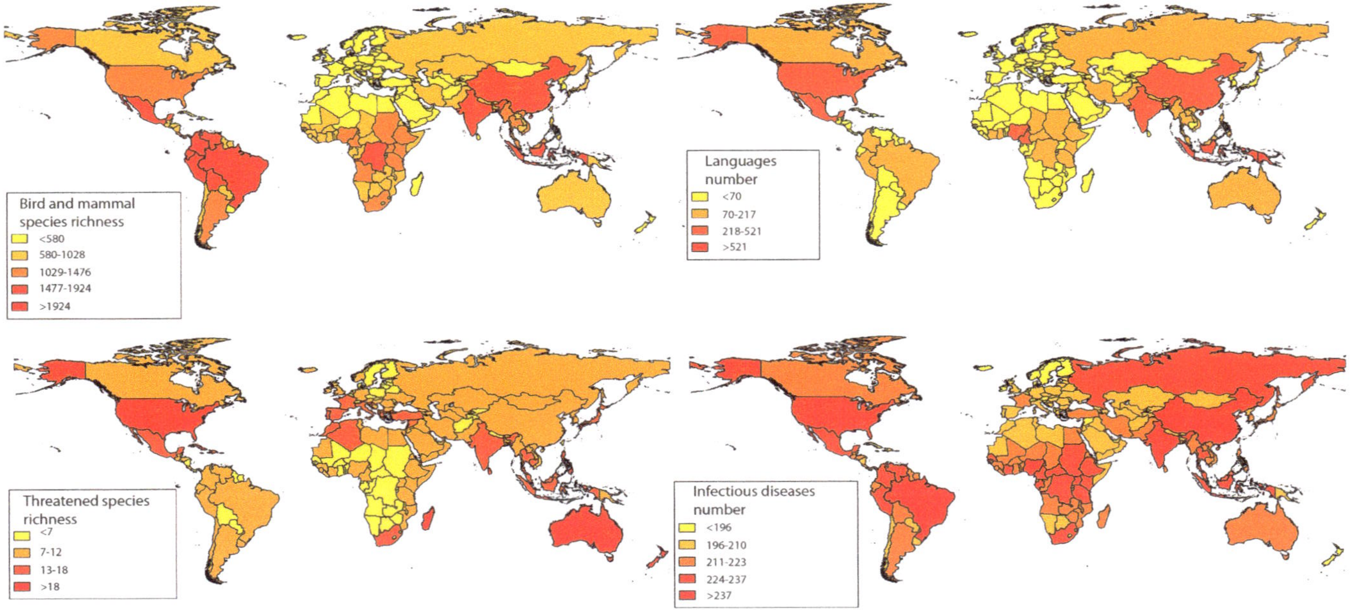

Fig. 3.1 Richness maps of bird and mammal species, languages, vertebrate species at threat, and human infectious diseases (the data are from Bird Life International, the Ethnologue, IUCN and GIDEON, see paragraph 3.3 for details)

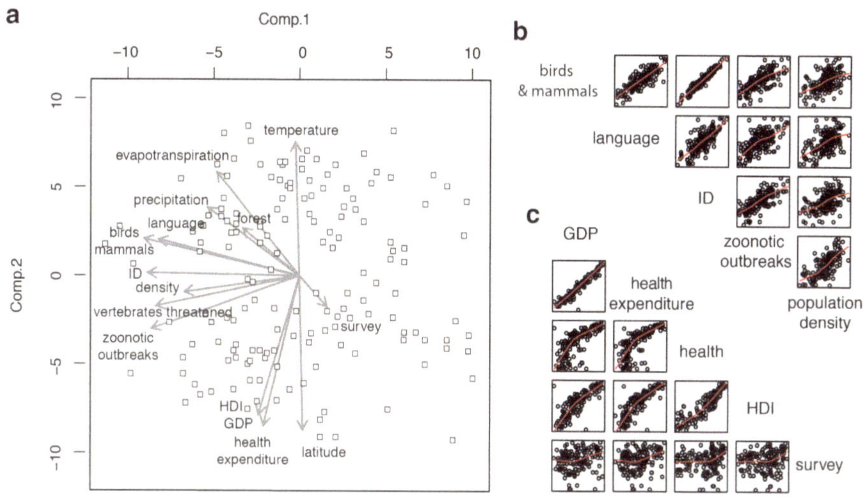

Fig. 3.2 (**a**) Principal Component Analysis on environmental, biodiversity, and socio-economic variables potentially linked with the infectious diseases and their outbreaks. Variables correlated with nation size were controlled by using their ratio to nation size (i.e. expressed as a density). See Sect. 3.3 for the sources of variables. (**b**) Cross correlation among variables concerning richness or number: birds and mammals, languages, infectious diseases, zoonotic outbreaks, and threaten vertebrates. (**c**) Cross correlation among socio-economic variables: GPD, health expenditure, health development index, human development index, and number of surveys on infectious diseases. See Sect. 3.3 for the sources of variables

evaporation and through transpiration) estimated using remote census methods, and GDP as potential explanatory variables. The best model, selected using Akaike's Information Criterion (AIC), showed that bird and mammal species richness (independent of nation size) is linked to human population density and evapotranspiration (Table 3.1). High productivity areas, using evapotranspiration as a proxy, support high biodiversity. Interestingly, human population density appears to be positively correlated with biodiversity independent of nation area, in agreement with previous studies showing that biodiversity hotspots are situated in high human? population density areas (Cincotta et al. 2000).

3.2.3 Main Variables Linked to the Number of Human Infectious Diseases

The best model showed that the number of infectious diseases (controlled for nation size) is statistically associated with bird and mammal richness, climate (precipitation and temperature) and GDP (Table 3.2). It appears that a country with high biodiversity (in terms of birds and mammals) relative to its size also has a greater diversity of infectious diseases (Fig. 3.3a), confirming the pattern already found at the global level by Dunn et al. (2010).

Table 3.1 Summarized results for a generalized linear model (GLM) explaining the richness density of bird and mammal species (expressed as the ratio of a country's richness to its area)

Dependent variable	Explanatory variables	Effect (P)	ANOVA F (P)	F, R^2,P
Species richness density in birds and mammals	Human population density	0.46 (<0.0001)	58.4 (<0.0001)	
	Evapotranspiration	0.002 (<0.0001)	25.5 (<0.0001)	$F_{2,152} = 41.93$, 0.36, <0.00001

Initial variables were population density, mean temperature, mean precipitation, evapotranspiration, and *per capita* gross domestic product (GDP). Selection of the model was done using Akaike's Information Criterion (AIC)

Table 3.2 Summarized results for GLMs explaining the number of infectious diseases (expressed as the ratio of a country's infectious disease richness to its area) and of zoonotic diseases with outbreaks (expressed as the ratio of the number of zoonotic diseases with outbreaks in a country to its area)

Dependent variable	Explanatory variables	Effect (P)	ANOVA F (P)	F, R^2,P
Number of infectious diseases (ratio to nation size)	Richness density in birds and mammals	1.18 (<0.0001)	3,552.2 (<0.0001)	
	Temperature	−0.01 (<0.0001)	82.9 (<0.0001)	
	Precipitation	−0.26 (<0.0001)	76.4 (<0.0001)	
	GDP	0.04 (0.02)	5.4 (0.02)	$F_{4,150} = 879.3$, 0.96, <0.00001
Number of zoonotic diseases with outbreaks (ratio to nation size)	Richness of infectious diseases (ratio to nation size)	0.63 (<0.0001)	589.3 (<0.0001)	
	Human population density	0.24 (<0.0001	138.3 (<0.0001)	
	Temperature	−0.02 (<0.0001)	65.1 (<0.0001)	
	Precipitation	0.23 (<0.0001)	28.2 (<0.0001)	
	GDP	0.16 (<0.0001)	70.9 (<0.0001)	
	Survey	0.23 (<0.0001)	36.4 (<0.0001)	$F_{6,148} = 154.7$, 0.86, <0.0001

Initial variables were human population density, language diversity, bird plus mammal species richness, mean temperature, mean precipitation, evapotranspiration, surveys, GDP and *per capita* health expenditure, and for the zoonotic disease outbreak model, number of infectious diseases. Model selection was done using the AIC criterion

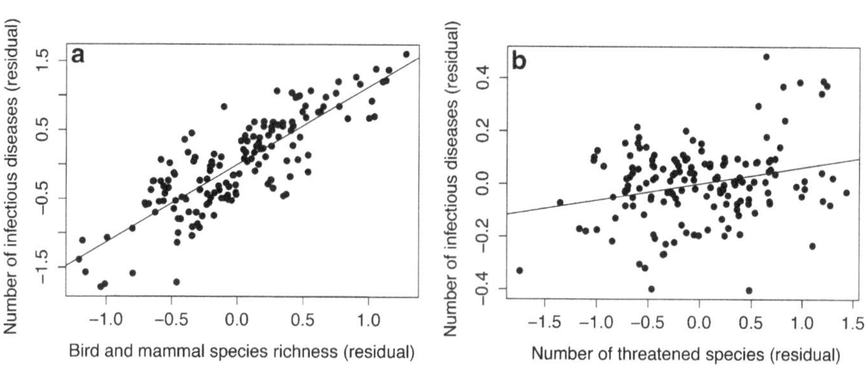

Fig. 3.3 (**a**) Relationship between the number in infectious diseases (residuals of GLM in Table 3.2) and the richness in birds and mammal species (residuals of GLM from Table 3.1). (**b**) Relationship between the number of infectious diseases (residuals of GLM in Table) and the number of vertebrate species at threat (residuals of GLM from Table 3.2)

3.2.4 Main Variables Linked to Zoonotic Outbreaks

Over the period from 1950 to 2008, there was a dramatic increase in both the number of outbreaks and the number of diseases causing outbreaks through time (Fig. 3.4) taken into account reporting and surveillance, which is in accordance with the global trend for EID (Jones et al. 2008). In the same manner as above, a GLM was used to identify explanatory factors of the number of zoonotic outbreaks, expressed as the ratio of the number of diseases with outbreaks to the nation area.

The number of zoonotic infectious diseases with outbreaks was correlated positively with infectious diseases, human population density, climate (positively with precipitation and negatively with temperature), GDP, and surveys (Table 3.2b). Population density is a well-known factor driving disease dynamics (Anderson and May 1978), and we find that increasing density increases the number of diseases with outbreaks. The inclusion of precipitation and temperature underscore the importance of climate in the occurrence of epidemics. Economic wealth, i.e. GDP, was also found positively correlated with the total number of diseases with outbreaks. Economic development potentially impacts the epidemiological environment in a variety of ways. One is that with increased economic resources, diseases are simply detected more frequently, which may be the case here since the number of zoonotic disease outbreaks is also positively correlated with surveys.

3.2.5 Main Variables Linked to Species at Threat and Human Infectious Diseases

We performed a GLM on the number of species at threat, expressed as a ratio to nation area, as dependent on the following factors: human population density, bird and mammal species richness, language diversity, number of infectious diseases,

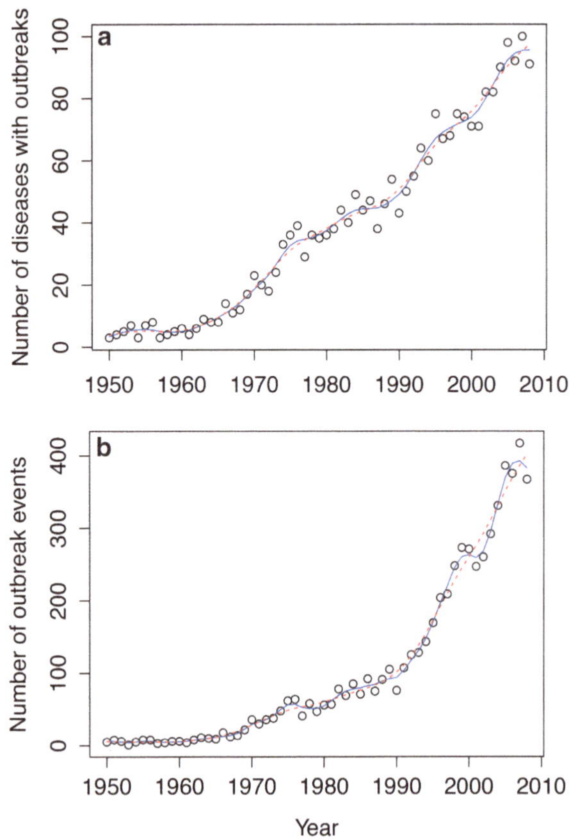

Fig. 3.4 (**a**) Changes in the number of diseases with at least one outbreak from 1950 to 2009 (data from GIDEON). (**b**) Changes in the number of total reported outbreaks from 1950 to 2009 (data from GIDEON)

number of zoonotic disease outbreaks, temperature, precipitation, evapotranspiration, forest cover and GDP. The best model explaining the number of species at threat selected three variables, all positively correlated with the number of threatened species: population density, number of human infectious diseases, and GDP (Table 3.3). Here, both population density and economic development are linked with the increasing threat to vertebrate species (see also Fig. 3.2a).

3.2.6 Capturing the Complexity of Human Zoonotic Outbreaks: Human Population Density and Development Matter

We performed a structured equation modelling (SEM) using the results of the GLMs presented in Tables 3.1, 3.2, and 3.3. SEM is a useful modelling technique to propose a causal chain of correlation. By fitting structural equation models, we show that population size has a strong effect on bird and mammal richness,

Table 3.3 Summarized results for a GLM explaining the number of vertebrate species at threat (ratio to the nation size)

Dependent variable	Explanatory variables	Effect (P)	F (P)	F, R^2,P
Number of species at threat (ratio to nation size)	Population density	0.39 (<0.0001	206.7 (<0.0001)	
	Number of infectious diseases (ratio to nation size)	0.75 (<0.0001)	198.4	
	GDP	0.19 (=0.0001)	15.7 (<0.0001)	$F_{3,151} = 140.3$, 0.73, <0.0001

Initial variables were human population density, language diversity, bird plus mammal richness, number of infectious diseases, number of zoonotic diseases with outbreaks, forest cover, mean temperature, mean precipitation, GDP and per capita health expenditure. Model selection was done using AIC criterion

vertebrate species at threat and zoonotic disease outbreaks (Fig. 3.5). The only climate variable significant in the SEM is precipitation, which has an effect on both bird and mammal species richness and zoonotic outbreaks. GDP *per capita* impacts the number of zoonotic outbreaks but also the number of threatened vertebrate species. GDP also has a positive effect on the number of surveys, which in turn helps at better recording outbreak events.

A simple causal chain seems to emerge from this analysis with increased population density and GDP threatening biodiversity and population density favouring epidemics. Indeed, positive population density effect on disease invisibility and spread has been theoretically formalized and empirically observed for numerous infectious diseases (Anderson and May 1978, 1991).

3.3 Mechanisms of Biodiversity as Insurance for Health

Among the many factors expected to affect parasitic transmission under natural conditions, abiotic factors such as climate (Munson et al. 2008; Shamam et al. 2010) and biotic factors including sex, age, behaviour, density, and community structure of hosts or parasites (Altizer et al 2003; Keesing et al. 2006; Jones et al. 2008; Ezenwa et al 2010) have all been stressed as important determinants of exposure and/or susceptibility to parasites. Importantly, anthropogenic drivers are more and more frequently identified as strong drivers of emerging diseases through mechanisms that favour parasitic persistence or transmission (Rohr et al. 2008; Acevedo-Whitehouse and Duffus 2009). Another factor potentially influencing parasitic diseases is host diversity, which was first presented as a buffer for parasite infection with the pioneering work of Elton (1958) who proposed the "disease diversity hypothesis" in plants. This hypothesis has found later echoes with the "dilution effect" in animals.

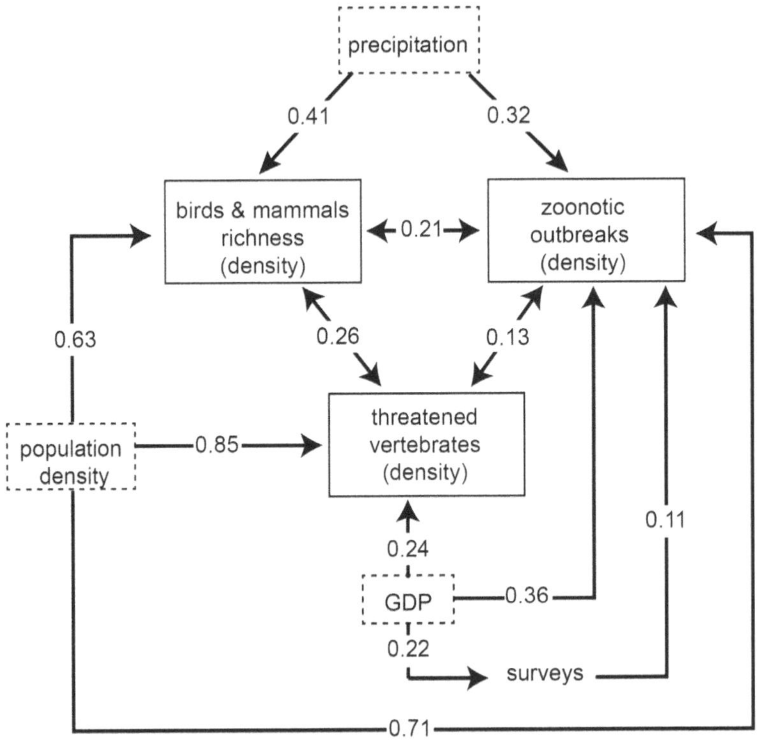

Fig. 3.5 Structured equation modeling (SEM) illustrating standardized coefficients for pathways among variables retained within the general linear modeling of Tables 3.1, 3.2, and 3.3 (all coefficients were significant at $P < 0.001$). Precipitation, population density and GDP appear at the principal correlates of zoonotic outbreaks

The expected links between host diversity and parasite transmission may be quite complex. On one hand, a higher diversity of hosts has been linked to a larger pool of novel pathogens. This positive link is often named the "amplification effect" (Ogden and Tsao 2009). On the other hand, biodiversity can be linked to reduced pathogen transmission through a "dilution effect" (Keesing et al. 2006, 2010). A dilution effect may occur when the addition of one or more host species to a host community is linked to "wasted transmissions" which negatively affect pathogen persistence, despite the presence of highly competent reservoir host species (Keesing et al. 2010). The initial proposed mechanism was a "direct" dilution effect with a lost of infective stages in "wrong" arthropod vectors (i.e. Lyme disease, West Nile fever). Other mechanisms were proposed for directly transmitted diseases, often named "indirect dilution effects". For instance, higher host diversity may lead to a reduction of the susceptible host density via interspecific competition or predator–prey interactions, which in turn will decrease the transmission (Begon et al. 2002).

Table 3.4 Examples of studies showing a link between a decrease of biodiversity and an increase of parasite/pathogen transmission (prevalence), and are therefore in agreement with the "dilution effect" hypothesis

Host species	Locality	Parasite/pathogen	References
Rodents	USA	Hantavirus (Sin Nombre Virus)	Suzán et al. (2009)
Rodents	USA	Hantavirus (Sin Nombre Virus)	Dizney and Ruedas (2009)
Rodents	Southeast Asia	Hantavirus (Seoul Virus)	Blasdell et al. (2011)
Rodents	USA	Hantavirus (Sin Nombre Virus)	Clay et al. (2009)
Rodents	Belgium	Hantavirus (Puumala Virus)	Tersago et al. (2008)
Rodents	USA	*Bartonella* spp (bacteria)	Bai et al. (2009)
Rodents and other mammal species	USA	*Borrelia burgdoferi* (Lyme disease)	Ostfeld and Keesing (2000), Lo Giudice et al. (2008), Keesing et al. (2006)
Small mammals	USA	*Anaplasma phagocytophilum* (bacteria)	Foley et al. (2008)
Birds	USA	West Nile virus	Ezenwa et al. (2006)
Birds	USA	West Nile virus	Allan et al. (2003)
Birds	USA	West Nile virus	Swaddle and Calos (2008)
Birds	Germany	*Himasthla elongata* (helminth)	Thieltges et al. (2008)
Amphibians	USA	*Batrachochytrium dendrobatidis* (fungi)	Searle et al. (2011)
Amphibians	USA	*Ribeiroia ondatrae* (helminth)	Johnson et al. (2013)
Snails	Laboratory	*Schistosoma mansoni* (helminth)	Johnson et al. (2009)
Snails	New Zealand	*Microphallus* sp. (helminth)	Kopp and Jokela (2007)

Several studies found strong support for a dilution effect (direct or indirect) (Table 3.4). This strong pattern encourages some scientists to infer that a consistent picture has emerged: biodiversity loss tends to increase pathogen transmission and disease incidence (Keesing et al. 2010). The local conditions of host species richness and composition in reservoir hosts may be the determining factors as to whether an increased number of potential hosts will divert transmission from highly competent reservoir host species and results in fewer new infections or not.

3.4 Discussion

3.4.1 Biodiversity and Infectious Diseases

Several studies have emphasized the need to preserve vertebrate biodiversity and community composition in order to significantly reduce the risk of infectious disease emergence (LoGiudice et al 2003; Keesing et al 2010). The results presented here agree with previous analyses (Dunn et al. 2010), that show biodiversity appears to be a main explanatory factor for the diversity of human infectious diseases. A country with high bird and mammal diversity, independently of its area, likely also harbours a high number of vectors and reservoirs, which constitute the essential elements for the transmission of infectious diseases.

Dunn et al. (2010) analyzed the diversity of human parasites by nation. Their results indicated that the diversity of human pathogens in a nation increased with human population size and the diversity of birds and mammals in a nation. We found the same result for biodiversity, but instead of population size, we looked at population density, which appears not to be a determinant of the diversity of pathogens but instead a determinant of the number of zoonotic disease outbreaks. We found that the density of bird and mammal species is positively linked with population density. Hence, the relationship between high pathogen richness and high biodiversity may be due to the fact that a nation with high biodiversity is generally related to high population density (Cincotta et al. 2000).

We encountered here a classical pitfall of correlative studies where two variables may be linked only because they are in fact linked to a third one. From this perspective, the pattern that pathogens are more abundant in the tropics could be linked not to higher biodiversity in tropical areas but rather to higher human population densities encountered in those parts of the Earth. As host density is linked with disease spread (Anderson and May 1978) and is also a strong determinant of parasite species richness across mammal species (Bordes and Morand 2011) these results seem to be rather related.

The increase in the number of outbreaks, both total number and number of diseases presenting outbreaks over the last 60 years, is in agreement with results from studies carried out worldwide (Morand and Waret-Szkuta 2012; Morand et al 2013). Several explanatory factors have been identified for the increase in outbreaks, including growing urbanization, forest alteration, agriculture intensification and global trade. It is important to note, however, that the increase in epidemic diseases is strongly correlated with improvements in disease detection and increased healthcare expenditures, reflected in part by the positive effect of GDP, which is strongly correlated with *per capita* health expenditure. The results presented here, particularly the SEM, strongly suggest that increases in population and wealth (GDP) both threaten biodiversity, whereas increases in population favour epidemics (Anderson and May 1991). This is an important qualification to some recent analysis showing positive impacts of development (and globalization) on health (Martens et al. 2010).

3.4.2 *Implications for Human Well-Being*

The ecosystem approach has been endorsed by the Millennium Ecosystem Assessment (MEA 2005) as a conceptual framework. This approach defines ecosystem services as the benefits that people obtain from healthy ecosystems. According to the Millennium Ecosystem Assessment (MEA 2005), one such ecosystem service is the regulation of human and animal diseases, a benefit largely due to positive effects of biodiversity on disease regulation. The MEA also notes that any changes in ecosystems can directly affect human pathogens.

This and the results presented in this chapter suggest that although biodiversity is a source of pathogens, well-preserved biodiversity may act as insurance against outbreaks. However, much work is necessary to confirm these results at regional and local levels where land planning can integrate the management of biodiversity as a way to improve human health.

Combining results of these studies on species richness, emergence, and incidence, we see that biodiversity has a complex relationship with human disease, serving both as a source of pathogens and a regulating factor that keeps incidence down.

Acknowledgment This work was part of the BiodivHealthSEA project (http://www.biodivhealthsea.org) funded by the French ANR programme CP&ES (grant number ANR 11CPEL 002). We thank also the ATPd "Emergence and Risques Sanitaires" of CIRAD.

References

Acevedo-Whitehouse K, Duffus ALJ (2009) Effects of environmental change on wildlife health. Philos Trans R Soc Lond B Biol Sci 364:3429–3438

Allan BF, Keesing F, Ostfeld RS (2003) Effect of forest fragmentation on Lyme disease risk. Conserv Biol 17:267–272

Altizer S, Nunn CL, Thrall PH et al (2003) Social organization and parasite risk in mammals: integrating theory and empirical studies. Annu Rev Ecol Evol Syst 34:517–547

Anderson RM, May RM (1978) Regulation and stability of host-parasite population interactions: I. Regulatory processes. J Anim Ecol 47:219–247

Anderson RM, May RM (1991) Infectious diseases of humans: dynamics and control. Oxford University Press, Oxford

Bai Y, Kosoy MY, Calisher CH et al (2009) Effects of rodent community diversity and composition on prevalence of an endemic bacterial pathogen *Bartonella*. Biodiversity 10:3–9

Begon D, Benett M, Bowers RG et al (2002) A clarification of transmission terms in host-microparasite models: numbers, densities and areas. Epidemiol Infect 129:147–153

Blasdell KR, Cosson JF, Chaval Y et al (2011) Rodent-borne hantaviruses in Cambodia, Lao PDR, and Thailand. Ecohealth 8:432–443

Bordes F, Morand S (2011) The impacts of multiple infections on wild animal hosts: a review. Inf Ecol Epidem 1:7346

Bordes F, Guégan JF, Morand S (2011) Microparasite species richness in rodents is higher at lower latitudes and is associated with reduced litter size. Oikos 120:1889–1896

Chivian E (2003) Biodiversity: its importance to human health. Harvard Med School, Boston

Chivian E, Bernstein AS (2004) Embedded in nature: human health and biodiversity. Environ Health Perspect 112:12–13
Cincotta RP, Wisnewski J, Engelman R (2000) Human population in the biodiversity hotspots. Nature 404:990–992
Clay C, Lehmer EM, St Jeor S et al (2009) Sin Nombre virus and rodent species diversity: a test of the dilution and amplification hypotheses. PLoS One 4:e6467
Collard IF, Foley RA (2002) Latitudinal patterns and environmental determinants of recent human cultural diversity: do humans follow biogeographical rules? Evol Ecol Res 4:371–383
Daszak P, Cunningham AA, Hyatt AD (2000) Emerging infectious diseases of wildlife: threats to biodiversity and human health. Science 287:433–449
Dizney LJ, Ruedas LA (2009) Increased host species diversity and decreased prevalence of Sin Nombre virus. Emerg Infect Dis 15:1012–1018
Dunn RR, Davies TJ, Harris NC et al (2010) Global drivers of human pathogen richness and prevalence. Proc R Soc Biol Sci B 277:2587–2595
Elton CS (1958) The ecology of invasions by animals and plants. Methuen, London
Ezenwa VO, Godsey MS, King RJ et al (2006) Avian diversity and West Nile virus: testing associations between biodiversity and infectious disease risk. Proc R Soc Lond B 273:109–117
Ezenwa VO, Etienne RS, Luikart G et al (2010) Hidden consequences of living in a wormy world: nematode- induced immune suppression facilitates tuberculosis invasion in African Buffalo. Am Nat 176:613–624
Foley JE, Nieto NC, Adjemian J et al (2008) *Anaplasma phagocytophilum* infection in small mammal hosts of ixodes ticks, Western United States. Emerg Infect Dis 14:1147–1150
Gallup JL, Mellinger AD, Sachs JD (1999) Geography datasets. In. Murray Research Archive
Guernier V, Hochberg ME, Guégan JF (2004) Ecology drives the worldwide distribution of human diseases. PLoS Biol 2:740–746
Harmon D, Maffi L (2002) Are linguistic and biological diversity linked? Conserv Biol Pract 3:26–27
Hijmans RJ, Cameron SE, Parra JL et al (2005) Very high resolution interpolated climate surfaces for global land areas. Int J Climat 25:1965–1978
Johnson PT, Lund PJ, Hartson RB et al (2009) Diversity reduces *Schistosoma mansoni* transmission and human infection risk. Proc R Soc Lond B 276:1657–1663
Johnson PT, Preston DL, Hoverman JT et al (2013) Biodiversity decreases disease through predictable changes in host community competence. Nature 494:230–234
Jones KE, Patel NG, Levy MA et al (2008) Global trends in emerging infectious diseases. Nature 451:990–994
Keesing F, Holt RD, Ostfeld RS (2006) Effects of species diversity on disease risk. Ecol Lett 9:485–498
Keesing F, Belden LK, Daszak P et al (2010) Impacts of biodiversity on the emergence and transmission of infectious diseases. Nature 468:647–652
Kopp K, Jokela J (2007) Resistant invaders can convey benefits to native species. Oikos 116:295–301
Lafferty KD (2009) The ecology of climate change and infectious diseases. Ecology 90:888–900
Lo Giudice K, Duerr STK, Newhouse MJ et al (2008) Impact of host community composition on Lyme disease risk. Ecology 89:2841–2849
LoGiudice K, Ostfeld RS, Schmidt KA et al (2003) The ecology of infectious disease: effects of host diversity and community composition on Lyme disease risk. Proc Natl Acad Sci U S A 100:567–571
Maffi L (2005) Linguistic, cultural, and biological diversity. Annu Rev Anthropol 29:599–617
Martens P, Akin S-M, Huynen M, Mohsin R (2010) Is globalization healthy: a statistical indicator analysis of the impacts of globalization on health. Global Health 6:16
MEA (2005) Millenium ecosystem assessment. Ecosystems and human well-being: biodiversity synthesis. Island Press, Washigton DC

Morand S, Waret-Szkuta A (2012) Determinants of human infectious diseases in Europe: biodiversity and climate variability influences. Bull Epidém Hebd 12–13:156–159
Morand S, Owers K, Waret-Szkuta A, McIntyre K et al (2013) Climate variability and outbreaks of infectious diseases in Europe. Sci Rep 3:1774
Munson L, Terio K, Kock R et al (2008) Climate extremes promote fatal co-infections during canine distemper epidemics in African lions. PLoS One 3(6):e2545
Ogden NH, Tsao JI (2009) Biodiversity and Lyme disease: dilution or amplification? Epidemics 1:196–206
Ostfeld RS, Keesing F (2000) Biodiversity and disease risk: the case of Lyme disease. Conserv Biol 14:722–728
Patz JA, Daszak P, Tabor GM et al (2004) Unhealthy landscapes: policy recommendations on land use change and infectious disease emergence. Environ Health Perspect 112:1092–1098
Rohr J, Schotthoefer AM, Raffel TR et al (2008) Agrochemicals increase trematode infections in a declining amphibian species. Nature 455:1235–1239
Schipper J, Chanson JS, Chiozza F et al (2008) The status of the world's land and marine mammals: diversity, threat, and knowledge. Science 322:225–230
Searle C, Biga LM, Spatafora JW et al (2011) A dilution effect in the emerging amphibian pathogen *Batrachochytrium dendrobatidis*. Proc Natl Acad Sci U S A 108:16322–16326
Shamam J, Pitzer VE, Viboud C, Grenfell BT, Lipsitch M (2010) Absolute humidity and the seasonal onset of influenza in the continental United States. PLoS Biol 8(2):e1000316
Sutherland WJ (2003) Parallel extinction risk and global distribution of languages and species. Nature 423:276–277
Suzán G, Marcé E, Giermakowski JT et al (2009) Experimental evidence for reduced rodent diversity causing increased hantavirus prevalence. PLoS One 4:e5461
Swaddle JP, Calos SE (2008) Increased avian diversity is associated with lower incidence of human West Nile infection: observation of the dilution effect. PLoS One 3:e2488
Tersago K, Schreurs A, Linard C et al (2008) Population, environmental and community effects on local bank vole (*Myodes glareolus*) Puumala virus infection in an area with low human incidence. Vector Borne Zoonotic Dis 8:235–244
Thieltges DW, Bordalo MD, Hernández AC et al (2008) Ambient fauna impairs parasite transmission in a marine parasite-host system. Parasitology 135:1111–1116
Wilcox BA, Colwell RR (2005) Emerging and reemerging infectious diseases: biocomplexity as an interdisciplinary paradigm. Ecohealth 2:244–257
Wilcox BA, Gubler DJ (2005) Disease ecology and the global emergence of zoonotic pathogens. Environ Health Prev Med 10:263–272
Wolfe ND, Dunavan CP, Diamond J (2007) Origins of major human infectious diseases. Nature 447(7142):279–283

Part II
Understanding Zoonotic Diseases Through a One Health Perspective

Chapter 4
Hantaviruses

Thomas M. Yuill and James N. Mills

Abstract This chapter focuses on hantaviruses of known human health importance. In humans, New World hantaviruses cause hantavirus pulmonary syndrome (HPS) and Old World hantaviruses cause hemorrhagic fever with renal syndrome (HFRS). Hantaviruses are distributed world-wide and are maintained primarily in rodent hosts.

Rodent hosts are persistently infected, transmitting the viruses among themselves, through fighting or contamination of their surroundings, the latter being the most frequent source of human infection. A variety of ecological factors influence the population dynamics of these rodent hosts and, in turn, the dynamics of transmission among them with corresponding risk of human infection.

Keywords Diagnosis • Disease • Hantaviruses • Prevention • Rodents • Transmission

4.1 Background

Hantaviruses and their diseases have interesting histories. These viruses are distributed nearly world-wide and are maintained primarily in rodent hosts (Fig. 4.1). An increasing number of hantaviruses have recently been discovered in association with insectivore (order Soricomorpha) species, but none have been associated with human disease and they will not be discussed here. The hantaviruses of the Old World cause hemorrhagic fever with renal syndrome (HFRS) (Tables 4.1). Hantaviruses are a cause of HFRS throughout much of Europe and Asia (Vapalahti et al. 2003). In 1934, a mild form of HFRS, much later found to be due to Puumala virus (PUUV), was reported in Finland. During World War II, Japanese troops in

T.M. Yuill (✉)
Department of Pathobiological Sciences, School of Veterinary Medicine,
University of Wisconsin-Madison, Madison, WI, USA
e-mail: tmyuill@wisc.edu

J.N. Mills
Population Biology, Ecology, and Evolution, Graduate Division of Biological and Biomedical Sciences, Emory University, Atlanta, GA, USA
e-mail: WildlifeDisease@gmail.com

A. Yamada et al. (eds.), *Confronting Emerging Zoonoses*,
DOI 10.1007/978-4-431-55120-1_4

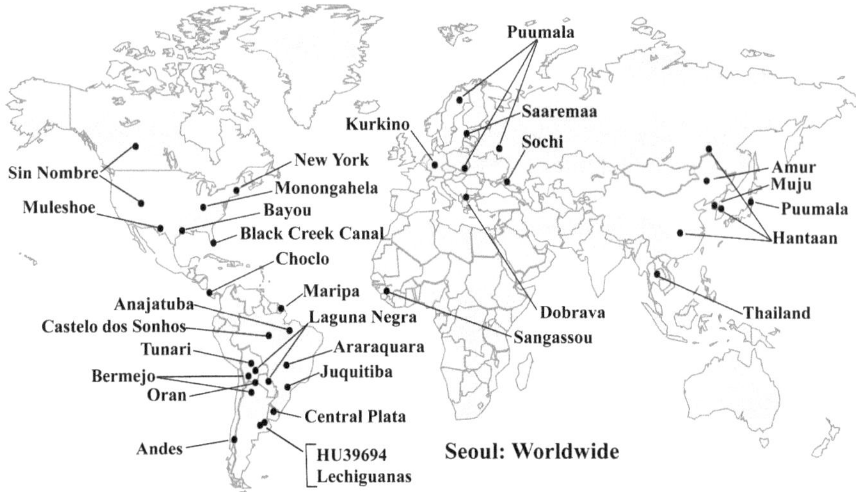

Fig. 4.1 Named hantavirus genotypes that have been associated with human disease

Manchuria succumbed to a more severe disease. Hantaan virus (HTNV) infections causing HFRS with a 10 % case fatality rate were first recognized in 1951 in United Nations troops in Korea, when the disease was termed Korean hemorrhagic fever (KHF). The disease doubtless occurred in Asia perhaps for centuries. From 1951 to 1954, approximately 3,200 troops developed KHF (Schmaljohn and Hjelle 1997). Fatality rates declined gradually, from 4.9 % in 1982 to 2.7 % in 1991, then to 1 % in 1995 through 2007 (Zhang et al. 2010). Many years after the clinical syndrome was recognized, HTNV was first isolated by Lee et al. (1982) and later propagated in cell culture (French et al. 1981).

The isolation of Dobrava virus from a yellow-necked mouse (*Apodemus flavicollis*) was reported in 1992 (Avsic-Zupanc et al. 1992). The isolation of Belgrade virus from a severe human case of HFRS was reported that same year (Gligic et al. 1992). Subsequently, these viruses were shown to be identical and the International Committee for Taxonomy of Viruses (ICTV 2013) designated the name as Dobrava-Belgrade virus (DOBV). Following the isolation of the virus, DOBV was shown to be widely distributed in Europe and eastward in Russia (Tables 4.1), with *A. flavicollis* as the main rodent host. DOBV was isolated from a sibling rodent species, Black Sea field mice (*A. ponticus*), in the Black sea region of European Russia (cited in Klempa et al. 2013). The Saarema isolate of DOBV was found in striped field mice (*A. agrarius*) in two Estonian islands (Plyusnin et al. 1997). A Central European DOBV also was isolated from a striped field mouse (Klempa et al. 2005) Other molecularly distinct DOBV isolates were made from striped field mice from the Kurkino region of Russia (Plyusnin et al. 1999) and from Slovakia (Sibold et al. 1999, 2001). Interestingly, in several European countries where both striped and yellow-necked mice are present in the same areas, their viruses in each are from distinct genetic lineages. Because of the difference in lineages, Klempa et al. (2013) proposed that DOBV be subdivided into four

Table 4.1 Named hantavirus genotypes, their putative rodent host, approximate recognized distribution, and disease for those hantaviruses known or suspected to be pathogenic for humans

Virus	Host scientific name[a]	Host common name[a]	Host family: subfamily	Known distribution of virus	Disease	References
Europe						
Dobrava-Belgrade[b] (DOBV)	*Apodemus flavicollis*	Yellow-necked field mouse	Muridae: Murinae	Balkans N to Slovakia, Czech Republic, Turkey	Moderate/severe HFRS	Avsic-Zupanc et al. (1992), Olsson et al. (2010)
Saaremaa[b] (SAAV)	*Apodemus agrarius (agrarius)*	Striped field mouse	Muridae: Murinae	Saaremaa Island, Estonia	Subclinical?[b]	Olsson et al. (2010), Plyusnin (2002)
Kurkino[b]	*Apodemus agrarius (agrarius)*	Striped field mouse	Muridae: Murinae	Germany, Balkans, N to Estonia, Russia	Mild/moderate HFRS	Klempa et al. (2013)
Sochi[b]	*Apodemus ponticus*	Caucasus field mouse	Muridae: Murinae	SE Russia	Moderate/severe HFRS	Klempa et al. (2008, 2013)
Puumala (PUUV)	*Myodes [Clethrionomys] glareolus*	Bank vole	Cricetidae: Arvicolinae	Europe, Scandinavia, Russia, Balkans,	Mild HFRS (Nephropathia epidemica)	Brummer-Korvenkontio et al. (1980), Olsson et al. (2010)
Asia						
Hantaan (HTNV)	*Apodemus agrarius (mantchuricus)*	Striped field mouse	Muridae: Murinae	Far Eastern Russia, N Asia, Balkans	Severe HFRS	Lee et al. (1978a, b)
Amur[c] (AMRV)	*Apodemus peninsulae*	Korean field mouse	Muridae: Murinae	Far Eastern Russia, Korea[c], China[c]	HFRS	Lokugamage et al. (2004)
Thailand (THAIV)	*Bandicota indica*	Greater bandicoot rat	Muridae: Murinae	Thailand	HFRS	Elwell et al. (1985), Hugot et al. (2006), Pattamadilok et al. (2006)

(continued)

Table 4.1 (continued)

Virus	Host scientific name[a]	Host common name[a]	Host family: subfamily	Known distribution of virus	Disease	References
Seoul[d] (SEOV)	*Rattus norvegicus*[d]	Brown rat	Muridae: Murinae	Nearly Worldwide	Mild/Moderate HFRS	Lee et al. (1982)
Muju (MUJV)	*Myodes regulus*	Korean red-backed vole	Cricetidae: Arvicolinae	South Korea	HFRS	Song et al. (2007)
Africa						
Sangassou (SANGV)	*Hylomyscus alleni* [*simus*]	Allen's hylomyscus	Muridae: Murinae	Guinea	Febrile syndrome	Klempa et al. (2010)
North America						
Sin Nombre (SNV)	*Peromyscus maniculatus*	North American deermouse	Cricetidae: Neotominae	North America	HPS	Childs et al. (1994)
Monongahela[e] (MGLV)	*Peromyscus maniculatus* (*nubiterrae*)	North American deermouse	Cricetidae: Neotominae	E USA and Canada	HPS	Song et al. (1996)
New York[f] (NYV)	*Peromyscus leucopus*[e]	White-footed deermouse	Cricetidae: Neotominae	NE USA	HPS	Hjelle et al. (1995)
Bayou (BAYV)	*Oryzomys palustris*	Marsh oryzomys	Cricetidae: Sigmodontinae	SE USA	HPS	Ksiazek et al. (1997)
Black Creek Canal (BCCV)	*Sigmodon hispidus* (*spadicipygus*)	Hispid cotton rat	Cricetidae: Sigmodontinae	S Florida, USA	HPS	Rollin et al. (1995)
Latin America (Americas South of USA)						
Río Mamoré[g] (RIOMV)	*Oligoryzomys microtis*	Small-eared colilargo	Cricetidae: Sigmodontinae	Bolivia, Peru	None recognized[g]	Hjelle et al. (1996)
Maripa[g]	Unknown		Unknown	French Guiana	HPS	Matheus et al. (2010)

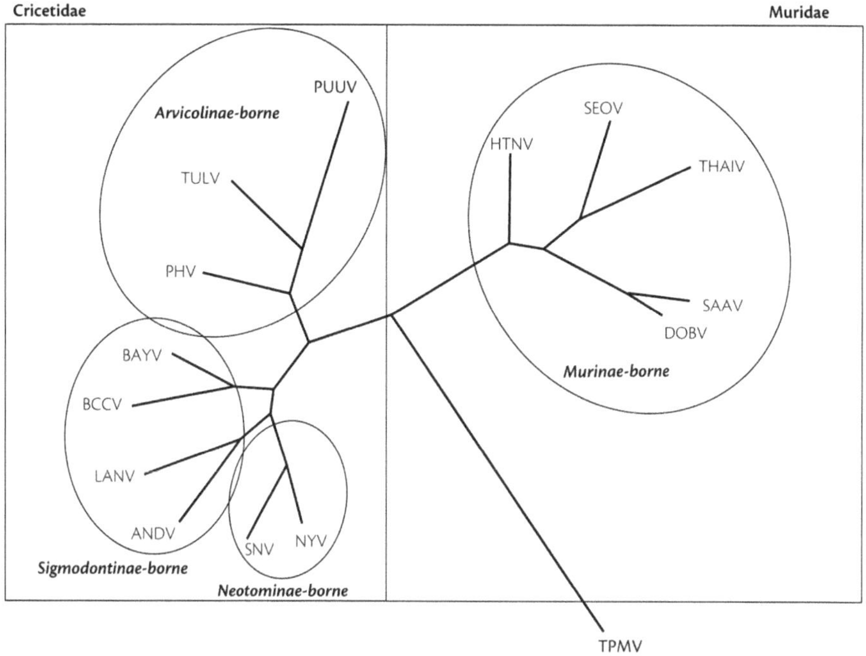

Fig. 4.2 Phylogeny and host relationships of some major hantaviruses. Thottapalayam virus is nonpathogenic for humans and is hosted by an insectivore (order Soricomorpha). All other viruses are hosted by rodents (order Rodentia) of the superfamily Muroidea, families Muridae (Europe and Asia) and Cricetidae. Three subfamilies of Cricetidae host hantaviruses: Sigmodontinae (South America and southern North America) Neotominae (North America), and Arvicolinae (circumboreal). The phylogram is constructed using the complete S segment. *PUUV* Puumala virus, *TULV* Tula virus, *PHV* Prospect Hill virus, *BAYV* Bayou virus, *BCCV* Black Creek Canal virus, *LANV* Laguna Negra virus, *ANDV* Andes virus, *SNV* Sin Nombre virus, *NYV* New York virus, *HTNV* Hantaan virus, *SEOV* Seoul virus, *THAIV* Thai virus, *SAAV* Saaremaa virus, *DOBV* Dobra virus. TULV and PHV are not known to be human pathogens and are not included in Tables 4.1. From: Vaheri et al. (2011)

different hantaviruses or between lineages of the same hantavirus. The evolutionary radiation of selected hantaviruses and association with their rodent hosts is illustrated in Fig. 4.2.

4.3 Pathogenesis in Human Disease

The hantaviruses are responsible for human disease in the Western Hemisphere, Europe, and Asia. Both disease syndromes (HPS and HFRS) involve capillary leakage (Peters et al. 1999). There is evidence that pathogenesis is due to immunopathology, rather than direct cell and tissue damage (Borges et al. 2006;

Kilpatrick et al. 2004; Terajima et al. 2004). It has been proposed (Terajima et al. 2007) that impairment of endothelial cells' defense mechanisms against cytotoxic CD8+ T cells is the mechanism resulting in capillary leakage in HPS and HFRS. Tumor necrosis factor-α, interleukin-6 and -10 and γ-interferon play a role in pathogenesis (Makela et al. 2002; Mori et al. 1999). Studies of 21 HPS patients indicated that IL-6 may have an important role in the pathogenesis of HPS and was associated with fatal outcome. A mixed Th1/Th2 immune response occurred during the course of HPS and the magnitude of Th1 response effector cytokines was correlated to HPS severity. Reduced levels of TGF-beta in HPS patients suggested immunoregulatory damage (Borges et al. 2008).

Clinically, HPS is a severe disease with a case fatality rate of 36 % in the USA. The case definition of HPS (CDC 1997) is a febrile illness characterized by bilateral interstitial pulmonary infiltrates and respiratory compromise usually requiring supplemental oxygen and clinically resembling acute respiratory disease syndrome. Hematological and pulmonary histopathological features were described for a series of fatal HPS cases (Zaki et al. 1995). The incubation period is 1–5 weeks after exposure. Four to ten days after the initial phase of illness, coughing and shortness of breath may occur. The typical prodrome consists of fever, chills, myalgia, headache, and gastrointestinal symptoms with thrombocytopenia progressing rapidly to acute pulmonary edema, hypoxia, respiratory insufficiency, hypotension, and cardiogenic shock. There is hemoconcentration, left shift in the white blood cell count, neutrophilic leukocytosis, thrombocytopenia, and circulating immunoblasts. Because hypoxia induces pulmonary epithelial and endothelial cells to secrete vascular endothelial growth factor it can also increase capillary permeability and edema locally (Duchin et al. 1994; Gavrilovskaya et al. 2012). SNV-specific CD8(+) cells contribute to the severity of HPS disease outcome (Kilpatrick et al. 2004).

In general, HFRS symptoms usually develop 1–2 weeks after infection, but may take up to 8 weeks to appear. Initial symptoms begin suddenly and include intense headaches, back and abdominal pain, fever, chills, nausea, and blurred vision. There may be flushing of the face, inflammation or redness of the eyes, or a rash. Symptoms can progress, including low blood pressure, acute shock, vascular leakage, and acute kidney failure, which can cause severe fluid overload. The severity of the disease varies depending upon the virus causing the infection. Hantaan and Dobrava virus infections usually cause severe symptoms, while Seoul is intermediate, and Saaremaa and Puumala virus infections are usually more moderate (Plyusnin et al. 2006). Complete recovery may take months (CDC 2013b; Muranyi et al. 2005).

Nephropathia epidemica (NE) is a clinically milder form of HFRS. There is sudden onset with high fever, headache, backache, abdominal pain, and transient thrombocytopenia in the early phase of the disease. Three days later, there may be conjunctival hemorrhages and palatine and truncal petechiae. Usually, only 1 % of patients develop severe neurologic signs including seizures or bladder paralysis. Hemorrhages develop with oliguria, azotemia, proteinuria, and hematuria. The rash disappears after the third day and polyuria develops. The convalescent phase lasts for several weeks, usually without sequelae. Fatality from acute renal failure ranges from 0.1 to 1 % (Beers and Berkow 2005). Patients with neurological disease were seen in a large study of HFRS patients infected by Puumala virus. The common symptoms included headache, blurred vision, and vomiting and some of the patients had all three. Nine patients had severe neurological manifestations: meningism and cerebral hemorrhage during the first week of illness, and epileptiform seizures and urinary bladder paralysis during the second week (Alexeyev and Morozov 1995). Impaired pulmonary function has been documented in some HFRS patients (Linderholm et al. 1997; Mustonen et al. 1996) and more than half of 70 PUUV-infected, hospitalized patients had abnormal cardiac findings (Makela et al. 2009). However, one patient with Puumala virus infection who developed a severe clinical syndrome typical of hantavirus pulmonary syndrome (Gizzi et al. 2013).

Severe HFRS disease occurs in 10–15 % of cases. HFRS involves capillaries and venules, causing hemorrhage circulation disorders. Acute renal failure may result from interstitial hemorrhage and infiltrates (Sirotin and Keiser 2001).

The clinical course of severe HFRS has five phases: febrile, hypotensive, oliguric, diuretic, and convalescent. The onset lasts for 3–4 days with high fever, backache, abdominal pain, chills, myalgia, malaise, and bradycardia. Photophobia, pharynx enanthema, and a diffuse flushing of the face occur. Petechia develop on the palate and conjunctival hemorrhages may appear with temporary vision impairment. There is hematuria with gross proteinuria. Hypotension occurs 3–6 days after onset of fever and there may be shock. Laboratory findings include leukocytosis and thrombocytopenia. Renal damage occurs, including acute tubulointerstitial nephritis, necrotizing glomerulonephritis, and IgA nephropathy (Cosgriff 1991). The oliguric phase starts and hemorrhage may become severe. The diuretic phase usually begins on day 11. Convalescence usually begins 3 weeks to 6 months after the acute phase. Sequelae are unusual, but when they occur may include hypertension and chronic renal failure (Beers and Berkow 2005).

4.4 Zoonotic Hantaviruses and Their Rodent Hosts

Hantaviruses establish persistent infection in their rodent hosts without apparent disease (Botten et al. 2000, 2002). Studies suggest that TGF-β1-expressing regulatory T cells may play an important role in limiting immunopathology in the natural reservoir host, but may interfere with viral clearance (Schountz et al. 2007). This response may have arisen as a mutually beneficial, coadaptive, evolutionary process between hantaviruses and their natural rodent reservoirs, limiting disease from the infection while allowing virus persistence (Schountz et al. 2007). Perhaps there has been historic host switching followed by long-term adaptation (Ramsden et al. 2009), with persistent infections and minimal host damage. However, that strict concept has been challenged with new information indicating that host switching can occur currently (Lin et al. 2012; Vapalahti et al. 1996).

Natural and human-induced environmental changes can affect the prevalence and transmission of hantaviruses among their rodent hosts and spillover of the viruses to humans. It has been suggested that regulators of hantavirus prevalence and transmission can be categorized into five major classes (Mills 2005). (1) Environmental regulators (weather and food availability) affect transmission rates through their effect on reproductive success and on population densities. (2) Anthropogenic factors, such as ecosystem disturbance, may simplify those systems, decreasing biodiversity and favoring opportunistic species that may be hantavirus reservoirs. (3) Genetic factors may influence susceptibility of mice to infection or for chronic shedding. (4) Behavior, such as fighting, can increase risk of hantavirus transmission, causing different patterns of infection between males and females. In temperate regions, communal nesting may result in overwinter transmission. (5) Physiologic factors can control host response to infection and length of time the host remains infectious. Thus, risk prediction is difficult because these several regulators often interact. The relative importance of each factor varies according to the status of the host species, season, year, and geographic location.

There is evidence that regulators 1 and 2 above may be interacting synergistically to produce counterintuitive effects on the incidence of HPS. Although high biodiversity has been associated with high risk of emergence of infectious diseases (Jones et al. 2008; Wolfe et al. 2005), outbreaks of HPS and of the rodent borne viral hemorrhagic fevers have been associated with areas with severe anthropogenic disturbance and extremely low biodiversity (Mills 2006). There are several possible explanations for this apparent paradox.

Anthropogenic disturbance is known to simplify natural ecosystems, decreasing biodiversity and resulting in local extirpation of some species. As a result of global climate change, which is increasingly blamed on human factors (Stocker and Qin 2013), numerous species, including rodent hosts of zoonotic diseases, are moving their ranges poleward and upward in altitude (Hickling et al. 2006; Jannett et al. 2007; Moritz et al. 2008). Although a similar migration occurred at the end of the last ice age (Wright et al. 1993); migration is now complicated by massive barriers created through mechanized agriculture, massive deforestation, building of towns and cities and highways. Many species, especially nonvolant specialists, will not be able to migrate (Root and Schneider 2002). The Intergovernmental Panel on Climate Change (IPCC IV) predicts that 20–30 % of species are likely to at increased risk of extinction as global average temperature exceeds 1.5–2.5 °C above pre-industrial levels. Species, however, will not go extinct randomly. Specialist species with narrow habitat and diet requirements will disappear quickly, while more adaptable opportunistic species take their place, rapidly reproducing to high population densities in the absence of their specialist competitors (Dahlberg 1992; Peters 1997; Ruedas et al. 2004). Although opportunistic species constitute a minority of total Cricetid species, they form the majority of known hantavirus hosts in the Americas. Thus, climate change and anthropogenic disturbance may be acting synergistically to create conditions where opportunistic hosts of hantaviruses thrive, increasing risk to humans inhabiting those disturbed habitats.

A phenomenon labeled the "dilution effect" is hypothesized to result in higher risk of vector-borne (Ostfeld and Keesing 2000) and, apparently, directly transmitted zoonotic diseases including HPS (Clay et al. 2009; Mills 2006; Suzan et al. 2009) to humans in ecosystems with lower potential host diversity. Studies of the dilution effect are providing increasing evidence that human perturbation of ecosystems in the form of direct disturbance and indirectly through climate change are resulting in biodiversity loss which leads to greater incidence of HPS and other zoonotic diseases.

4.4.1 The Old World Zoonotic Hantaviruses and Their Rodent Hosts

Three hantaviruses in Europe, Dobrava, Seoul, and Puumala viruses, cause HFRS (Olsson et al. 2010).

4.4.1.1 Hantaan Virus (HTNV)

HTNV and its Asian variants are harbored in mice of the genus *Apodemus* (Lee et al. 1978a, b, 1981a, b). The reservoir of HTNV is *A. agrarius koreae* in eastern Asia including China. In China, the density of *A. agrarius*, the water-level difference in the Huai River and crop production were correlated with the incidence of HFRS (Bi et al. 1998). In another study employing a landscape epidemiologic approach, along with geographic information system and remote sensing techniques, multivariate logistic regression analysis showed that incidence of HFRS from HTNV infection was associated with elevation, a vegetation index, precipitation, annual cumulative air temperature, semihydromorphic soils, timber forests, and orchards (Yan et al. 2007). In farming areas of the Anhui Province of China, where cases of HFRS occurred, there was higher risk of HTNV infection among those who had slept on the ground or had been engaged in heavy farm work (Xu et al. 1985; Chen et al. 1986). HTNV variants Amur and Soochong viruses, both human pathogens, are harbored by *A. peninsulae* (Baek et al. 2006; Zhang et al. 2007). In laboratory studies, laboratory rats (*R. norvegicus*) infected with HTNV shed virus in saliva and feces but longest in urine (Lee et al. 1981a). A study of HFRS related hantaviruses in the Russian Far East has shown the presence of several different genotypes of viruses related to, but with S segments somewhat different from, HTNV (Yashina et al. 2000).

4.4.1.2 Puumala Virus (PUUV)

The bank vole *Myodes glareolus*, the reservoir host of PUUV is widely distributed in northern and central continental Europe (Bernshtein et al. 1999; Brummer-Korvenkontio et al. 1980). As with SNV in deer mice, PUUV infection in adults is associated with wounds in the fall, i.e., at the end of the breeding season, but not in spring. In addition, in a study in southern Belgium, sexually active animals were significantly more often wounded and positive for infection. Wounds resulting from biting or scratching were observed mainly in adult rodents. PUUV infection was associated with higher mobility in juvenile and subadult males (Escutenaire et al. 2002).

HFRS epidemics due to PUUV can be predicted based on the population dynamics of bank voles, the reservoir hosts (Kallio et al. 2009; Olsson et al. 2009). A recently developed model (Sauvage et al. 2007) indicated that only vole populations with multi-annual fluctuations resulted in simultaneously high numbers of infected voles with a high proportion of voles in the acute virus excretion phase as the population increase was ending. This results in a brief peak of exceptionally high virus concentrations in the environment with an increased risk of human exposure. In assessing factors influencing the association of increased PUUV infection in bank voles in a northern Sweden area endemic for the virus, four factors together predicted 80 % of the model outcome: age, body mass index, population abundance (increasing, peak, or declining/low), and sex . Specifically, voles that were in the oldest age class, of higher body mass index, in the peak stage of abundance and of male sex were most likely to be infected (Olsson et al. 2002). In this area in northern Sweden, particular environmental characteristics associated with old-growth moist forests dominated by *Alectoria spp.*, *Picea abies*, fallen wood, and *Vaccinium myrtillus* were associated with increased bank vole abundance and, hence, the number of PUUV-infected voles, whereas dry forests did not have this association (Olsson et al. 2005).

Efficient spread and transmission of PUUV depends on high bank vole population densities that increase as food availability becomes abundant during good mast (edible shrub and tree seeds or nuts) years in central Europe (Schwarz et al. 2009; Tersago et al. 2009). Climate change can affect hantavirus occurrence in rodent hosts and risk of human PUUV infection. Climatic conditions favoring increased mast production have resulted in an increase in HFRS cases a year later (Clement et al. 2009). Increased average temperatures in West-Central Europe have been associated with more frequent PUUV outbreaks with increased bank vole population densities due to increased mast production. Warmer climate, perhaps coupled with increased surveillance, have resulted in an increase in documented cases of HFRS from PUUV infections (Heyman et al. 2009; Schwarz et al. 2009). In contrast, warm winters in Scandinavia have led to a decline in vole populations as a result of reduced protective snow cover with increased predation (Klempa 2009). Similar to findings with deermice and SNV discussed below, a study of wild populations of bank voles in Finland showed that PUUV infected voles had significantly lower overwinter survival (Kallio et al. 2007), slowed maturation, and reduced survival and reproduction compared to uninfected voles (Tersago et al. 2012).

4.4.1.3 Dobrava-Belgrade Virus (DOBV)

DOBV is a genetically distinct hantavirus (Xiao et al. 1993). In Europe, *Apodemus flavicollis*, the yellow-necked forest mouse is the reservoir species of DOBV (Avsic-Zupanc et al. 1992; Taller et al. 1993). Sequence comparison and phylogenetic analysis of the Slovenian wild type DOBV from *A. flavicollis* strains were closely related whereas the strain harbored by *A. agrarius* had a high level of genetic diversity from other Slovenian DOBV strains and clustered together on phylogenetic trees with other DOBV strains harbored by *A. agrarius* from Russia, Estonia, and Slovakia. These findings suggest that the DOBV variants harbored by the two species of *Apodemus* in Europe represent distinct genetic lineages (Avsic-Zupanc et al. 2000). DOBV was recently reported to occur in *A. uralensis* captured on the Black Sea coast of Turkey (Oktem et al. 2014).

4.4.1.4 Seoul Virus (SEOV)

The brown rat (*Rattus norvegicus*) is the main reservoir host of SEOV and is doubtless responsible for moving it around the world. Lin et al. (2012) hypothesized that an ancestor of phylo-group A SEOV variants was first exported from China to Europe and then spread through the New World following the migration of brown rats. Phylogenetic analysis carried out by these authors that included 136 novel Chinese strains, indicated existence of four distinct phylogroups. All non-Chinese SEOV strains and most of the Chinese variants fell into the phylogroup A, while the Chinese strains originating from mountainous areas clustered into three other distinct groups (B, C, and D).

SEOV has also been isolated from wild (Lee et al. 1982) and domestic pet *Rattus rattus* (Jameson et al. 2013). SEOV variants Geo and Serang viruses are associated with *R. rattus* in China and *R. tanezumi* in Indonesia respectively (Leduc 1987; Plyusnin et al. 2009). Transmission between brown rats can occur through exposure to SEOV-containing excrement or by wounds inflicted through fighting (Glass et al. 1988; Hinson et al. 2004). Infection of newborn laboratory *R. norvegicus* resulted in persistent infection, but infection was transitory when adult rats were infected (Kariwa et al. 1996). SEOV was detected in pet rats (*R. norvegicus*) in England (Jameson et al. 2013) and in Sweden (Lundkvist et al. 2013) with a risk of human infection to humans in contact with these pets.

4.4.2 The New World Zoonotic Hantaviruses and Their Rodent Hosts

HPS due to SNV infection was first recognized in 1993 in the Four Corners area of the USA. SNV was shown to be harbored by the North American deermouse, *Peromyscus maniculatus*, a species widely distributed in North America that occurs in a variety of habitats (Childs et al. 1994; Nichol et al. 1993). Subsequently, presence of SNV antibody indicated that deermice were infected throughout their range in the United States, Canada, and Mexico (Mills et al. 1998; Monroe et al. 1999). Studies during the initial outbreak indicated that cases clustered seasonally and temporally by biome type and geographic location. Areas of human exposure and areas of highest prevalence in host populations were most often found in pinyon-juniper woodlands, grasslands, and Great Basin desert scrub lands, at elevations of 1,800–2,500 m (Engelthaler et al. 1999; Mills et al. 1997). Elevation, as well as habitat evaluation accrued from satellite data, showed an association between environmental conditions and HPS risk the following year (Glass et al. 2000). It was hypothesized that the 1991–1992 El Niño-southern oscillation (ENSO) caused increased precipitation, increasing the food supply that led to greater rodent population densities with enhanced risk of transmission to humans. The result of this series of events may have been responsible for the 1993–1994 outbreak of the disease in the Four Corners states. A second strong ENSO occurred in 1997–1998. Researchers who were monitoring rainfall, habitat quality, and deermouse population density in the southwestern US at the time observed increased rainfall, green-up of vegetation, increased food supplies, and an abrupt spike in deermouse population density. The peak in rodent density was followed by a similar peak in HPS cases in the Four Corners states in 1998–1999 (Mills 2005; Yates et al. 2002). Most of the 1998–1999 patients reported indoor exposure to deermice (Hjelle and Glass 2000). Several of the initial (1993) cases in the Four Corners area occurred in Native Americans. Interestingly, the oral history of local American Indian healers describes clusters of similar deaths occurring over three cycles during the twentieth century in association with specific ecological changes (Chapman and Khabbaz 1994).

Subsequent field studies indicated that SNV occurred in an extensive geographic area where *P. maniculatus* is found. However, infections in deermouse populations were focal, being common in some areas and absent in others. Areas of high SNV antibody prevalence in deermouse populations were determined and satellite imagery was used to identify environmental characteristics associated with transmission within its reservoir population. At least 2 years of high-risk conditions were needed for prevalence to increase. Areas with persistently high-risk environmental conditions may serve as maintenance foci for the survival of SNV in local deermouse populations (Glass et al. 2002). Despite very low rodent population densities at some sites, hantavirus infection persisted at a low level, perhaps because of chronic infection and shedding in a few long-lived individuals or by periodic virus reintroduction from neighboring populations. Hantavirus antibody prevalence was seasonal with multiyear patterns suggesting a delayed density-dependent relationship between prevalence and population density (Mills et al. 1999). SNV infection was more prevalent in males and individuals of larger body mass (Douglass et al. 2001; Mills et al. 1997). In deermice, IgG antibody prevalence was positively associated with delayed population density. Virus transmission among deermice, as indicated by seroconversion, occurred primarily during the breeding season (spring through summer) and primarily affected males (Douglass et al. 2007). In some sites, there was a second peak in the in the winter that affected males and females equally (Calisher et al. 1999). This might be explained in that breeding season transmission may involve male-to-male aggression while mid-winter transmission may be associated with communal huddling that involves both sexes. Serological results suggested that the longer deermice live, the greater the probability they will be infected with SNV (Calisher et al. 1999). At numerous sites in the southwestern US, SNV antibody prevalence in deermice was inversely correlated with rodent species richness (number of species) and diversity (a measure combining richness and proportional representation of species) (Calisher et al. 2002; Clay et al. 2009; Mills 2006).

Higher prevalence of SNV infections in male deermice related to virus transmission from aggression was demonstrated in several studies. In longitudinal studies in Colorado and Montana, there was a positive association between wounds and SNV antibody in adult deermice suggesting that when infected rodents fight with uninfected individuals, virus transmission occurs (Calisher et al. 1999; Douglass et al. 2001). As in other studies, male rodents comprised a larger percentage of the total seropositive mice suggesting that male mice contribute more to the SNV epizootic cycle than female mice.

Antibody prevalence and occurrence of SNV RNA in blood have been used as indications of presence of SNV in deermouse populations. However, the antibody response to SNV infection in naturally infected deermice was shown to be highly variable. Presence of SNV RNA has been used as an indicator of relative risk of transmission. Blood levels of viral RNA varied as much as 100-fold, even in individuals infected with identical strains of virus. Deermice that infected other susceptible individuals tended to have a higher SNV RNA levels than those that did not infect other individuals (Bagamian et al. 2012).

Although hantavirus infections have been considered to have little adverse effect on their natural rodent hosts, a field study in Montana showed that recently infected male deermice gained less weight over the 1-month period following seroconversion than did those that did not acquire antibody, suggesting that SNV infection may have negatively impacted the health of infected individuals (Douglass et al. 2007) and decreased survival time (Luis et al. 2012). SNV infected deermice had smaller movement areas on trapping grids than uninfected individuals (Amman et al. 2013), another effect that can influence population dynamics and the maintenance of the virus in nature.

The ecological setting can influence the prevalence of SNV infections in deer mice populations. Studies of SNV infections in *P. maniculatus* populations in California and Nevada characterized the vegetation type and density, elevation, slope, and hydrologic features of study sites using remote sensing and geographic information systems data. The data retroactively predicted infection status of deermice with up to 80 % accuracy (Boone et al. 2000), with potential use as a determinant of risk of human infection.

Numerous other rodents of the families Cricetidae, Muridae, and even Sciuridae (squirrels) have been occasionally found with antibody to SNV, or even SNV RNA, as a result of spillover infection, especially during periods of very high deermouse population density (Childs et al. 1994). It is likely that similar spillover occurs for all hantaviruses and such events in the past may have led to host-switching, coadaptation and the evolution of new hantaviral lineages.

Among the New World hantaviruses, SNV epidemiology is the best studied. After its discovery, other hantaviruses that cause HPS were discovered in various areas of North, Central and South America (Tables 4.1). At least 13 named hantavirus genotypes have been described in 12 species of rodent hosts of the family Cricetidae, subfamily Sigmodontinae (New World rats and mice) in the four countries of the Southern Cone of South America (Palma et al. 2012).

Andes virus is endemic in the Andean region of Chile and Argentina. Andes virus HPS in Chile and Argentina occurs where there is contact with the pygmy rice rat (*Oligoryzomys longicaudatus*) and its habitat in rural areas. The habitat is characterized by shrub vegetation, and forest edge with bamboo (colihue cane, *Chusquea quila*) and brushy peridomestic areas. Population irruptions of these rodents can occur when colihue cane flowers, producing seeds that are a preferred source of food for them (Palma et al. 2012). Population densities of these rodents during population irruptions may reach as much as 100 individuals per ha (Jimenez et al. 1992; Gallardo and Mercado 1999). Interestingly, ANDV antibody prevalence in the long-tailed pygmy rice rats was seasonally highest in the spring, when the population was comprised mainly of adults, than in the autumn, when the populations, although of higher densities, consisted mainly of juvenile individuals. The highest relative seroprevalence of *O. longicaudatus* was found in the Mediterranean ecoregion. Torres-Pérez et al. (2010) postulated that spatial features such as landscape structure and habitat fragmentation are major components in differences in ANDV antibody prevalence in this rodent in Chile. This relationship suggests delayed density dependent effects on antibody prevalence (Mills et al. 2007). The peridomestic habitat places them in close juxtaposition with human dwellings.

In Panama, cases of HPS occur in the Azuero Peninsula where contact with the Costa Rican pygmy rice rat (*Oligoryzomys fulvescens costaricensis*), the host of Choclo virus, and its habitat occur. This rodent is found in anthropogenically disturbed habitats characterized by several crops including corn (maze), watermelon, beans, coffee and sugar cane (Salazar-Bravo et al. 2004).

Other hantaviruses associated with human disease in North, Central, and South America are listed in Table 4.1. The rodent hosts for most of these hantaviruses have been identified, although there is incomplete understanding of host and virus taxonomy.

4.5 Transmission to Humans

Risk of transmission to humans usually depends on presence of viable virus in the environment. Several hantaviruses have been shown to be relatively stable under environmental conditions. Transmission of hantaviruses to humans most frequently occurs through breathing of aerosols of virus-contaminated rodent excreta with exposure often occurring when entering or cleaning rodent-infested structures (Armstrong et al. 1995). Large quantities of infectious virus may be excreted in the urine, saliva, and feces of the infected rodents. Voles infected with PUUV have shown that the shed virus is surprisingly stable, and thereby infectious over long periods outside the rodent host (Kallio et al. 2006). In laboratory studies, cell-cultured HTNV on glass or plastic surfaces remained infectious for 1–3 days at room temperature (Schmaljohn et al. 1998). Similar laboratory experiments with dried PUUV from cell culture gave similar results with all virus becoming noninfectious after 24 h. In contrast, naïve bank voles that were exposed to bedding (wood shavings) contaminated by laboratory infected "donor" voles became infected when exposed to the contaminated bedding up to 15 days after the donor voles had been removed (Kallio et al. 2006). In cotton rats (*Sigmodon hispidus*) infected experimentally with Black Creek Canal virus, virus was sporadically isolated from wet bedding and consistently from dried feces, strongly suggesting that the virus was somewhat stable in the environment (Hutchinson et al. 1998).

These experiments were conducted in the laboratory environment. Exposure to sunlight would likely have rendered the virus noninfectious in a few hours, if not minutes. Virus may be shed by infected hosts at least intermittently for the life of the rodent. The quantity of virus shed may be orders of magnitude higher during the first 2–8 weeks following infection (Lee et al. 1981a; Hutchinson et al. 1998). Peak shedding of PUUV occurred at 11–28, 14–21, and 11–28 days postinfection for saliva, urine, and feces, respectively (Hardestam et al. 2008).

4.6 Diagnosis and Treatment of Hantavirus Infections

4.6.1 Diagnosis

Clinically, the patterns of symptoms, the rate of development of signs and symptoms, and specific clinical laboratory parameters discriminated between individuals having SNV infections and developing HPS versus those suspected of having the infections but later were ruled out (Chapman et al. 1999).

Laboratory diagnosis can be made by serology (detection of immunoglobulin M or rising titers of immunoglobulin G or A) (Padula et al. 2000), by demonstration of hantavirus-specific RNA by polymerase chain reaction (PCR) (Giebel et al. 1990) or of hantavirus antigens by immunohistochemistry (CDC 1997). The requirement for a high level of biological containment for infectious hantaviruses has been a significant constraint in the development of serological diagnostic tests. Use of an expression vector for the nucleocapsid protein gene of provided specific serological reactivity with immune sera (Schmaljohn et al. 1983). The low level of biological containment required for production of this protein offers a significant advantage over live virus antigens for serological diagnosis of hantavirus infections. As a field test, a strip immunoblot assay bearing four immobilized antigens for SNV showed promise for the detection of SNV antibodies early in the course of HPS (Hjelle et al. 1997). Presumptive diagnosis may be made by examination of a peripheral blood smear after the onset of the cardiopulmonary phase of HPS (Mertz et al. 2006). A rapid and easy test for PUUV and ANDV employed an immunochromatographic assay and was useful for the diagnosis of nephrophathy from PUUV infection in Europe. Recently, multiparametric indirect immunofluorescence assays (IFA) based on biochip mosaics were developed to detect serum antibodies against clinically important Old and New World hantaviruses simultaneously. The multiparametric IFA provided highly sensitive and specific serological diagnosis of HTNV, PUUV, SEOV, SAAV, DOBV, SNV or ANDV (Lederer et al. 2013). With both human and animal sera, antibody responses fell into two groups: those that reacted with HTNV, SEOV, and DOBV and those reacting with SNV and PUUV (Elgh et al. 1997).

A real-time reverse transcriptase (RT) PCR method developed for detection of PUUV RNA had a detection threshold for PUUV cDNA of two copies per reaction. A two-step qualitative RT-PCR to detect PUUV RNA showed 100 % agreement with the real-time RT-PCR assay (Evander et al. 2007).

4.6.2 Treatment

Critical care management for HPS includes the avoidance of fluid overload and cardiac output maintained. In severe SNV infections, patients experience extreme pulmonary edema and cardiogenic shock and may require mechanical ventilation and occasionally extracorporeal membrane oxygenation therapy (Koster and Jenison 1997). About 5 % of hospitalized PUUV and 16–48 % of DOBV patients require dialysis and prolonged intensive-care treatment (Vaheri et al. 2013).

Treatment with intravenous ribavirin is probably not effective when initiated during the cardiopulmonary phase of HPS (Mertz et al. 2006). However, ribavirin has been used successfully to reduce mortality and the severity of disease in HFRS patients

4.6.3 Prevention and Control

Vaccines are used to prevent HTNV infections in Asia. To control and prevent HFRS in China, a comprehensive preventive strategy has been implemented and includes public health education and promotion, rodent control, surveillance, and vaccination (Zhang et al. 2010). Approximately two million doses of inactivated rodent brain- or cell culture-derived HFRS vaccines are given annually in China. Although vaccination, along with public education and rodent control measures, have coincided with a reduction in HFRS cases to less than 20,000 per year, China still has the highest number of HFRS cases and deaths in the world (Zhang et al. 2010). A rodent brain-derived inactivated HFRS vaccine has also been used in the Republic of Korea since the early 1990s and has similarly resulted in reduced numbers of HFRS cases (Cho et al. 2002). This vaccine has been used in Korea for military personnel and high-risk rural residents (Cho and Howard 1999; Lee et al. 1990; Sohn et al. 2001). Schmaljohn et al. (1990) found that passively transferred neutralizing monoclonal antibodies prevented infection, suggesting that an antibody response alone can prevent infection.

There are no commercially available vaccines for New World hantaviruses. In experiments designed as a proof of concept, Rhesus macaques were vaccinated with an expression plasmid that contained HTNV and ANDV M genome segments (Custer et al. 2003). The animals developed high levels of neutralizing antibodies that not only neutralized ANDV but also cross-neutralized other HPS-associated hantaviruses, including SNV (Hooper et al. 2006). However, the high cost of development of commercial vaccines is prohibitive considering the relatively small number of HPS cases in the Americas.

Prevention is the best approach to avoid becoming infected with hantaviruses. Although the CDC recommendations are focused on North American hantaviruses, (CDC 2013a) the same measures apply to avoidance of hantavirus infection in other geographic areas. The CDC preventive measures, as well as those of most national and local health agencies include:

- Eliminate or minimize contact with rodents in the house, workplace, or campsite.
- Seal up holes and gaps in the house and garage.
- Place traps in and around the house to decrease rodent infestation.
- Eliminate available rodent food sources.
- Ventilate closed areas that have been unoccupied for a long period of time that may be inhabited by rodents.
- Avoid making dust. Wet down floors and other surfaces with bleach before sweeping.
- Keep food and garbage in rodent-proof thick plastic or metal containers with tight lids.
- Keep outside cooking areas and grills clean.
- Put pet food away after use and do not leave pet-food out overnight.
- Keep garbage in a rodent-proof thick plastic or metal garbage can with a tight lid.
- Keep bird feeders away from the house and utilize squirrel guards to limit access to the feeder by rodents.
- Keep compost bins as far away from the house as possible (30 m or more).

- Keep grains and animal feed in rodent-proof thick plastic or metal containers with tight lids. In the evening, uneaten animal feed should be returned to containers with lids.
- Eliminate possible rodent nesting sites outside the house. Elevate hay, woodpiles, and garbage cans at least 30 cm off the ground.
- Move woodpiles far away from the house (30 m or more).
- Get rid of old trucks, cars, and old tires that could house mice and rats.
- Keep grass cut short and shrubbery within 30 m of the home well trimmed.

References

Alexeyev OA, Morozov VG (1995) Neurological manifestations of hemorrhagic fever with renal syndrome caused by Puumala virus: review of 811 cases. Clin Infect Dis 20(2):255–258

Amman BR, Manangan AP, Flietstra TD, Calisher CH, Carroll DS, Wagoner KD, Mills JN (2013) Association between movement and Sin Nombre virus (*Bunyaviridae*: *Hantavirus*) infection in North American deermice (*Peromyscus maniculatus*) in Colorado. J Wildl Dis 49(1):132–142

Armien B, Pascale JM, Muñoz C, Mariñas J, Núñez H, Herrera M, Trujillo J, Sánchez D, Mendoza Y, Hjelle B, Koster F (2013) Hantavirus fever without pulmonary syndrome in Panama. Am J Trop Med Hyg 89:489–494

Armstrong LR, Zaki SR, Goldoft MJ, Todd RL, Khan AS, Khabbaz RF, Ksiazek TG, Peters CJ (1995) Hantavirus pulmonary syndrome associated with entering or cleaning rarely used, rodent-infested structures [letter]. J Infect Dis 172(4):1166

Avsic-Zupanc T, Xiao SY, Stojanovic R, Gligic A, Vandergroen G, Leduc JW (1992) Characterization of Dobrava virus: a hantavirus from Slovenia, Yugoslavia. J Med Virol 38:132–137

Avsic-Zupanc T, Nemirov K, Petrovec M, Trilar T, Poljak M, Vaheri A, Plyusnin A (2000) Genetic analysis of wild-type Dobrava hantavirus in Slovenia: co-existence of two distinct genetic lineages within the same natural focus. J Gen Virol 81(7):1747–1755

Baek LJ, Kariwa H, Lokugamage K, Yoshimatsu K, Arikawa J, Takashima I, Kang JI, Moon SS, Chung SY, Kim EJ et al (2006) Soochong virus: an antigenically and genetically distinct hantavirus isolated from Apodemus peninsulae in Korea. J Med Virol 78(2):290–297

Bagamian KH, Towner JS, Kuenzi AJ, Douglass RJ, Rollin PE, Waller LA, Mills JN (2012) Transmission ecology of Sin Nombre hantavirus in naturally infected North American deermouse populations in outdoor enclosures. PLoS One 7(10):e47731

Beers MH, Berkow R (2005) Infectious diseases; viral diseases. The merck manual of diagnosis and therapy, 17th edn. Wiley, Indianapolis

Bernshtein AD, Apekina NS, Mikhailova TV, Myasnikov YA, Khlyap LA, Korotkov YS, Gavrilovskaya IN (1999) Dynamics of Puumala hantavirus infection in naturally infected bank voles (*Clethrinomys glareolus*). Arch Virol 144(12):2415–2428

Bi P, Wu X, Zhang F, Parton KA, Tong S (1998) Seasonal rainfall variability, the incidence of hemorrhagic fever with renal syndrome, and prediction of the disease in low-lying areas of China. Am J Epidemiol 148(3):276–281

Bi Z, Formenty PB, Roth CE (2008) Hantavirus infection: a review and global update. J Infect Dev Ctries 2(1):3–23

Boone JD, Mcgwire KC, Otteson EW, Debaca RS, Kuhn EA, Villard P, Brussard PF, St Jeor SC (2000) Remote sensing and geographic information systems: charting Sin Nombre virus infections in deer mice. Emerg Infect Dis 6(3):248–258

Borges AA, Campos GM, Moreli ML, Souza RL, Aquino VH, Saggioro FP, Figueiredo LT (2006) Hantavirus cardiopulmonary syndrome: immune response and pathogenesis. Microbes Infect 8 (8):2324–2330

Borges AA, Campos GM, Moreli ML, Moro Souza RL, Saggioro FP, Figueiredo GG, Livonesi MC, Moraes Figueiredo LT (2008) Role of mixed Th1 and Th2 serum cytokines on pathogenesis and prognosis of hantavirus pulmonary syndrome. Microbes Infect 10(10–11):1150–1157

Botten J, Mirowsky K, Kusewitt D, Bharadwaj M, Yee J, Ricci R, Fedderson RM, Hjelle B (2000) Experimental infection model for Sin Nombre hantavirus in the deer mouse (*Peromyscus maniculatus*). Proc Natl Acad Sci U S A Biol Sci 97(19):10578–10583

Botten J, Mirowsky K, Ye C, Gottlieb K, Saavedra M, Ponce L, Hjelle B (2002) Shedding and intracage transmission of Sin Nombre hantavirus in the deer mouse (Peromyscus maniculatus) model. J Virol 76(15):7587–7594

Brummer-Korvenkontio M, Vaheri A, Hovi T, Von Bonsdorff CH, Vuorimies J, Manni T, Penttinen K, Oker-Blom N, Lähdevirta J (1980) Nephropathia epidemica: detection of antigen in bank voles and serologic diagnosis of human infection. J Infect Dis 141:131–134

Calisher CH, Sweeney W, Mills JN, Beaty BJ (1999) Natural history of Sin Nombre virus in western Colorado. Emerg Infect Dis 5(1):126–134

Calisher CH, Root JJ, Mills JN, Beaty BJ (2002) Assessment of ecologic and biologic factors leading to hantavirus pulmonary syndrome, Colorado, U.S.A. Croat Med J 43(3):330–337

CDC (1997) Case definitions for infectious conditions under public health surveillance. Centers for disease control and prevention. MMWR Recomm Rep 46(RR-10):1–55

CDC (2013a) Hantavirus. http://www.cdc.gov/hantavirus. Accessed 20 May 2013

CDC (2013b) Hemorrhagic fever with renal syndrome. http://www.cdc.gov/hantavirus/hfrs/index.html. Accessed 15 June 2013

CDC (2014) Reported cases of HPS. http://www.cdc.gov/hantavirus/surveillance/index.html. Accessed 14 August 2014

Chapman LE, Khabbaz RF (1994) Etiology and epidemiology of the Four Corners hantavirus outbreak. Infect Agents Dis 3(5):234–244

Chapman LE, Mertz GJ, Peters CJ, Jolson HM, Khan AS, Ksiazek TG, Koster FT, Baum KF, Rollin PE, Pavia AT et al (1999) Intravenous ribavirin for hantavirus pulmonary syndrome: a safety and tolerance during 1 year of open-label experience. Antivir Ther 4:211–219

Chen H-X, Qiu F-X, Dong B-J, Ji S-Z, Li Y-T, Wang Y et al (1986) Epiedmiologic studies on hemorrhagic fever with renal syndrome in China. J Infect Dis 154:394–398

Childs JE, Ksiazek TG, Spiropoulou CF, Krebs JW, Morzunov S, Maupin GO, Rollin PE, Sarisky J, Enscore RE, Frey JK et al (1994) Serologic and genetic identification of *Peromyscus maniculatus* as the primary rodent reservoir for a new hantavirus in the southwestern United States. J Infect Dis 169:1271–1280

Cho HW, Howard CR (1999) Antibody responses in humans to an inactivated hantavirus vaccine (Hantavax). Vaccine 17(20–21):2569–2575

Cho HW, Howard CR, Lee HW (2002) Review of an inactivated vaccine against hantaviruses. Intervirology 45:328–333

Clay CA, Lehmer EM, St Jeor S, Dearing MD (2009) Sin Nombre virus and rodent species diversity: a test of the dilution and amplification hypotheses. PLoS One 4:e6467

Clement J, Vercauteren J, Verstraeten WW, Ducoffre G, Barrios JM, Vandamme AM, Maes P, Van RM (2009) Relating increasing hantavirus incidences to the changing climate: the mast connection. Int J Health Geogr 8:1

Cosgriff TM (1991) Mechanisms of disease in hantavirus infection: pathophysiology of hemorrhagic fever with renal syndrome. Rev Infect Dis 13(1):97–107

Cruz CD, Forshey BM, Vallejo E, Agudo R, Vargas J, Blazes DL, Guevara C, Laguna-Torres VA, Halsey ES, Kochel TJ (2012) Novel strain of Andes virus associated with fatal human infection, central Bolivia. Emerg Infect Dis 18(5):750–757

Cueto GR, Cavia R, Bellomo C, Padula PJ, Suarez OV (2008) Prevalence of hantavirus infection in wild *Rattus norvegicus* and *R. rattus* populations of Buenos Aires City, Argentina. Trop Med Int Health 13(1):46–51

Custer DM, Thompson E, Schmaljohn CS, Ksiazek TG, Hooper JW (2003) Active and passive vaccination against hantavirus pulmonary syndrome with Andes virus M genome segment-based DNA vaccine. J Virol 77(18):9894–9905

Dahlberg KA (1992) The conservation of biological diversity and U.S. agriculture: goals, institutions, and policies. Agric Ecosyst Environ 42:177–193
Delfraro A, Clara M, Tome L, Achaval F, Levis S, Calderón GE, Enría D, Lozano M, Russi J, Arbiza J (2003) Yellow pygmy rice rat (*Oligoryzomys flavescens*) and hantavirus pulmonary syndrome in Uruguay. Emerg Infect Dis 9(7):846–852
Delfraro A, Tome L, D'elia G, Clara M, Achaval F, Russi JC, Rodonz JR (2008) Juquitiba-like hantavirus from 2 nonrelated rodent species, Uruguay. Emerg Infect Dis 14(9):1447–1451
Douglass RJ, Wilson T, Semmens WJ, Zanto SN, Bond CW, Van Horn RC, Mills JN (2001) Longitudinal studies of Sin Nombre virus in deer mouse dominated ecosystems of Montana. Am J Trop Med Hyg 65(1):33–41
Douglass RJ, Calisher CH, Wagoner KD, Mills JN (2007) Sin Nombre virus infection of deer mice in Montana: characteristics of newly infected mice, incidence, and temporal pattern of infection. J Wildl Dis 43(1):12–22
Duchin JS, Koster FT, Peters CJ, Simpson GL, Tempest B, Zaki SR, Rollin PE, Nichol S, Umland ET, Group HS (1994) Hantavirus pulmonary syndrome: a clinical description of 17 patients with a newly recognized disease. New Engl J Med 330:949–955
Elgh F, Lundkvist A, Alexeyev OA, Stenlund H, Vsic-Zupanc T, Hjelle B, Lee HW, Smith KJ, Vainionpaa R, Wiger D et al (1997) Serological diagnosis of hantavirus infections by an enzyme-linked immunosorbent assay based on detection of immunoglobulin G and M responses to recombinant nucleocapsid proteins to five viral serotypes. J Clin Microbiol 35 (5):1122–1130
Elwell MR, Ward GS, Tingpalapong M, Leduc JW (1985) Serologic evidence of Hantaan-like virus in rodents and man in Thailand. Southeast Asian J Trop Med Public Health 16:349–354
Engelthaler DM, Mosley DG, Cheek JE, Levy CE, Komatsu KK, Ettestad P, Davis T, Tanda DT, Miller L, Frampton JW et al (1999) Climatic and environmental patterns associated with hantavirus pulmonary syndrome, Four Corners region, United States. Emerg Infect Dis 5 (1):87–94
Enria D, Padula P, Segura EL, Pini N, Edelstein A, Posse CR, Weissenbacher MC (1996) Hantavirus pulmonary syndrome in Argentina Possibility of person to person transmission. Medicina (B Aires) 56:709–711
Escutenaire S, Chalon P, De Jaegere F, Karelle-Bui L, Mees G, Brochier B, Rozenfeld F, Pastoret PP (2002) Behavioral, physiologic, and habitat influences on the dynamics of Puumala virus infection in bank voles (Clethrionomys glareolus). Emerg Infect Dis 8:930–936
Evander M, Eriksson I, Pettersson L, Juto P, Ahlm C, Olsson GE, Bucht G, Allard A (2007) Puumala hantavirus viremia diagnosed by real-time reverse transcriptase PCR using samples from patients with hemorrhagic fever and renal syndrome. J Clin Microbiol 45(8):2491–2497
Firth C, Tokarz R, Simith DB, Nunes MR, Bhat M, Rosa ES, Medeiros DB, Palacios G, Vasconcelos PF, Lipkin WI (2012) Diversity and distribution of hantaviruses in South America. J Virol 86(24):13756–13766
French G, Foulke R, Brand O, Eddy G (1981) Korean hemorrhagic fever: propagation of the etiologic agent in a cell line of human origin. Science 211:1046–1048
Gallardo MH, Mercado CL (1999) Mast seeding of bamboo shrubs and mouse outbreaks in southern Chile. Mastozool Neotrop 6:103–111
Gavrilovskaya I, Gorbunova E, Koster F, Mackow E (2012) Elevated VEGF levels in pulmonary edema fluid and PBMCs from patients with acute hantavirus pulmonary syndrome. Adv Virol 2012:674360
Giebel LB, Zoller L, Bautz EKF, Darai G (1990) Rapid detection of genomic variations in different strains of hantaviruses by polymerase chain reaction techniques and nucleotide sequence analysis. Virus Res 16:127–136
Gizzi M, Delaere B, Weynand B, Clement J, Maes P, Vergote V, Laenen L, Hjelle B, Verroken A, Dive A, Michaux I, Evrard P, Creytens D (2013) Bulpa P (2013) Another case of "European hantavirus pulmonary syndrome" with severe lung, prior to kidney, involvement, and diagnosed by viral inclusions in lung macrophages. Eur J Clin Microbiol Infect Dis 32:1341–1345

Glass GE, Childs JE, Korch GW, Leduc JW (1988) Association of intraspecific wounding with hantaviral infection in wild rats (*Rattus norvegicus*). Epidemiol Infect 101:459–472

Glass GE, Cheek JE, Patz JA, Shields TM, Doyle TJ, Thoroughman DA, Hunt DK, Enscore RE, Gage KL, Irland C et al (2000) Using remotely sensed data to identify areas at risk for hantavirus pulmonary syndrome. Emerg Infect Dis 6(3):238–247

Glass GE, Yates TL, Fine JB, Shields TM, Kendall JB, Hope AG, Parmenter CA, Peters CJ, Ksiazek TG, Li CS et al (2002) Satellite imagery characterizes local animal reservoir populations of Sin Nombre virus in the southwestern United States. Proc Natl Acad Sci U S A 99(26):16817–16822

Gligic A, Dimkovic N, Xiao SY, Buckle GJ, Jovanovic D, Velimirovic D, Stojanovic R, Obradovic M, Diglisic G, Micic J et al (1992) Belgrade virus: a new hantavirus causing severe hemorrhagic fever with renal syndrome in Yugoslavia. J Infect Dis 166:113–120

Hang CS, Song G, Qiu XZ, Du YL, Zhao JN, Liao HX et al (1982) Investigation of the agent causing mild type of hemorrhagic fever [in Chinese]. Zhonghua Liu Xing Bing Xue Za Zhi 3:204–205

Hardestam J, Karlsson M, Falk KI, Olsson G, Klingstrom J, Lundkvist A (2008) Puumala hantavirus excretion kinetics in bank voles (*Myodes glareolus*). Emerg Infect Dis 14 (8):1209–1215

Hepojoki J, Strandin T, Lankinen H, Vaheri A (2012) Hantavirus structure–molecular interactions behind the scene. J Gen Virol 93(Pt 8):1631–1644

Heyman P, Vaheri A, Lundkvist A, Vsic-Zupanc T (2009) Hantavirus infections in Europe: from virus carriers to a major public-health problem. Expert Rev Anti Infect Ther 7(2):205–217

Hickling R, Roy DB, Hill JK, Fox R, Thomas CD (2006) The distributions of a wide range of taxonomic groups are expanding polewards. Glob Change Biol 12(3):450–455

Hinson ER, Shone SM, Zink MC, Glass GE, Klein SL (2004) Wounding: the primary mode of Seoul virus transmission among male Norway rats. Am J Trop Med Hyg 70(3):310–317

Hjelle B, Glass GE (2000) Outbreak of hantavirus infection in the Four Corners region of the United States in the wake of the 1997–1998 El Niño-Southern oscillation. J Infect Dis 181 (5):1569–1573

Hjelle B, Jenison S, Torrez-Martinez N, Yamada T, Nolte K, Zumwalt R, Myers G (1994) A novel hantavirus associated with an outbreak of fatal respiratory disease in the southwestern United States: evolutionary relationships to known hantaviruses. J Virol 68:592–596

Hjelle B, Krolikowski J, Torrez-Martinez N, Chavez-Giles F, Vanner C, Laposata E (1995) Phylogenetically distinct hantavirus implicated in a case of hantavirus pulmonary syndrome in the northeastern United States. J Med Virol 46(1):21–27

Hjelle B, Torrez-Martinez N, Koster FT (1996) Hantavirus pulmonary syndrome-related virus from Bolivia [letter]. Lancet 347:57

Hjelle B, Jenison S, Torrez-Martinez N, Herring B, Quan S, Polito A, Pichuantes S, Yamada T, Morris C, Elgh F et al (1997) Rapid and specific detection of Sin Nombre virus antibodies in patients with hantavirus pulmonary syndrome by a strip immunoblot assay suitable for field diagnosis. J Clin Microbiol 35(3):600–608

Hooper JW, Custer DM, Smith J, Wahl-Jensen V (2006) Hantaan/Andes virus DNA vaccine elicits a broadly cross-reactive neutralizing antibody response in nonhuman primates. Virology 347 (1):208–216

Hugot JP, Plyusnina A, Herbreteau V, Nemirov K, Laakkonen J, Lundkvist A, Supputthamongkol Y, Henttonen H, Plyusin A (2006) Genetic analysis of Thailand hantavirus in *Bandicota indica* trapped in Thailand. Virol J 3(72)

Hutchinson K, Rollin P, Peters CJ (1998) Pathogenesis of a North American hantavirus, Black Creek Canal virus, in experimentally infected *Sigmodon hispidus*. Am J Trop Med Hyg 59:58–65

ICTV (2013) Virus taxonomy: 2012 release. http://ictvonline.org/virusTaxonomy.asp. Accessed 6 Apr 2013

Jameson LJ, Taori SK, Atkinson B, Levick P, Featherstone CA, Van Der Burgt G, Mccarthy N, Hart J, Osborne JC, Walsh AL et al (2013) Pet rats as a source of hantavirus in England and Wales, 2013. Euro Surveill 18(9)

Jannett FJ, Broschart MR, Grim LH, Schaberl JP (2007) Northerly range extensions of mammalian species in Minnesota. Am Midl Nat 158(1):168–176

Jimenez JE, Feinsinger P, Jaksic FM (1992) Spatiotemporal patterns of an irruption and decline of small mammals in northcentral Chile. J Mammal 173:356–364

Johnson AM, Bowen MD, Ksiazek TG, Williams RJ, Bryan RT, Mills JN, Peters CJ, Nichol ST (1997) Laguna Negra virus associated with HPS in western Paraguay and Bolivia. Virology 238(1):115–127

Johnson AM, De Souza LTM, Ferreira IB, Pereira LE, Ksiazek TG, Rollin PE, Peters CJ, Nichol ST (1999) Genetic investigation of novel hantaviruses causing fatal HPS in Brazil. J Med Virol 59:527–535

Jones KE, Patel NG, Levy MA, Storeygard A, Balk D, Gittleman JL, Daszak P (2008) Global trends in emerging infectious diseases. Nature 451(7181):990–993

Kallio ER, Klingstrom J, Gustafsson E, Manni T, Vaheri A, Henttonen H, Vapalahti O, Lundkvist A (2006) Prolonged survival of Puumala hantavirus outside the host: evidence for indirect transmission via the environment. J Gen Virol 87:2127–2134

Kallio ER, Voutilainen L, Vapalahti O, Vaheri A, Henttonen H, Koskela E, Mappes T (2007) Endemic hantavirus infection impairs the winter survival of its rodent host. Ecology 88 (8):1911–1916

Kallio ER, Begon M, Henttonen H, Koskela E, Mappes T, Vaheri A, Vapalahti O (2009) Cyclic hantavirus epidemics in humans–predicted by rodent host dynamics. Epidemics 1(2):101–107

Kariwa H, Kimura M, Yoshizumi S, Arikawa J, Yoshimatsu K, Takashima I, Hashimoto N (1996) Modes of Seoul virus infections: persistency in newborn rats and transiency in adult rats. Arch Virol 141(12):2327–2338

Kilpatrick ED, Terajima M, Koster FT, Catalina MD, Cruz J, Ennis FA (2004) Role of specific CD8+ T cells in the severity of a fulminant zoonotic viral hemorrhagic fever, hantavirus pulmonary syndrome. J Immunol 172(5):3297–3304

Klempa B (2009) Hantaviruses and climate change. Clin Microbiol Infect 15(6):518–523

Klempa B, Schmidt HA, Ulrich R, Kaluz S, Labuda M, Meisel H, Hjelle B, Kruger DH (2003) Genetic interaction between distinct Dobrava hantavirus subtypes in *Apodemus agrarius* and *A. flavicollis* in nature. J Virol 77(1):804–809

Klempa B, Stanko M, Labuda M, Ulrich R, Meisel H, Kruger DH (2005) Central European Dobrava hantavirus isolate from striped field mouse, Apodemus agrarius. J Clin Microbiol 43:2756–2763

Klempa B, Tkachenko EA, Dzagurova TK, Yunicheva YV, Morozov VG, Okulova NM, Slyusareva GP, Smirnov A, Kruger DH (2008) Hemorrhagic fever with renal syndrome cause by 2 lineages of Dobrava hantavirus, Russia. Emerg Infect Dis 14(4):617–625

Klempa B, Koivogui L, Sylla O, Koulemou K, Auste B, Kruger DH, Ter Meulen J (2010) Serological evidence of human hantavirus infections in Guinea, West Africa. J Infect Dis 201(7):1031–1034

Klempa B, Avsic-Zupanc T, Clement J, Dzagurova TK, Henttonen H, Heyman P, Jakab F, Kruger DH, Maes P, Papa A et al (2013) Complex evolution and epidemiology of Dobrava-Belgrade hantavirus: definition of genotypes and their characteristics. Arch Virol 158(3):521–529

Koster FT, Jenison SA (1997) Hantaviruses. In: Gorbach SL, Bartlett JG, Blacklow NR (eds) Infectious diseases. Saunders, Philadelphia

Kraus AA, Priemer C, Heider H, Kruger DH, Ulrich R (2005) Inactivation of Hantaan virus-containing samples for subsequent investigations outside biosafety level 3 facilities. Intervirology 48(4):255–261

Ksiazek T, Peters CJ, Rollin P, Zaki S, Nichol S, Spiropoulou C (1995) Identification of a new North American hantavirus that causes acute pulmonary insuffieiency. Am J Trop Med Hyg 52:117–123

Ksiazek TG, Nichol ST, Mills JN, Groves MG, Wozniak A, Mcadams S, Monroe MC, Johnson AM, Martin ML, Peters CJ et al (1997) Isolation, genetic diversity, and geographic distribution of Bayou virus (*Bunyaviridae*: *Hantavirus*). Am J Trop Med Hyg 57(4):445–448

Lederer S, Lattwein E, Hanke M, Sonnenberg K, Stoecker W, Lundkvist A, Vaheri A, Vapalahti O, Chan PK, Feldmann H et al (2013) Indirect immunofluorescence assay for the simultaneous detection of antibodies against clinically important old and new world hantaviruses. PLoS Negl Trop Dis 7(4):e2157

Leduc JW (1987) Epidemiology of Hantaan and related viruses. Lab Anim Sci 37:413–418

Leduc JW, Smith GA, Childs JE, Pinheiro FP, Maiztegui JI, Niklasson B, Antoniades A, Robinson DM, Khin M, Shortridge KF (1986) Global survey of antibody to Hantaan-related viruses among peridomestic rodents. Bull World Health Organ 64:139–144

Lee HW, Lee PW, Tamura M, Tamura T, Oku-no Y (1979) Etiological relation between Korean hemorrhagic fever and epidemic hemorrhagic fever in Japan. Biken J 22:41–45

Lee HW, Lee PW, Johnson KM (1978b) Isolation of the etiologic agent of Korean hemorrhagic fever. J Infect Dis 137:298–308

Lee HW, Bark DH, Baek LJ, Choi KS, Whang YN, Woo MS (1980) Korean hemorrhagic fever patients in urban areas of Seoul. Korean J Virol 10:1–6

Lee HW, French GR, Lee PW, Baek LJ, Tsuchiya K, Foulke RS (1981a) Observations on natural and laboratory infection of rodents with the etiologic agent of Korean hemorrhagic fever. Am J Trop Med Hyg 30(2):477–482

Lee HW, Lee PW, Baek LJ, Song CK, Seong IW (1981b) Intraspecific transmission of Hantaan virus, etiologic agent of Korean hemorrhagic fever, in the rodent *Apodemus agrarius*. Am J Trop Med Hyg 30:1106–1112

Lee HW, Baek LJ, Johnson KM (1982) Isolation of Hantaan virus, the etiologic agent of Korean hemorrhagic fever, from wild urban rats. J Infect Dis 146:638–644

Lee HW, Ahn CN, Song JW, Baek LJ, Seo TJ, Park SC (1990) Field trial of an inactivated vaccine against hemorrhagic fever with renal syndrome. Arch Virol (Suppl 1):35–47

Levis S, Morzunov SP, Rowe JE, Enría D, Pini N, Calderón GE, Sabattini M, Jeor SC (1998) Genetic diversity and epidemiology of hantaviruses in Argentina. J Infect Dis 177(3):529–538

Liang M, Li D, Xiao SY, Hang C, Rossi CA, Schmaljohn CS (1994) Antigenic and molecular characterization of hantavirus isolates from China. Virus Res 31:219–233

Lin XD, Wang W, Guo WP, Zhang XH, Xing JG, Chen SZ, Li MH, Chen Y, Xu J, Plyusnin A et al (2012) Cross-species transmission in the speciation of the currently known murinae-associated hantaviruses. J Virol 86(20):11171–11182

Linderholm M, Sandstrom T, Rinnstrom O, Groth S, Blomberg A, Tarnvik A (1997) Impaired pulmonary function in patients with hemorrhagic fever with renal syndrome. Clin Infect Dis 25 (5):1084–1089

Liu PQ, Liao HX, Fu JL, Hang CS, Song G (1984) Isolation of epidemic hemorrhagic fever virus from Rattus losea and Rattus confucianus and their antigenic identification. Bull Jiangxi Med Coll 3:1–7

Lokugamage K, Kariwa H, Lokugamage N, Miyamoto H, Iwasa M, Hagiya T, Araki K, Tachi A, Mizutani T, Yoshimatsu K et al (2004) Genetic and antigenic characterization of the Amur virus associated with hemorrhagic fever with renal syndrome. Virus Res 101(2):127–134

Lopez N, Padula P, Rossi C, Lazaro ME, Franze-Fernandez MT (1996) Genetic identification of a new hantavirus causing severe pulmonary syndrome. Virology 220:223–226

Luis AD, Douglass RJ, Hudson PJ, Mills JN, Bjornstad ON (2012) Sin Nombre hantavirus decreases survival of male deer mice. Oecologia 169(2):431–439

Lundkvist A, Verner-Carlsson J, Plyusnina A, Forslund L, Feinstein R, Plyusnin A (2013) Pet rat harbouring Seoul hantavirus in Sweden. Euro Surveill 18(27):pii = 20521. http://www.eurosurveillance.org/ViewArticle.aspx?ArticleId=20521

Makela S, Mustonen J, La-Houhala I, Hurme M, Partanen J, Vapalahti O, Vaheri A, Pasternack A (2002) Human leukocyte antigen-B8-DR3 is a more important risk factor for severe Puumala hantavirus infection than the tumor necrosis factor-alpha(−308) G/A polymorphism. J Infect Dis 186(6):843–846

Makela S, Kokkonen L, La-Houhala I, Groundstroem K, Harmoinen A, Huhtala H, Hurme M, Paakkala A, Porsti I, Virtanen V et al (2009) More than half of the patients with acute Puumala hantavirus infection have abnormal cardiac findings. Scand J Infect Dis 41(1):57–62

Martin ML, Lindsey Regnery H, Sasso DR, Mccormick JB, Palmer E (1985) Distinction between *Bunyaviridae* genera by surface structure and comparison with Hantaan virus using negative stain electron microscopy. Arch Virol 86:17–28

Matheus S, Djossou F, Moua D, Bourbigot AM, Hommel D, Lacoste V, Dussart P, Lavergne A (2010) Hantavirus pulmonary syndrome, French Guiana. Emerg Infect Dis 16(4):739–741

Mertz GJ, Hjelle B, Crowley M, Iwamoto G, Tomicic V, Vial PA (2006) Diagnosis and treatment of new world hantavirus infections. Curr Opin Infect Dis 19(5):437–442

Mills JN (2005) Regulation of rodent-borne viruses in the natural host: implications for human disease. Arch Virol 19:45–57

Mills JN (2006) Biodiversity loss and emerging infectious disease: an example from the rodent-borne hemorrhagic fevers. Biodiversity 7(1):9–17

Mills JN, Ksiazek TG, Ellis BA, Rollin PE, Nichol ST, Yates TL, Gannon WL, Levy CE, Engelthaler DM, Davis T et al (1997) Patterns of association with host and habitat: antibody reactive with Sin Nombre virus in small mammals in the major biotic communities of the southwestern United States. Am J Trop Med Hyg 56(3):273–284

Mills JN, Johnson JM, Ksiazek TG, Ellis BA, Rollin PE, Yates TL, Mann MO, Johnson RM, Campbell ML, Miyashiro J et al (1998) A survey of hantavirus antibody in small-mammal populations in selected U.S. National Parks. Am J Trop Med Hyg 58(4):525–532

Mills JN, Ksiazek TG, Peters CJ, Childs JE (1999) Long-term studies of hantavirus reservoir populations in the southwestern United States: a synthesis. Emerg Infect Dis 5(1):135–142

Mills JN, Ellis BA, Calderón G, Enría DA, Siazek TG (2007) A longitudinal study of hantavirus infection in three sympatric reservoir species in agroecosystems on the Argentina Pampa. Vector Born Zoon Dis 7:229–240

Monroe MC, Morzunov SP, Johnson AM, Bowen MD, Artsob H, Yates T, Peters CJ, Rollin PE, Ksiazek TG, Nichol ST (1999) Genetic diversity and distribution of *Peromyscus*-borne hantaviruses in North America. Emerg Infect Dis 5(1):75–86

Mori M, Rothman AL, Kurane I, Montoya JM, Nolte KB, Norman JE, Waite DC, Koster FT, Ennis FA (1999) High levels of cytokine-producing cells in the lung tissues of patients with fatal hantavirus pulmonary syndrome. J Infect Dis 179(2):295–302

Moritz C, Patton JL, Conroy CJ, Parra JL, White GC, Beissinger SR (2008) Impact of a century of climate change on small-mammal communities in Yosemite National Park, USA. Science 322 (5899):261–264

Muranyi W, Bahr U, Zeier M, Van Der Woude FJ (2005) Hantavirus infection. J Am Soc Nephrol 16(12):3669–3679

Mustonen J, Partanen J, Kanerva M, Pietila K, Vapalahti O, Pasternack A, Vaheri N (1996) Genetic susceptibility to severe course of nephropathia epidemica caused by Puumala hantavirus. Kidney Int 49(1):217–221

Nichol ST, Spiropoulou CF, Morzunov S, Rollin PE, Ksiazek TG, Feldmann H, Sanchez A, Childs JE, Zaki S, Peters CJ (1993) Genetic identification of a hantavirus associated with an outbreak of acute respiratory illness. Science 262:914–917

Oktem IMA, Uyar Y, Dincer E, Gozalan A, Schlegel M, Babur C, Celebi B, Sozen M, Karatis A et al (2014) Dobrava-Belgrade virus in *Apodemus flavicollis* and *A. uralensis* mice, Turkey. Emerg Infect Dis 20:121–130

Olsson GE, White N, Ahlm C, Elgh F, Verlemyr AC, Juto P, Palo RT (2002) Demographic factors associated with hantavirus infection in bank voles (*Clethrionomys glareolus*). Emerg Infect Dis 8(9):924–929

Olsson GE, White N, Hjalten J, Ahlm C (2005) Habitat factors associated with bank voles (*Clethrionomys glareolus*) and concomitant hantavirus in northern Sweden. Vector Borne Zoonotic Dis 5(4):315–323

Olsson GE, Hjertqvist M, Lundkvist A, Hornfeldt B (2009) Predicting high risk for human hantavirus infections, Sweden. Emerg Infect Dis 15(1):104–106

Olsson GE, Leirs H, Henttonen H (2010) Hantaviruses and their hosts in Europe: reservoirs here and there, but not everywhere? Vector Borne Zoon Dis 10(6):549–561

Ostfeld RS, Keesing F (2000) The function of biodiversity in the ecology of vector-borne zoonotic diseases. Can J Zool 78(2061):2078

Padula PJ, Edelstein A, Miguel SDL, Lopez NM, Rossi CM, Rabinovich RD (1998) Hantavirus pulmonary syndrome outbreak in Argentina: molecular evidence for person-to-person transmission of Andes virus. Virology 241:323–330

Padula PJ, Rossi CM, La Valle MO, Martinez PV, Colavecchia SB, Edelstein A, Miguel SD, Rabinovich RD, Segura EL (2000) Development and evaluation of a solid-phase enzyme immunoassay based on Andes hantavirus recombinant nucleoprotein. J Med Microbiol 49 (2):149–155

Padula P, La Valle MG, Alai MG, Cortada P, Villagra M, Gianella A (2002) Andes virus and first case report of Bermejo virus causing fatal pulmonary syndrome. Emerg Infect Dis 8(4):437–439

Palma RE, Polop JJ, Owen RD, Mills JN (2012) Ecology of rodent-associated hantaviruses in the Southern Cone of South America: Argentina, Chile, Paraguay, and Uruguay. J Wildl Dis 48 (2):267–281

Pattamadilok S, Lee BH, Kumperasart S, Yoshimatsu K, Okumura M, Nakamura I, Araki K, Khoprasert Y, Dangsupa P, Panlar P et al (2006) Geographical distribution of hantaviruses in Thailand and potential human health significance of Thailand virus. Am J Trop Med Hyg 75 (5):994–1002

Peters CM (1997) Sustainable use of biodiversity: myths, realities, and potential. In: Grifo F, Rosenthal J (eds) Biodiversity and human health. Island Press, Washington DC, pp 312–333

Peters CJ, Simpson GL, Levy H (1999) Spectrum of hantavirus infection: hemorrhagic fever with renal syndrome and hantavirus pulmonary syndrome. Annu Rev Med 50:531–545

Plyusnin A (2002) Genetics of hantaviruses: implications to taxonomy. Arch Virol 147:665–682

Plyusnin A, Vapalahti O, Vasilenko V, Henttonen H, Vaheri A (1997) Dobrava virus in Estonia: does the virus exist throughout Europe? Lancet 349:1369–1370

Plyusnin A, Nemirov K, Apekina N, Plyusnina A, Lunkdvist Å, Vaheri A (1999) Dobrava hantavirus in Russia. Lancet 353:207

Plyusnin A, Vaheri A, Lundkvist A (2006) Saaremaa hantavirus should not be confused with its dangerous relative, Dobrava virus. J Clin Microbiol 44(4):1608–1611

Plyusnin A, Ibrahim IN, Plyusnin A (2009) A newly recognized hantavirus in the Asian house rat (*Rattus tanezumi*) in Indonesia. J Gen Virol 90(Pt 1):205–209

Ramsden C, Holmes EC, Charleston MA (2009) Hantavirus evolution in relation to its rodent and insectivore hosts: no evidence for codivergence. Mol Biol Evol 26:143–153

Razzauti M, Plyusnina A, Sironen T, Henttonen H, Plyusnin A (2009) Analysis of Puumala hantavirus in a bank vole population in northern Finland: evidence for co-circulation of two genetic lineages and frequent reassortment between strains. J Gen Virol 90(Pt 8):1923–1931

Rivera PC, Ittig REG, Fraire HJR, Levis S, Gardenal CN (2007) Molecular identification and phylogenetic relationships among the species of the genus *Oligoryzomys* (Rodentia, Cricetidae) present in Argentina, putative reservoirs of hantaviruses. Zool Scr 36(3):231–239

Rodriguez LL, Owens JH, Peters CJ, Nichol ST (1998) Genetic reassortment among viruses causing hantavirus pulmonary syndrome. Virology 242:99–106

Rollin PE, Ksiazek TG, Elliott LH, Ravkov EV, Martin ML, Morzunov S, Livingstone W, Monroe M, Glass G, Ruo S et al (1995) Isolation of Black Creek Canal virus, a new hantavirus from *Sigmodon hispidus* in Florida. J Med Virol 46(1):35–39

Root TL, Schneider SH (2002) Climate change: overview and implications for wildlife. In: Schneider SH, Root TL (eds) Wildlife responses to climate change. Island Press, Washington DC, pp 1–56

Ruedas LA, Salazar-Bravo J, Tinnin DS, Armien B, Caceres L, Garcia A, Diaz MA, Gracia F, Suzan G, Peters CJ et al (2004) Community ecology of small mammal populations in Panama following an outbreak of hantavirus pulmonary syndrome. J Vector Ecol 29(1):177–191

Salazar-Bravo J, Armién B, Suzán G, Armién A, Ruedas L, Avila M, Zaldívar Y, Pascale JM, Gracia F, Yates TL (2004) Serosurveyof wild rodents for hantaviruses in Panama, 2000–2002. J Wildl Dis 40:103–109

Sauvage F, Langlais M, Pontier D (2007) Predicting the emergence of human hantavirus disease using a combination of viral dynamics and rodent demographic patterns. Epidemiol Infect 135 (1):46–56

Schmaljohn CS, Dalrymple JM (1983) Analysis of Hantaan virus RNA: evidence for a new genus of *Bunyaviridae*. Virology 131:482–491

Schmaljohn C, Hjelle B (1997) Hantaviruses: a global disease problem. Emerg Infect Dis 3:97–103

Schmaljohn CS, Hasty SE, Harrison SA, Dalrymple JM (1983) Characterization of Hantaan virions, the prototype virus of hemorrhagic fever with renal syndrome. J Infect Dis 148:1005–1012

Schmaljohn CS, Hasty SE, Dalrymple JM, Leduc JW, Lee HW, Von Bonsdorff CH, Brummer-Korvenkontio M, Vaheri A, Tsai TF, Regnery HL et al (1985) Antigenic and genetic properties of viruses linked to hemorrhagic fever with renal syndrome. Science 227:1041–1044

Schmaljohn CS, Jennings GB, Hay J, Dalrymple JM (1986) Coding strategy of the S genome segment of Hantaan virus. Virology 155:633–643

Schmaljohn CS, Chu YK, Schmaljohn AL, Dalrymple JM (1990) Antigenic subunits of Hantaan virus expressed by baculovirus and vaccinia virus recombinants. J Virol 64:3162–3170

Schmaljohn C, Huggins J, Calisher CH (1998) Laboratory and field safety (Chapter XII) In: Lee HW, Calisher CH, Schmaljohn C (eds) Manual of hemorrhagic fever with renal syndrome and hantavirus pulmonary syndrome. WHO Collaborating Center for Virus Reference and Research, Asan Institute for Life Sciences, Seoul, Korea, pp 191–198

Schountz T, Prescott J, Cogswell AC, Oko L, Mirowsky-Garcia K, Galvez AP, Hjelle B (2007) Regulatory T cell-like responses in deer mice persistently infected with Sin Nombre virus. Proc Natl Acad Sci U S A 104(39):15496–15501

Schwarz AC, Ranft U, Piechotowski I, Childs JE, Brockmann SO (2009) Risk factors for human infection with Puumala virus, southwestern Germany. Emerg Infect Dis 15(7):1032–1039

Sibold C, Meisel H, Lunkdvist Å, Cifire F, Ulrich R, Kozuch O, Labuda M, Kruger DH (1999) Simultaneous occurrence of Dobrava, Puumala, and Tula hantaviruses in Slovakia. Am J Trop Med Hyg 61:409–411

Sibold C, Ulrich R, Labuda M, Lundkvist Å, Schütt M, Gerke P, Leitmeyer K, Meisel H, Kruger DH (2001) Dobrava hantavirus causes hemorrhagic fever with renal syndrome in central Europe and is carried by two different Apodemus mice species. J Med Virol 63:158–167

Sirotin BZ, Keiser NP (2001) On the history of the study of haemorrhagic fever with renal syndrome in eastern Russia. Nephrol Dial Transplant 16(6):1288–1290

Sohn YM, Rho HO, Park MS, Kim JS, Summers PL (2001) Primary humoral immune responses to formalin inactivated hemorrhagic fever with renal syndrome vaccine (Hantavax): consideration of active immunization in South Korea. Yonsei Med J 42(3):278–284

Song G (1999) Epidemiological progresses of hemorrhagic fever with renal syndrome in China. Chin Med J 112(5):472–477

Song JW, Baek LJ, Nagle JW, Schlitter D, Yanagihara R (1996) Genetic and phylogenetic analyses of hantaviral sequences amplified from archival tissues of deer mice (*Peromyscus maniculatus nubiterrae*) captured in the eastern United States. Arch Virol 141(5):959–967

Song KJ, Baek LJ, Moon S, Ha SJ, Kim SH, Park KS, Klein TA, Sames W, Kim HC, Lee JS et al (2007) Muju virus, a novel hantavirus harboured by the arvicolid rodent *Myodes regulus* in Korea. J Gen Virol 88(Pt 11):3121–3129

Stocker TF, Qin D (2013) Climate change 2013: the physical science basis. In: Working Group I Contribution to the fifth assessment report of the intergovernmental panel on climate change:

summary for policymakers. http://www.ipcc.ch/report/ar5/wg1/docs/WGIAR5_SPM_brochure_en.pdf
Suzan G, Marce E, Giermakowski JT, Mills JN, Ceballos G, Ostfeld RS, Armien B, Pascale JM, Yates TL (2009) Experimental evidence for reduced rodent diversity causing increased hantavirus prevalence. PLoS One 4(5):e5461
Suzuki A, Bisordi I, Levis S, Garcia J, Pereira LE, Souza RP, Sugahara TKN, Pini N, Enría D, Souza LTM (2004) Identifying rodent hantavirus reservoirs. Brazil Emerg Infect Dis 10 (12):2127–2134
Taller AM, Xiao SY, Godec MS, Gligic A, Vsic-Zupanc T, Goldfarb LG, Asher DM (1993) Belgrade virus, a cause of hemorrhagic fever with renal syndrome in the Balkans, is closely related to Dobrava virus of field mice. J Infect Dis 168:750–753
Terajima M, Vapalahti O, Van Epps HL, Vaheri A, Ennis FA (2004) Immune responses to Puumala virus infection and the pathogenesis of nephropathia epidemica. Microbes Infect 6 (2):238–245
Terajima M, Hayasaka D, Maeda K, Ennis FA (2007) Immunopathogenesis of hantavirus pulmonary syndrome and hemorrhagic fever with renal syndrome: Do CD8+ T cells trigger capillary leakage in viral hemorrhagic fevers? Immunol Lett 113(2):117–120
Tersago K, Verhagen R, Servais A, Heyman P, Ducoffre G, Leirs H (2009) Hantavirus disease (nephropathia epidemica) in Belgium: effects of tree seed production and climate. Epidemiol Infect 137(2):250–256
Tersago K, Crespin L, Verhagen R, Leirs H (2012) Impact of Puumala virus infection on maturation and survival in bank voles: a capture-mark-recapture analysis. J Wildl Dis 48 (1):148–156
Toro J, Vega JD, Khan AS, Mills JN, Padula P, Terry W, Yadon Z, Valderrama R, Ellis BA, Pavletic C et al (1998) An outbreak of hantavirus pulmonary syndrome, Chile, 1997. Emerg Infect Dis 4(4):687–694
Torres-Pérez F, Navarrete-Droguett J, Aldunate R, Yates TL, Mertz GJ, Vial PA, Ferres M, Marquet PA, Palma RE (2004) Peridomestic small mammals associated with confirmed cases of human hantavirus disease on southcentral Chile. Am J Trop Med Hyg 70:305–309
Torres-Pérez F, Palma RE, Hjelle B, Ferrés M, Cook JA (2010) Andes virus infections in the rodent reservoir and in humans vary across contrasting landscapes in Chile. Infect Genet Evol 10:820–825
Travassos Da Rosa ES, Mills JN, Padula PJ, Elkhoury MR, Ksiazek TG, Mendes WS, Santos ED, Aruojo GC, Martinez VP, Edelstein A (2005) Newly recognized hantaviruses associated with hantavirus pulmonary syndrome in northern Brazil: Partial genetic characterization of viruses and serologic implication of likely reservoirs. Vector-Borne Zoon Dis 5(1):11–19
Travassos Da Rosa ES, Medeiros DB, Nunes MR, Simith DB, De Souza PA, Elkhoury MR, Lavocat M, Marques AA, Via AV, D'andrea P et al (2011) Pygmy rice rat as potential host of Castelo dos Sonhos Hantavirus. Emerg Infect Dis 17(8):1527–1530
Umenai T, Lee PW, Toyoda T, Yoshinaga K, Horiuchi T, Lee HW, Saito T, Hongo M, Nobonaga T, Ishida N (1979) Korean hemorrhagic fever in an animal laboratory. Lancet 1 (8130):1314–1316
Vaheri A, Mills JN, Christina F, Spiropoulou F, Hjelle B (2011) Hantaviruses. In: Palmer SR, Soulsby L, Torgerson T, Brown DWG (eds) Oxford textbook of zoonoses. Oxford University Press, Oxford, pp 307–322 (Chap. 28 Fig. 28.2)
Vaheri A, Henttonen H, Voutilainen L, Mustonen J, Sironen T, Vapalahti O (2013) Hantavirus infections in Europe and their impact on public health. Rev Med Virol 23(1):35–49
Vapalahti O, Lundkvist A, Kukkonen SKJ, Cheng Y, Gilljam M, Kanerva M, Manni T, Pejcoch M, Niemimaa J, Kaikusalo A et al (1996) Isolation and characterization of Tula virus, a distinct serotype in the genus *Hantavirus*, family *Bunyaviridae*. J Gen Virol 77(Pt 12):3063–3067
Vapalahti O, Mustonen J, Lundkvist A, Henttonen H, Plyusnin A, Vaheri A (2003) Hantavirus infections in Europe. Lancet Infect Dis 3(10):653–661

Vincent MJ, Quiroz E, Gracias F, Sanchez A, Ksiazek T, Kitzutani P, Ruedas LA, Tinnin D, Caceres L, Garcia A et al (2000) Hantavirus pulmonary syndrome in Panama: identification of novel hantaviruses and their likely reservoirs. Virology 277:14–19

Wang H, Yoshimatsu K, Ebihara H, Ogino M, Araki K, Kariwa H, Wang Z, Luo Z, Li D, Hang C et al (2000) Genetic diversity of hantaviruses isolated in China and characterization of novel hantaviruses isolated from *Niviventer confucianus* and *Rattus rattus*. Virology 278(2):332–345

Wilson DE, Reeder DM (2005) Mammal species of the world. Johns Hopkins University Press, Baltimore

Wolfe ND, Daszak P, Kilpatrick AM, Burke DS (2005) Bushmeat hunting, deforestation, and prediction of zoonoses emergence. Emerg Infect Dis 11:1822–1827

Wright HE, Kutzbach JE, Webb T, Ruddiman WE, Street-Perrott FA, Bartlein PJ (1993) Global climates since the last glacial maximum. University of Minnesota Press, Minneapolis

Xiao SY, Diglisic G, Vsic-Zupanc T, Leduc JW (1993) Dobrava virus as a new Hantavirus: evidenced by comparative sequence analysis. J Med Virol 39:152–155

Xu ZY, Gou CS, Wu YL, Zhang XW, Liu K (1985) Epidemiological studies of hemorrhagic fever with renal syndrome: analysis of risk factors and of transmission. J Infect Dis 152:137–144

Yan L, Fang LQ, Huang HG, Zhang LQ, Feng D, Zhao WJ, Zhang WY, Li XW, Cao WC (2007) Landscape elements and Hantaan virus-related hemorrhagic fever with renal syndrome, People's Republic of China. Emerg Infect Dis 13(9):1301–1306

Yashina LN, Patrushev NA, Ivanov LI, Slonova RA, Mishin VP, Kompanez GG, Zdanovskaya NI, Kuzina II, Safronov PF, Chizhikov VE et al (2000) Genetic diversity of hantaviruses associated with hemorrhagic fever with renal syndrome in the far east of Russia. Virus Res 70(1–2):31–44

Yates TL, Mills JN, Parmenter CA, Ksiazek TG, Parmenter RR, Vande Castle JR, Calisher CH, Nichol ST, Abbott KD, Young JC et al (2002) The ecology and evolutionary history of an emergent disease: hantavirus pulmonary syndrome. Bioscience 52(11):989–998

Zaki SR, Greer PW, Coffield LM, Goldsmith CS, Nolte KB, Foucar K, Feddersen RM, Zumwalt RE, Miller GL, Khan AS et al (1995) Hantavirus pulmonary syndrome: pathogenesis of an emerging infectious disease. Am J Pathol 146:552–579

Zhang YZ, Zou Y, Yao LS, Hu GW, Du ZS, Jin LZ, Liu YY, Wang HX, Chen X, Chen HX et al (2007) Isolation and characterization of hantavirus carried by *Apodemus peninsulae* in Jilin, China. J Gen Virol 88(Pt 4):1295–1301

Zhang YZ, Zou Y, Fu ZF, Plyusnin A (2010) Hantavirus infections in humans and animals, China. Emerg Infect Dis 16(8):1195–1203

Zou Y, Hu J, Wang ZX, Wang DM, Yu C, Zhou JZ, Fu ZF, Zhang YZ (2008) Genetic characterization of hantaviruses isolated from Guizhou, China: evidence for spillover and reassortment in nature. J Med Virol 80(6):1033–1041

Chapter 5
Enterohemorrhagic *E. coli* Infections

Larry I. Lutwick

Abstract

...there is always some little thing that is too big for us

Don Marquis, Archy and Mehitabel, 1929

Enterohemorrhagic *E. coli* (EHEC) represent a classic example for the role of the One Health paradigm. A silent barnyard colonizer of ruminants, this zoonotic group of organisms was recognized to have jumped the species barrier to humans barely three decades ago and now is an human enteric infection which is the most common cause of acute kidney failure in children. Transmitted primarily through a breakdown in the food chain that facilitates contamination by some of the world's overwhelming amount of bovine manure, EHEC demonstrate a great deal of genetic diversity and clearly show the ability to develop into more virulent strains. EHEC also have interactions with a spate of other life forms including viruses, fungi, protozoa and plants.

Keywords *E. coli* • EHEC • Hemolytic uremic syndrome • Shiga toxin • Zoonosis

5.1 Introduction

Escherichia coli, *E. coli* for short, by far the most common facultative Gram negative bacillus in the human intestinal tract, starts to colonize the gut within hours of birth. Within its niche in the intestinal tract, the organism is part of the normal bacterial flora with many functions to help its human host including preventing pathogen colonization, degrading dietary and in situ-produced compounds, and producing nutrients. The genetic makeup of these symbiotic *E. coli* is varied and quite dynamic. Lawrence and Ochman (1998) estimated that almost a fifth of its genome has been transferred horizontally from other organisms. Although pathogenicity of these normal flora strains of the organism occurs when the commensal enters sterile parts of the body including the peritoneum, urinary bladder and blood stream, some *E. coli* have acquired virulence factors to become

L.I. Lutwick, MD (✉)
Medicine and Biomedical Sciences, Homer Stryker MD School of Medicine, Western Michigan University, 1000 Oakland Drive, Kalamazoo, MI 49008, USA
e-mail: larry.lutwick@med.wmich.edu

A. Yamada et al. (eds.), *Confronting Emerging Zoonoses*,
DOI 10.1007/978-4-431-55120-1_5

Table 5.1 *E. coli* enteropaths

Type	Disease
Enterotoxigenic *E. coli* (ETEC)	Weanling diarrhea
	Traveler's diarrhea
Enteropathogenic *E. coli* (EPEC)	Acute/chronic diarrhea in children <2 years of age
Enteroinvasive *E. coli* (EIEC)	Bacillary dysentery
	Bacteremia
Diffusely adhering *E. coli* (DAEC)	Childhood diarrhea
Enteroaggregative *E. coli* (EAEC)	Acute/chronic diarrhea
	Traveler's diarrhea
Enterohemorrhagic *E. coli* (EHEC)	Bloody diarrhea
	Hemolytic Uremic Syndrome

enteropathogenic for man. There are currently six enteropaths of the bacterium described (Table 5.1), one being the zoonotic EHEC.

These *E. coli* enteropaths, as well as all other *Enterobacteriaceae*, can be classified based on surface antigenic serotypes, the somatic (O antigen) and flagellar antigens (H antigen). This typing system was derived from studies with *Proteus vulgaris*. Some sera directed against *Proteus* would suppress the swarming motility of the organism and some would not (Weil and Felix 1916). The swarming quality referred to as *hauch*, the German word for "breath", is now known to correspond to the flagellar antigen. Non-motility associated antigenicity was referred to as *ohne hauch*, German for "without breath", corresponding to the somatic antigen. Now, the modified Kauffman serotyping scheme is used to classify each *E. coli* by its individual O and H antigens, for example, *E. coli* O157:H7. Some bacilli are non-motile and "NM" or "H-" replaces the H antigen in the designation. These antigens are organism-specific so that *E. coli* O1 is not O antigenically the same as *Vibrio cholerae* O1.

EHEC serotype O157:H7 literally burst on the microbiological scene during the first half of 1982 when found to be associated with two outbreaks of hemorrhagic colitis in Oregon and Michigan linked to undercooked beef (Riley et al. 1983). The serotype was rare, having been characterized only once before in the USA, in 1975 from a California woman with bloody diarrhea, by the Centers for Disease Control and Prevention (CDC). Isolated cases of a similar illness had been described in Japan and USA, usually associated with antimicrobial use, but no organism was identified as the cause (Sakurai et al. 1979; Toffler et al. 1978; Pittman et al. 1974).

An association of EHEC with hemolytic uremic syndrome (HUS) came soon afterwards with the finding of a toxic factor (Shiga toxin) produced by *E. coli* O157:H7 having a cytotoxic effect very similar to that of *Shigella dysenteriae* serotype 1 (O'Brien et al. 1983) and the observation that serotype O157:H7 infection could be followed by HUS (Karmali et al. 1983). Two decades previously, Barnard and Kibel (1965) suggested that *E. coli* infection could result in HUS and a decade before intrafamily clustering of HUS had been observed (Kaplan et al. 1975). Prior to the 1982 outbreak, as well, Konowalchuk et al. (1977) had found that some

E. coli strains produced a toxin that could kill Vero cells derived from African green monkey kidneys, referring to it as Verotoxin. Verotoxin and Shiga toxin were subsequently shown to be identical.

5.2 The EHEC Organisms

EHEC are types of *E. coli* that can produce Shiga toxin and contain the genes to effectively deliver the toxin, thereby causing disease in man. Not all *E. coli* containing one or both Shiga toxin genes (stx1 and stx2) can do so. Indeed, there are more than 380 different Shiga toxin producing O:H serotypes that have been characterized but only a limited number are linked to toxin-related human disease (Karmali et al. 2003). The canonical EHEC organism is *E. coli* O157:H7 but non-O157 serotypes that can cause an indistinguishable Shiga toxin-linked disease include O serogroups 26, 103, 111, 121, and 145. Among serotype O157:H7 isolates, single nucleotide polymorphism (SNP) analysis (Manning et al. 2008) has grouped clinical isolates into nine clades. Among them, clade 8 is more frequently linked to more severe disease.

5.2.1 EHEC Population Genetics

The clonal group of serotype O157:H7 is only distantly related to EHEC of other serotypes and it seems to have developed from an enteropathogenic progenitor *E. coli* O55:H7, a cause of infantile diarrhea worldwide (Whittam et al. 1988, 1993). Whittam and colleagues (Feng et al. 1998; Wick et al. 2005) have set forth an evolutionary model in which O157:H7 emerged from the progenitor through a series of genetic changes starting with acquiring the phage carrying the gene for Shiga toxin 2 and followed by the EHEC plasmid pO157 and the phage for Shiga toxin 1, and losing both sorbitol fermentation and beta-glucuronidase activity. A sorbitol fermenting, non-motile clone (O157:H-) also developed which was initially seen primarily in Europe (Karch and Bielaszewska 2001) and is characterized by a higher risk of HUS progression (Buvens et al. 2009). The evolution of non-O157 EHEC has been felt to begin with a serotype O26:H11 that acquired one or both Shiga toxin phages and other virulence factors (Donnenberg and Whittam 2001). Bacteriophage gain and loss play a very dominant role in bacterial genomic change with acquisition of genes by horizontal transfer or loss by deletion occurring at a rate 140 times greater than that for point mutations in standard housekeeping genes (Wick et al. 2005) whose open reading frame backbones tend to be remarkably well preserved.

5.2.2 Shiga Toxins

Evolutionarily, bacterial acquisition of protein exotoxins such as the Shiga toxin is unlikely to be related to effects of the toxin on humans, as man is likely neither the original nor primary target of such toxins (Matz and Kjelleberg 2005). Since a substantial degree of bacterial mortality is related to predation by single celled eukaryotic protozoa, toxin release before or after ingestion could be a major defense mechanism. Indeed, Shiga toxin can kill *Tetrahymena thermophila*, a protozoan bacterial predator (Lainhart et al. 2009; Stolfa and Koudelka 2013). Therefore, EHEC disease in man can be thought of as collateral damage in *E. coli*'s fight for survival.

EHEC carry copies of one or two toxin-converting lambda-like bacteriophages. Bacteriophages, accounting for around 10^{31} individuals, are as a group the most abundant organism on our planet (Whitman et al. 1998). The toxins are two-thirds of the Shiga toxin family, prototypically represented by the *Shigella dysenteriae* Shiga toxin. Shiga toxin 1 (Stx1) differs from the *S. dysenteriae* toxin by only one amino acid whereas Shiga toxin 2 (Stx2) is only 57 % identical. There is no antitoxin cross-neutralization between the EHEC toxins. Variant toxins exist, a through d for Stx1 and a through g for Stx2 (Bergan et al. 2012). Stx2c seems particularly linked to increased toxicity and overall Stx2 is several hundred times more toxic than Stx1 in animal models.

Stxs consist of an enzymatically active A subunit with five non-toxic B subunits arranged in a pentameric ring around the A moiety. The B subunits facilitate binding to target cells, each having three separate binding sites, producing high affinity adherence to the trisaccharide component of the receptor globotriaosyl-ceramide (Gb3). The degree of binding is a major part of the differences in toxicity of Stx variants (Shimizu et al. 2007). Stx2 binding is not solely to the trisaccharide part of the molecule but also involves other surface components of the ceramide portion of Gb3 (Gallegos et al. 2012). These and other differences in receptor binding not only are involved in differential targeting of susceptible tissues but also affect intracellular trafficking, impacting upon toxicity. Clade 8 of *E. coli* O157:H7 overexpresses Stx2 (Neupane et al. 2011) and other virulence factors (Abu-Ali et al. 2010), contributing to the more severe disease associated with this clade.

After receptor binding, Stx enters cells by endocytosis. Cells resistant to Shiga toxin, like those of the bovine intestinal and vascular tissues, are able to degrade the protein in lysosomes (Hoey et al. 2003). In toxin-sensitive cells, however, a retrograde transport route occurs, bypassing the late endocytic pathway by trafficking from the early/recycling endosomes directly to the *trans*-Golgi network and then to the endoplasmic reticulum where it can be transferred to the cytosol. Once in the cytosol, the A subunit, functioning as an RNA N-glysosidase, removes a single adenine from the 28S rRNA inhibiting elongation factor-1-dependent amino-acyl-tRNA binding to ribosomes and with cessation of protein synthesis and subsequent cell death (Obrig et al. 1987). Interestingly, the plant toxin ricin, derived from castor beans of the *Ricinus communis* plant, removes the same adenine, illustrating evolutionary convergence in quite divergent organisms (Brigotti 2012). Although this inhibition of protein translation has been the classic cause for Stx injury, Marsden's laboratory (Petruzziello-Pellegrini and Marsden 2012) has shown that at levels of Stx that minimally affect protein synthesis, dramatic changes in gene expression occur altering the phenotype of susceptible vascular endothelial cells, promoting proinflammatory cytokines, upregulating chemokine receptors and their ligands, and cellular apoptosis (Colpoys et al. 2005; Petruzziello-Pelligrini et al. 2012). Shiga toxin, in a murine model, also causes complement activation, contributing to microvascular thrombosis (Morigi et al. 2011).

Once released into the intestinal milieu, the Shiga toxin phage infects other *E. coli* causing lytic cycles and releasing more toxin. In a murine model, intestinal flora susceptible to the phage augmented Stx2 production whereas fecal toxin levels were diminished if phage-resistant *E. coli* were present (Gamage et al. 2003). This process likely also plays a role in the degree of disease produced by the infection and can be enhanced by biofilms produced by EHEC (Solheim et al. 2013). As there are often multiple prophages in an EHEC (the Sakai strain had 18, many of which are defective), recombination between them can occur helping to disseminate virulence factors to other *E. coli* (Asadulghani et al. 2009).

Shiga toxin-negative strains of EHEC O26 can be isolated from patients who have had HUS (Bielaszewska et al. 2007b). During the course of HUS, a subset of EHEC may suffer excision of the toxin-converted phage. This particular O26:H11 Stx2a clone, which is sometimes H-, has become more prevalent in Europe and is linked to a higher risk of HUS. It is postulated by Bielaszewska, et al. (2013) that loss of the Stx phage can increase adaptability in the human gut, increasing in the number of organisms and then reacquiring the Stx gene to emerge highly virulent. Isolates losing the Stx prophage have been designated EHEC-LST (EHEC—Lost Shiga Toxin) (Bielaszewska et al. 2007a).

5.2.3 Other Virulence Factors

The locus of enterocyte effacement (LEE) is a chromosomally located pathogenicity island encoding for proteins involved in the formation of the attaching and effacing lesions by EPEC and EHEC including a regulator (Ler), activated by quorum sensing (Sperandio et al. 2001). Quorum sensing relates to gene expression in response to fluctuations in cell population density through hormone-like mediators called autoinducers. The process bestows upon bacteria some qualities of higher organisms and could have been an early step in the development of multicellularity (Miller and Bassler 2001). LEE encodes for a type III secretion system, intimin (the binding moiety) and effector proteins. The secretion system (abbreviated TTSS or T3SS) is an injectosome that serves to detect target organisms and introduce effector proteins into the host cell, including Tir (translocated intimin receptor that produces adherence between intimin and Tir).

Sorbitol fermenting O157 strains, which can have a higher risk of HUS, adhere to intestinal cells at higher levels than the standard nonfermenting strains. These strains, besides fermenting sorbitol, are nonmotile, do not encode for the hemolysin but contain a unique gene cluster not found in other EHEC or other *Enterobacteriaceae* encoding for novel fimbriae and mediating mannose-resistant hemagglutination (Brunder et al. 2001). Other molecular mechanisms related to EHEC colonization are reviewed by Farfan and Torres (2012).

5.2.4 Identification of EHEC strains

Much of the description of the EHEC bacteriology and clinical disease is based on *E. coli* O157:H7. EHEC, however, is a group of organisms heterogeneous in serotypic and other phenotypic markers, but like the group of bacilli that produce botulinum toxin, are tied together by the production of a toxin, in this case Shiga toxin which can causes hemorrhagic colitis and hemolytic uremic syndrome.

EHEC are non-invasive zoonotic, Gram negative bacilli that intermittently colonize cattle and other animals, essentially as commensals causing minimal damage, but when crossing species boundaries into man can precipitate significant disease. Serotype O157:H7 has been identified by culture on selective media, specifically sorbitol MacConkey agar with or without cefixime and tellurite. MacConkey agar was developed to identify non-lactose fermenting Gram negative enteric pathogens such as *Salmonella* and *Shigella* in stool samples. Fermentation of lactose (or in the EHEC media, sorbitol) will produce CO_2 in the local milieu causing a decrease in pH. Since the medium contains a neutral red indicator, the colony and sometimes the surrounding medium will become pink in color (Fig. 5.1a) and non-fermenting colonies will remain beige (Fig. 5.1b). On sorbitol MacConkey agar, nonfermentors can be identified as serotype O157 by agglutination with specific antisera. Non-O157 serotypes ferment sorbitol as do some O157 strains, so this detection modality does not identify these strains. To avoid missing these strains of EHEC, Shiga toxin can be assayed by immunoassay or the gene can be probed for by PCR. Usually, these assays are done on overnight enrichment broth cultures rather than the diarrhea directly.

In many diagnostic laboratories, screening for EHEC is performed only on bloody diarrheal stools or if specifically requested. Needless to say, not testing for EHEC makes diagnosis more difficult. To avoid this, the CDC now recommends routine testing of acute community-acquired diarrhea for EHEC using culture on selective media and with an assay to detect Shiga toxin (Gould 2012).

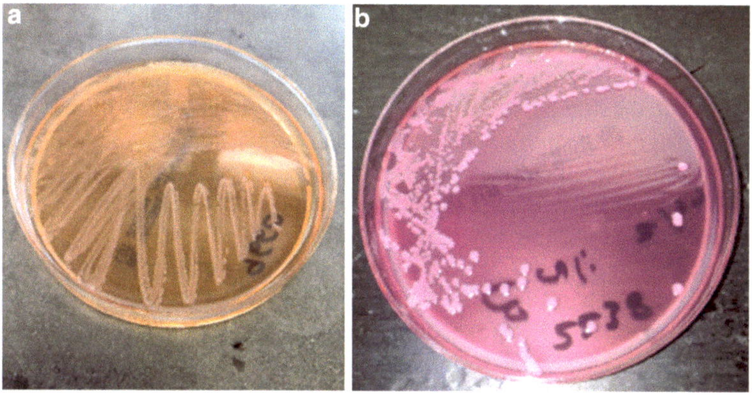

Fig. 5.1 A sorbitol MacConkey agar plate showing a fermenting *E. coli* (9.1a) and one showing a non-fermenting *E. coli* O157:H7 (9.1b)

5.3 Animal Reservoirs

Ferens and Hovde (2011) describe EHEC prevalence in livestock as a paradox where the bacteria occur worldwide but are found infrequently in many prevalence studies, reflecting issues such as the sporadic nature of carriage, the low numbers of bacteria present, insufficiently sensitive sampling or culturing approaches, and the poor correlation between fecal shedding and harboring the organism extraintestinally. As examples of the latter, EHEC has been noted to be more prevalent in cattle on the ears than in feces in one study from Sweden (Boqvist et al. 2009) and serotype O157 could be isolated from the oral cavity or hides of 130 of 139 feedlot cattle including 50 cattle that were fecal-negative (Keen and Elder 2002).

5.3.1 Cattle

The prototypical association of cattle and human EHEC infection could reflect the predominance of beef and dairy cattle among domesticated animals together with their voluminous output of bovine manure (Ferens and Hovde 2011). Disease does not occur in cattle and most livestock because of the lack of vascular receptors for the toxin although diarrhea may be caused in young calves (Kang et al. 2004). The organism is not entirely a commensal as it may cause mucosal hemorrhage in the intestines of colonized cattle that remain for several months after shedding ceases (Walle et al. 2012). EHEC is typically spread quickly when a carrier animal is placed in a pen of non-infected cattle. McGee et al. (2004) reported that within 2 days the organism contaminated the hides of many pen mates and soon after could be found in the stool of the previously uninfected. Shedding could be found in adjacent pens within 8 days (Shere et al. 2002). Disseminated among dairy cattle is more efficient than beef cattle, likely because a dairy herd is moved several times daily for milking (Ferens and Hovde 2011).

Most infected cattle pass EHEC in stool at low numbers (less than 100 colony forming units/gram of stool). Such low numbers can make detection difficult, often underestimating overall carriage prevalence (LeJeune et al. 2006) and excretion is often intermittent. Some animals, however, harbor quite high amounts of EHEC for prolonged amounts of time. One report (Omisakin et al. 2003) found that 9 % of carriers excreted over 96 % of total EHEC. It is these animals, termed supershedders, which are the major transmitters to other animals' gastrointestinal tracts and/or hides. Non-EHEC shedding cattle are five times more likely not to be housed with a supershedder (Cobbold et al. 2007). Supershedding is linked to colonization of the rectoanal junction of the animal where proteins other than LEE seem to contribute to adherence to the rectoanal junction stratified squamous epithelial cells (Kudva et al. 2012). It must be noted that the meat contamination may not necessarily be related to direct fecal contact at all but rather due to spread from contaminated hides (Arthur et al. 2010).

One feature of particular importance is the marked variability of EHEC prevalence in cattle. Prevalence depends on the time of the year with peaks between the late spring and early fall, seen particularly in temperate climates. As examples, in Nebraska, USA, feedlot cattle had a prevalence of serotype O157:H7 carriage of almost 100 % in the fall diminishing to less than 20 % by March (Arthur et al. 2009) and in Alberta, Canada, dairy herd positivity for EHEC was 15 times more common between June and September than during other periods (Stanford et al. 2005). The cause of this variability is unclear but may be related to increased day length impacting on bovine hormone production (Lahti et al. 2003; Edrington et al. 2006). Alternatively, warmer temperatures increase EHEC levels in contaminated drinking water available to the herd and represent a higher risk of spread (Gautam et al. 2011).

A variety of interventions have been studied in both dairy and beef cattle regarding reducing EHEC rates, often with heterogeneous results. Two issues will be mentioned: dietary manipulations and bovine immunization. Grain-fed cattle tend to carry higher amounts of EHEC than forage-fed cattle and by abruptly switching from grain to an all hay diet, EHEC populations can decline (Callaway et al. 2009). Finishing beef and lactating dairy cattle are commonly fed high grain rations, often distiller's grain, in order to maximize performance and production efficiency. Probiotics, however, may decrease EHEC prevalence. As an example, a combination of *Lactobacillus acidophilus* and *Proprionibacterium freudenreichii* reduced the prevalence of EHEC in feces and on hides as well as increasing the growth efficiency of cattle (Stephens et al. 2007). Cull et al. (2012), however, found no effect of the probiotic in a pen-level field trial of feedlot cattle as compared to controls.

The Cull study (2012) also reported that a serotype O157 outer membrane-based vaccine (siderophore receptor and porin proteins) reduced the overall prevalence of *E. coli* O157:H7 as well as the prevalence of supershedders.

This large study confirmed the observations of older, smaller studies. A meta-analysis of this vaccine and a second commercial vaccine targeting the type III secretion system proteins also suggested bovine vaccination as an effective modality in the control of EHEC (Varela et al. 2012). Snedeker et al. (2012) reported similar reductions of prevalence in their review but cautioned the interpretation because of heterogeneity in the results. Among vaccine candidates being evaluated are a fusion protein carrying a detoxified Stx2 A subunit together with a Stx1 B subunit monomer (Cai et al. 2011); an intranasally administered fusion protein with both Stx 1 and 2 B subunit monomers, the Tir protein and the zonula occludens toxin from a bacteriophage of *Vibrio cholerae* (Zhang et al. 2012); and a double repeated B subunit produced in transgenic lettuce designed for oral administration of the vegetable (Matsui et al. 2011). The potential use of edible transgenic plant vaccine technology (Lutwick 2008) can generate produce that may prevent EHEC infection.

5.3.2 Other Livestock

EHEC can also be found in bovine species other than cattle. With the increasing popularity of bison meat, it should be noted that the fecal prevalence of EHEC ranges from 17 to 83 %, averaging 42 % (Reinstein et al. 2007). High prevalence rates have also been reported in water buffalo in Turkey (Seker and Yardimci 2008), Bangladesh (Islam et al. 2008), and Italy (Lorusso et al. 2009; Borriello et al. 2012). The pathogen has also been isolated from raw yak milk and milk products (Bandyopadhyay et al. 2012).

Goats and sheep: The prevalence of EHEC in these small domestic ruminants is less well documented than in cattle and varies considerably (La Ragione et al. 2009). Prevalence in sheep has been reported to be as high as 31 % during the summer (Kudva et al. 1996). There is less documentation in goats but studies do suggest a similar or slightly lower prevalence. Sheep in particular have been found to harbor a variety of EHEC serogroups that can affect man. As an example, an outbreak of EHEC O103:H25 infection with HUS was reported in Norway traced to dry cured mutton sausage (Schimmer et al. 2008).

Pigs: Unlike most of the animals that can carry EHEC without symptoms, porcine hosts develop "edema disease" related to EHEC enteric infection with organisms that produce Shiga toxin 2e. In Japan, the same level of carriage of EHEC O157 exists in pigs as in cattle and such isolates can persist in swine. Ferens and Hovde (2011) postulate that swine may be an unlikely source of human EHEC infection because of their sensitivity to virulent Shiga toxin producing strains.

Fowl: Chickens appear to be rare carriers of EHEC but the bacterium may be present in turkeys. One report (Doane et al. 2007) found serotype O157:H7 in only 0.9 % of chickens but 7.5 % of turkeys. Experimentally, however, chickens inoculated with O157:H7 could become colonized, shed bacteria and contaminate eggs up to 11 months (Schoeni and Doyle 1994). The organism could be isolated from egg shells but unlike *Salmonella enterica* serotype Enteritidis, was not found inside intact eggs.

Camels: *E. coli* O157:H7 was isolated from camel meat (Hajian et al. 2011) and feces (Salehi et al. 2012) in Iran, but an older study of 400 camels from five East African countries found no evidence of the organism (El-Sayed et al. 2008).

Horses: Carriage of EHEC appears to be uncommon in horses (Lengacher et al. 2010) although a case report exists suggesting horse-to-man transmission of serotype O157 (Chalmers et al. 1997).

5.3.3 Other Animals

Deer: Prevalence of EHEC O157 in white-tailed deer has ranged from 0.25 to 2.4 % in various studies and non-O157 EHEC has been reported in 5 % of white-tailed deer feces (Rounds et al. 2012). In limited studies of deer meat from various species, non-O157 EHEC contamination ranged from 7.5 % in Germany to 22 % in Belgium (Miko et al. 2009).

Rabbits: In 2001, a cluster of EHEC O157 infections occurred in visitors to a wildlife park in the UK which was associated with wild rabbit feces. The rabbits were living in an adjacent field along with O157 colonized cattle (Scaife et al. 2006). Subsequently, Dutch Belted rabbits naturally infected with EHEC O153 were found to develop a hemolytic uremic syndrome-like disease (Garcia et al. 2006).

Dogs and cats: EHEC including serotype O157 have been isolated from dogs and cats (Kataoke et al. 2010; Franiek et al. 2012). Dog-to-human transmission has been reported to occur (ProMED-mail 2001)

5.4 The One Health Interface: Vehicles for Transmission

An infectious dose of EHEC is not entirely established and may vary for different strains and serotypes but given the various ways it can be transmitted including person-to-person spread, the ID_{50} (estimated number of organisms to infect 50 % of people) seems to be quite low. Studies (Paton and Paton 1998; Griffin et al. 1994) suggest that it is between 10 and 100 colony-forming units with the lower end occurring in the more susceptible: young children, the elderly and the immunocompromised.

5.4.1 Ground Beef

The seminal outbreak of EHEC O157:H7 (Riley et al. 1983) was associated with undercooked ground beef and this vehicle will remain prototypically associated with the infection. Commercial ground beef is not simply a piece of meat that is run through a grinder but rather it is usually produced in bulk using the parts of the cow remaining after primal cuts of beef (to produce steaks and roasts) are removed, usually from multiple cows and even different slaughterhouses. This lower grade product, called beef trim, tends to be cut from the parts of the cow more likely to have had contact with bovine sources of EHEC.

Prior to butchering, carcasses have antibacterial sprays or dips in an attempt to diminish pathogen burdens. These methods often include combinations of high-pressure ambient temperature water washing, hot water sprays, and warm lactic or acetic acid sprays. Although treatments decrease pathogen burden, care must be taken to avoid spreading of the contamination to other areas of the carcass (Castillo et al. 1998). Wolf et al. (2012) have reported that a 4.4 % lactic acid dip was superior to lactic acid spray for control in reducing pathogen counts, but Harris et al. (2012) emphasize that all these types of interventions require validation under industry conditions to ensure their effectiveness.

The US Department of Agriculture (USDA) Food Safety and Inspection Service (FSIS) has tested for *E. coli* O157:H7 in slaughter houses since 1996. Now FSIS tests for six other EHEC strains (O26, O45, O103, O111, O121, and O145). If any of these EHEC serotypes is detected, recall implementation is evaluated carefully considered. Primal and subprimal (subprimal cuts are meat larger than a steak or roast but smaller than a side of beef which are shipped to markets for final cutting to reduce processing costs and retard spoilage) cuts are not generally tested and may be surface contaminated, so home or retail store grinding is not recommended by the USDA. Even if EHEC-contaminated, adequate cooking to an internal temperature of at least 160 °F and appropriate kitchen hygiene (Table 5.2) essentially eliminates any risk of infection. Moreover, USDA-FSIS inspection of meat and poultry using Hazard Analysis Critical Control Point (HACCP), Standard Sanitation Operating Procedures and other techniques has helped significantly reduce EHEC (and other foodborne illness risks) in the farm to table food movement dynamic. In addition, more widespread use of food irradiation would likely eliminate more pathogenic microorganism contaminants from risk meat and poultry offered to consumers.

5.4.2 Non-ground Beef

When a roast or steak is contaminated on its cut surface, surface searing alone will eradicate pathogens. However, individual steaks are not necessarily intact per se. Many are sold nonintact, that is, after being mechanically and/or chemically

Table 5.2 Appropriate Hamburger Market and Kitchen Hygiene Practices

While shopping and packing, keep meat (and poultry) away from other foods
Refrigerate or freeze as quickly as possible
Defrost frozen meat in the refrigerator, not at room temperature
Wash hands with soap and hot water prior to preparing any foods
Do not taste the uncooked hamburger prior to adequate cooking
Wash hands with soap and hot water just after handling the raw meat
Cook until internal temperature of the hamburger is 160 °F
The internal color of the meat not being pink is NOT an adequate measure
It is preferable to flip the hamburger multiple times (Rhee et al. 2003)
After adequate cooking, do not place the meat (or other food) on the uncleaned cutting board or plate that held the uncooked meat to prevent cross-contamination

flavor enhanced by blade and needle tenderization, chemical injections, or hydrodynamic shock. These methodologies can transfer pathogens deeper in the meat, requiring more cooking to eradicate. Several outbreaks related to nonintact steaks have been reported (Laine et al. 2005; ProMED-mail 2007). In addition, EHEC has been transmitted via other beef vehicles including steak tartare (Greenland et al. 2009) and dried, fermented beef salami (MacDonald et al. 2004).

5.4.3 *Other Meat Reservoirs*

As noted in Sect. 5.3, EHEC has been isolated from a variety of non-cattle sources, for the most part there are only anecdotal reports of non-bovine meat as vehicles for transmission. EHEC contaminated bison meat has resulted in two separate recalls in the USA, at least one of the recalls being associated with human cases (ProMED-mail 2010). Some deer are EHEC colonized in the wild and human infection has been associated with both venison (Rounds et al. 2012) and jerky made from deer meat (Keene et al. 1997).

5.4.4 *Seafood*

A Japanese seafood outbreak of EHEC occurred in 1998 with 62 cases linked to salted salmon roe, where EHEC must have survived the salinity (Asai et al. 1999). Additionally, crab meat can be another vehicle for EHEC. In an outbreak in the UK during 2011, a statistically significant association between illness and the consumption of crab meat away from home and presumably contaminated was reported (ProMED-mail 2011a).

5.4.5 Direct Exposure to Animal Reservoirs

Direct exposure to livestock reservoirs of EHEC is closely linked to transmission, especially in children. Recommendations issued by the CDC (2011) to prevent enteric disease transmission associated with farm animal contact were originally prompted by two EHEC outbreaks occurring in the USA in 2000 (CDC 2001; Crump et al. 2002). Exposures can be directly at a farm or be associated with petting zoos such as those found at state fairs and other gatherings. Risk factors linked in most clusters include touching either the animals or their pens, feeding the animals and eating or drinking prior to hand washing. Developing a lather with soap and water while hand washing diminishes risk, demonstrating the value of thorough hand washing but drying one's hands on clothing can increase the risk by recolonization of the hands (CDC 2011).

Temporary animal exhibits including petting zoos, festivals and Christmas tree lots are particularly vulnerable to design flaws leading to increased transmission. One of the largest petting zoo outbreaks of *E. coli* O157:H7 was linked to the 2004 North Carolina State Fair (Goode et al. 2009). The outbreak involved 108 reported cases. Especially among children <5 years of age who visited the petting zoo, contact with animal manure and hand-to-mouth behaviors were strongly linked to disease. Genetically identical serotype O157:H7 was isolated from the environment. Of note, parental awareness of the risk of zoonotic diseases in petting zoo was protective but using alcohol-based hand sanitary rubs instead of soap and water was not. Tables 5.3 and 5.4 overview some of the important issues related to livestock in public settings.

Disease transmission in this environment may occur indirectly, even after the event itself has ended. The outbreak strain in a North Carolina State Fair petting zoo cluster was isolated from shoes and wood shavings stuck to a stroller in households of visitors (Goode et al. 2009). Additionally, in an Oregon county fair outbreak (Keene et al. 2004) the risk of illness was linked to certain environmental exposures, not directly to animals. Cases of EHEC illness have also occurred linked to buildings a week or more after a petting zoo had been in them with the outbreak strain still found in the environment 42 weeks after the event (Varma et al. 2003), underscoring the need for adequate decontamination of such sites. A puzzling Italian cluster of mixed EHEC serotypes was recorded over several months in 1993. No common food vehicles and no cattle exposures were found but the epidemiology of the outbreak suggested exposure to live poultry or chicken coops (Tozzi et al. 1994).

In almost all of the outbreaks related to livestock exposures, cases occur in visitors rather than a parallel group of adults and children with daily exposure to the colonized animals. This suggests that previous exposures to EHEC act to confer a degree of protection against symptomatic infection (Reymond et al. 1996; Wilson et al. 1996).

Table 5.3 Diminishing EHEC Risk in Livestock/Public Interaction in Non-Animal Areas (adapted from CDC (2011)

Non-Animal Areas
Do not permit animals, except service animals, in non-animal areas
Prepare, serve, and consume food and beverages only in non-animal areas
Provide soap and hand-washing facilities and display hand-washing signs where food or beverages are served
Transition Areas Between Non-Animal and Animal Areas
Entrance Transition Area
Post signs or otherwise notify visitors that they are entering an animal area and that there are risks linked to animal contact
Reinforce visitors not to eat, drink, smoke, place their hands in their mouths, or use bottles or pacifiers while in the animal area
Provide holding areas for strollers
Control visitor traffic to prevent overcrowding
A one way flow is preferred with separate entrance and exit points
Exit Transition Area
Post signs or otherwise notify visitors to adequately wash their hands when leaving the animal area
Provide accessible hand washing stations for all visitors, including children and persons with disabilities
Position staff venue members near exits to underscore the importance of hand washing with soap

Table 5.4 Diminishing EHEC Risk in Livestock/Public Interaction in Animal Areas (Adapted from CDC (2011))

No food or beverages allowed in animal areas
No toys, pacifiers, spill-proof cups, baby bottles, strollers or similar items allowed
No smoking or use of other tobacco products allowed
Discourage hand-to-mouth behaviors such as nail-biting and thumb sucking
Prevent contact with animal manure or soiled bedding
Do not allow children to play or sit in animal areas
Any overt hand soiling should have supervised hand washing immediately
Ensure regular animal food and water areas are not publicly accessible
Animal feeding allowed only if barriers present
Animal feed containers should never be ones eaten by humans (such as ice cream cones)
All manure and soiled animal bedding requires prompt removal from the area
Trained staff members should be present to prevent inappropriate interactions and reduce potential risks
Avoid transporting stored manure or soiled bedding through public areas
If feasible, disinfect animal areas at least once daily
Provide adequate ventilation for both animals and humans

5.4.6 Human-To-Human Transmission

Secondary cases of EHEC infection are reported in most outbreaks, particularly notable from one small child to another. In reviewing 90 serotype O157 outbreaks from several continents, Snedeker and colleagues (2009) reported that about 20 % of cases were secondary, especially in the household setting. A later study (Locking et al. 2011) reported that 11 % of total cases were secondary. Since transmission relates to the poor bathroom hygiene of small children, it has been suggested that promptly separating siblings, especially if the primary case is less than 5 years old, can limit secondary spread (Werber et al. 2008). Chains of transmission, however, seem uncommon.

Two other settings linked to EHEC transmission are infant daycare and the elderly adult nursing home. Both have in common a cohort of more susceptible individuals with poor sanitary habits. A Minnesota study (Belongia et al. 1993) reported person-to-person EHEC transmission in all nine daycare facilities examined with a median attack rate of 22 %. The median age of the primary case was 26 months. Introduction of EHEC into the daycare center is often from an asymptomatic infant who acquired the infection at home. After EHEC is introduced into this setting, screening policies should not be based solely upon reported symptoms since it is clear that asymptomatic shedding occurs. Because of this, in the presence of EHEC, screening of all children in attendance at the daycare setting with at least one stool examination for EHEC is recommended (Gilbert et al. 2008).

The geriatric nursing home is another ideal situation for EHEC spread. Introduction into the facility is more likely to be foodborne so the number of cases in the primary wave can be prominent. Carter et al. (1987) reported that older age and previous gastrectomy were risk factors for acquiring EHEC and higher mortality rates occur. In a review of nursing home-associated EHEC infections (Reiss et al. 2006), the authors underscored the need for considering EHEC in the differential diagnosis of any case of gastroenteritis or bloody stool in this setting because of the higher complication rate. As with daycare settings, infection can occur in workers at nursing homes but with less morbidity and mortality. Intrahospital spread of EHEC has occurred in the laboratory (Spina et al. 2005) as well as in the clinical setting to both patients and health care workers (Weightman and Kirby 2000). In this milieu, adequate hand hygiene is important in prevention.

5.4.7 Inanimate Interfaces

5.4.7.1 Water

EHEC has been transmitted via contaminated water from municipal or rural sources of drinking water as well as from water used for recreational or other purposes. EHEC contaminated drinking water is more likely to occur in small water systems or wells supplying rural townships or camps. In general, contamination occurs when cow manure seeps into shallow, unchlorinated, poorly protected water supplies after large rainstorms (Olsen et al. 2002). Examples include a 1998 outbreak, from contaminated well water, of EHEC O157:H7 (and some *Campylobacter*) causing 921 cases of diarrhea at a New York State county fair (CDC 1999), a 2000 outbreak involving an estimated 2,300 cases of EHEC O157:H7 in Walkerton, Ontario, Canada from the town's water supply (Bruce-Grey-Owen Sound Health Unit 2000) and a large 1992 outbreak of serotype O157:H- in southern Africa (Effler et al. 2001).

Contamination of recreational lakes has been linked with outbreaks of EHEC as well as other enteric pathogens, usually linked to domestic or wild animal fecal contamination or untreated sewage runoff (Bruce et al. 2003; McCarthy et al. 2001). Swimming pools can also be associated with EHEC (Verma et al. 2007), minimized by maintaining adequate chlorine levels. Contamination of the pool water is usually directly due to pool users.

5.4.7.2 Unpasteurized Dairy and Other Foods

The widespread adoption of pasteurization primarily for dairy products, starting whole scale in the early twentieth century, produced dramatic decreases in milk/dairy product-related infections, a milestone in the history of food safety (Tauxe and Esteban 2007). EHEC is only one of a multitude of infections that can be transmitted by raw milk including campylobacteriosis, salmonellosis, yersiniosis, listeriosis, brucellosis as well as tuberculosis due to *Mycobacterium bovis*. Some individuals believe that pasteurization removes nutritional factors present in raw milk but no added benefits from the raw product has been scientifically proven (Potter et al. 1984) and there any clear superiority of using "certified" raw milk.

In the USA, it is illegal to sell unpasteurized milk across state lines but each state has laws regarding its sale. Some ban the sale, some allow sale of "cow shares" where the consumer owns part of the cow, while others allow sale for pet consumption or for sale only at the farm. During the 20 years between 1973 and 1992, 40 (87 %) of the 46 outbreaks of unpasteurized milk-associated illness occurred in states where the sale of raw milk was legal (Headrick et al. 1998). The Program for Monitoring Emerging Diseases (ProMED) regularly reports on such outbreaks related to EHEC or other pathogens (www.promedmail.org). Internationally, unpasteurized cow's milk is much more available. Pasteurized milk can be cross-contaminated following pasteurization, especially if defects in the process occur. An outbreak of non-O157 EHEC infection linked to ice cream produced on a farm in Belgium (de Schrijver et al. 2008) and a UK outbreak from yogurt (Morgan et al. 1993) illustrate this point.

Since EHEC can be found in other milk-producing livestock, it would be expected that cases of this infection can be linked to such a vehicle as shown by a Czech Republic cluster of serotype O157:H7 associated with unpasteurized goat's milk (Bielaszewska et al. 1997). Fecal contamination of milk from sheep and goats is less likely than for cows as their feces tend to be more solid (Farrokh et al. 2012).

Fresh cheese made from unpasteurized milk can also serve as EHEC vehicles, as examples, cases have been reported with cheese from both raw caprine (Espie et al. 2006) and bovine milk (CDC 2000). Much raw bovine milk cheese produced is aged 60 or more days at an appropriate temperature which has been claimed to create conditions able to eradicate milkborne pathogens (FDA 21 CFR Part 133). Observations following epidemics of typhoid fever caused by cheddar cheese made from raw milk in the 1940s and survival data on *Brucella abortus* in cheddar led to the 60 day curing period in the USA (Johnson et al. 1990; D'Amico 2008).

Prior to the emergence of EHEC, it was already clear that enteric pathogens may survive beyond the 60 day curing period. Reitsma (1995) subsequently reported that EHEC O157 could survive the manufacture and curing of cheddar cheese and using a 5-strain inoculum of EHEC O157, Schlesser et al. (2006) also confirmed that 60 day aging could be inadequate to eliminate EHEC. Confirming these experimental findings, several outbreaks of EHEC O157 (Honish et al. 2005; McCollum et al. 2012) have been linked to aged raw milk cheese. The physiological responses of EHEC (Peng et al. 2011), as measured by stress responses to acid, osmotic pressure and heat shock contribute individually or in combination to survival of during raw milk cheese production.

Although foods with a pH of less than 4.6 are generally regarded as safe by the USA Food and Drug Administration (FDA), EHEC O157:H7 can survive such acidic environments in unpasteurized foods. Both unpasteurized apple juice (Cody et al. 1999) and apple cider (CDC 1997) have caused outbreaks of EHEC made with apples that were likely unwashed and fallen from the tree. Mayonnaise or mayonnaise-based dressings or sauces have also been implicated despite their acidic pH (Weagant et al. 1994). Experimentally, EHEC did not replicate but survived in commercial mayonnaise up to 55 days at refrigerator temperatures (Zhao and Doyle 1994).

5.4.7.3 Produce

Produce that is eaten with minimal or no cooking can transmit EHEC. Soiling occurs at any point in the food chain from field to market and beyond usually from inadvertent exposure to water contaminated by animal droppings due to inappropriate irrigation, flooding or direct exposure to EHEC-containing feces. A strawberry-associated EHEC outbreak was linked to contamination by wild deer droppings (ProMED-mail 2011b). Among produce associated with outbreaks of EHEC infection are spinach, various kinds of sprouts, lettuce, tomatoes, raw or lightly pickled cabbage, salad greens, green onions, cantaloupes, mangoes, and nuts including walnuts and hazelnuts. Cases have even been linked to raw flour made into commercial refrigerated cookie dough that had been eaten raw (Neil et al. 2012).

Sprouts are a very notable vehicle for EHEC. Sprouting generally involves germinating seeds (which may be contaminated) at temperatures higher than ambient temperature facilitating a higher bacterial inoculum. Sprouts have been the vehicle for two notable outbreaks. In July 1996, almost 10,000 cases of *E. coli* O157:H7 occurred primarily in schoolchildren in Sakai City, Osaka, Japan, transmitted by white radish sprouts (Michino et al. 1999). During 2011, an outbreak centered in northern Germany (Buchholz et al. 2011) involved a chimeric, non-zoonotic EHEC/EAEC *E. coli* O104:H4 spread by fenugreek sprouts and infecting almost 4,000 people. This unique outbreak produced one of the largest number of deaths ever reported from a Shiga toxin-producing *E. coli*, more than 50.

While washing produce with water before use (or in the field with water and/or antibacterial agents such as dilute bleach) is recommended to help remove any dirt, bacteria or residual pesticide, this procedure cannot always be successful in preventing EHEC or other pathogens from causing illness. Bacteria including EHEC are able to internalize into plants, survive and multiply (Deering et al. 2012). Internalization can occur in seedlings or adult plants from exposure to roots, stems, or leaves. Arbuscular mycorrhizal fungi, existing in symbiosis with plant root structures, appear to enhance the survival of EHEC in growing plants (Gurtler et al. 2013). These observations underscore the need to prevent the initial contamination of the produce at all stages of growth. Produce contamination can also be due to poor kitchen hygiene in either the home or in a commercial restaurant. In several chain restaurant-associated outbreaks of *E. coli* O157:H7, cross-contamination of multiple salad bar items occurred from meat (Jackson et al. 2000). Cross-contamination can also occur from using reusable grocery bags by not washing between uses (Williams et al. 2011).

5.5 EHEC Human Infection

Paralleling the peak time for EHEC excretion in cattle, both O157 and non-O157 EHEC infections show a summer and fall peak. In 2009, the US reported a total of 3,108 EHEC infections with a rate of 1.01 per 100,000 population (CDC 2012). States in the upper Midwest had the highest rates, whereas those in the South had the lowest. Only a small percentage of these illnesses are diagnosed as Scallan et al. (2011) estimated a total of about 231,000 EHEC illness in the US. In the European Union, 3,573 cases were reported for 2009, 0.75 cases/100,000 (EFSA 2012). In Japan, 3,878 cases were reported for 2009 for a rate of 3.1/100,000 (NIID 2010).

5.5.1 EHEC Clinical Illness

5.5.1.1 Gastrointestinal Symptoms

The incubation period of EHEC O157:H7 infection ranges from 2 to 12 days, so it may be substantially longer than other *E. coli* and different enteric infections with the exception of typhoid fever. Bell et al. (1994) has reported a mean incubation period of 3.7 days. In most cases, the initial diarrhea is non-bloody for up to 3 days, not associated with significant fever or abundant stool leukocytes. The peripheral blood leukocyte count is often elevated and the affected individuals may complain of substantial abdominal pain. The pathophysiology of this initial diarrhea may not be related to Shiga toxin as *E. coli* O157 strains that are inherently Shiga toxin negative can produce watery diarrhea (Allerberger et al. 2000).

Bloody diarrhea subsequently begins in as many as 90 % of symptomatic infections, sometimes with almost all blood and little diarrhea. Bloody stools are often the reason to present for medical care. Although normal human intestinal lining cells do not express the Gb3 receptor for Shiga toxin, the toxin does translocate to cause damage to the mucosa, resulting in bloody diarrhea and enters the bloodstream to reach susceptible vascular epithelium of target organs (Schüller 2011). The combination of afebrile patients with watery diarrhea and significant abdominal pain should alert the clinician to the possibility of EHEC infection, even before the onset of bloody diarrhea. Imaging studies of the colon may show a thumbprinting appearance and thickening of the colon wall suggestive of either colonic ischemia or pseudomembranous colitis. Indeed, in retrospect, some cases of colonic ischemia may have been caused by EHEC (Su et al. 1998). Vascular injury causing peripheral edema and decreased platelet counts are common during this phase of illness; neither sign predict the development of HUS. During the usual gastrointestinal phase, clinical signs of systemic inflammation are not present, the individuals are not febrile, hypotensive, or acidotic. Quite mild or asymptomatic infections with EHEC occur and may serve as a reservoir for secondary cases. Cases of HUS have been reported in patients without diarrhea but are found to have EHEC in their stool (Tarr et al. 2005). The length of EHEC shedding in the stool is typically 10–20 days, generally to be shorter in adults and asymptomatic individuals. Chronic carriage is rare but may be a cause of chronic diarrhea (Spacek et al. 2004).

Risk factors for the subsequent development of HUS compiled by Tarr and colleagues (2005) include children less than 10 years of age, elevated white blood cell counts, persistent low platelet counts without reversal, and the use of either antimicrobial agents or antimotility agents. The onset of definable HUS is seen about a week after the onset of diarrhea but ranges between 4 to 12 days after the onset of symptoms. High levels of stromal cell-derived factor-1 (SDF-1), also called chemokine ligand 12 (CXCL12), caused by activation of CXCR4/CXCR7 receptors, may distinguish those who subsequently develop HUS (Petruzziello-Pelligrini et al. 2012). This may also be true for increased urinary levels of vascular

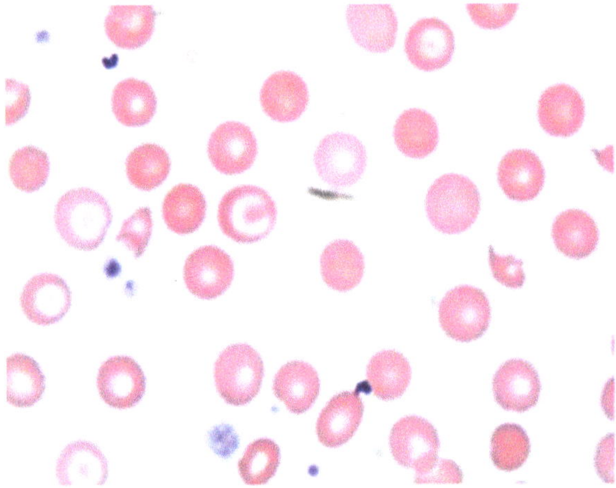

Fig. 5.2 A peripheral blood smear showing the misshapen schistocytes

endothelial growth factor (VEGF) (Stearns-Kurosawa et al. 2011). Extraintestinal infections with EHEC are uncommon but urinary tract infections have occurred and may (Starr et al. 1998) or may not (Gadea et al. 2012) be linked to HUS.

5.5.1.2 Thrombotic Microangiopathy

HUS and thrombotic thrombocytopenic purpura (TTP) are examples of the manifestations of microangiopathy. TTP per se is generally not associated with EHEC but the neurological signs and symptoms prominent in TTP do occur in HUS as poor prognostic features. HUS is a thrombotic illness caused primarily by the effects of Shiga toxin on the vascular endothelium of organs where the glycosphingolipid receptors for the toxin are expressed. The syndrome consists of the combination of prominent thrombocytopenia, intravascular hemolysis resulting in a prominent anemia with schistocytes (fragmented red blood cells) seen in the circulating blood (Fig. 5.2), and diminishing renal function that can result in overt kidney failure.

Shiga toxin induces dysfunction in the vascular endothelial cells of organs such as the kidney and brain with induction of proinflammatory mediators and widespread microvascular thrombosis. The coagulopathy differs from disseminated intravascular coagulation as coagulation factors are not consumed and markers of fibrinolysis are only mildly elevated. The thrombocytopenia is caused by direct activation of platelets causing aggregation and clearance by the reticuloendothelial system. Hemolytic anemia is related to mechanical injury of erythrocytes induced by passing through microthrombus-narrowed capillaries, especially in the renal vasculature. HUS targets the kidney vasculature particularly as the Gb3 toxin receptor is expressed at high density on glomerular endothelial cells (Obrig et al. 1993) and is the most common cause of acute renal failure in children.

Fibrin-rich microvascular thrombi occur in the renal glomeruli with endothelial swelling and detachment from the glomerular basement membrane, accompanied by subendothelial deposits and vessel lumen narrowing. A microvasculature-on-a-chip model has been developed to study the manifestations of HUS and could be useful in the understanding of the process and for drug discovery (Tsai et al. 2012).

Neurological involvement in diarrhea-associated HUS occurs primarily in those who develop renal failure, portending substantially higher mortality and sequelae (Nathanson et al. 2010). Manifestations include seizures which may be protracted, coma, cranial nerve abnormalities and stroke (Sheth et al. 1986; Hamano et al. 1993). Cardiac manifestations are uncommon, tending to occur during the acute phase of HUS related to fluid overload, hypertension, myocarditis or myocardial microangiopathic thrombosis, presenting as congestive heart failure (Poulton et al. 1987). Cardiac ischemia or infarction is rare.

As the microangiopathy resolves, renal function and thrombocytopenia improve. The hemolytic component of the process seems to be the last to resolve. In a mouse model, VEGF and angiopoietin 1 appear to play important roles in the regeneration of the endothelial injury (Ohara et al. 2012).

5.5.2 EHEC Management

5.5.2.1 Antimicrobials

The early observations regarding what was likely EHEC infection seemed to suggest that antimicrobial intervention had a deleterious effect by increasing HUS risk. Subsequent analyses for the most part underscored this risk of antimicrobial use. Wong et al. (2000) found that, in a multivariate analysis, antimicrobial administration was a substantial risk factor for HUS development with a relative risk of 17.3 and 95 % confidence interval 2.2 to 137. Smith et al. (2012) also reported an increased risk of HUS when bacteriocidal antimicrobials were used, within the initial either 3 or 7 days, after the onset of diarrhea. Not all studies showed such an association (Bell et al. 1997; Safdar et al. 2002) suggesting as the effect may be dependent on the particular strain of EHEC and the type of antimicrobial used.

Different classes of antimicrobials affect the amount of Shiga toxin released after exposure to subinhibitory concentrations of the drug. Using several different strains of EHEC O157, some which produced either Shiga toxin 1 or 2 or both, trimethoprim-sulfamethoxazole, trimethoprim, azithromycin and gentamicin markedly increased Shiga toxin levels in all strains, and cefixime, ceftriaxone, penicillin, ciprofloxacin and erythromycin increased toxin production in some (Grif et al. 1998). In another study using 5 different O157 strains (Walterspiel et al. 1992), significant differences between strains and different antimicrobials were found but consistently ciprofloxacin induced the largest increase in free Shiga toxin levels (as much as 436 %) with trimethoprim/sulfamethoxazole second followed by cefixime and tetracycline. Interestingly, Zhang et al. (2000) also reported in mice that, although both ciprofloxacin and fosfomycin reduced EHEC fecal levels, ciprofloxacin produced increased Shiga toxin levels and increased mortality in mice whereas fosfomycin did not. Ciprofloxacin was found to not only induce the toxin-related phage but also caused enhanced intraintestinal transfer of toxin prophage from one *E. coli* to another. The mechanism of increased Shiga toxin production with antimicrobial use is, in part, related to stimulation of the SOS stress reaction of the bacterium with the induction of the Shiga toxin prophage precipitating toxin production and cell lysis.

Some antimicrobials appear not to induce toxin production or phage-mediated lysis in EHEC including rifampin (Matar and Rahal 2003) and rifaximin (Ochoa et al. 2007). Additionally, imipenem, a carbapenem antimicrobial, does not appear to induce toxin increases (Takahashi et al. 1997). The latter antimicrobial, with broad spectrum antibacterial activity, could be used in the management of bacterial complications of EHEC infection such as bacteremia or colon perforation and even might be useful in treating EHEC. A rifamycin or fosfomycin might also have such a role in EHEC infection or in clearing colon colonization but, if possible, all antimicrobial use in EHEC infection should be avoided. Be that as it may, many EHEC strains, particularly non-O157 ones, can be multidrug resistant (Folster et al. 2011; Horoi et al. 2012).

5.5.2.2 Supportive Therapy

Few interventions have clearly had a modulating effect on preventing the progression of EHEC disease. Ake et al. (2005), however, reported that volume expansion using parenteral hydration with isotonic solutions, when begun in the first 4 days of illness, protected against the development of oligo- or anuric HUS. This therapy must be carefully monitored if HUS ensues. Supportive measures that should be avoided is the use of antimotility agents (including opiates) since they appear, when used in the first 3 days of illness, to prolong the duration of bloody diarrhea (Bell et al. 1997). Non-steroidal anti-inflammatory agents should also be avoided as this class of medications reduces renal blood flow. Hemodialysis is used if substantial kidney failure should occur and is continued until renal function begins to return. Plasmaphoresis, although pathologically rational in TTP or atypical, non-EHEC-associated HUS (Bu et al. 2012), it is not a first-line therapy for EHEC-linked HUS at any age (Tarr et al. 2012). Likewise, an orally administered Shiga toxin-binding agent given to children with HUS had no effect in diminishing the severity of disease (Trachtman et al. 2003). Once full blown HUS has developed, it may well be past the time for many interventions.

Table 5.5 Selected potential new modalities for EHEC treatment

Mechanism	Description	References
Circulating Stx Binding	Synthetic compounds with clustered trisaccharides can bind circulating Stx in the circulation	Nishikawa (2011)
Gb3 Downregulation	C-9, a specific inhibitor of glucosylceramide synthetase, decreased Gb3 expression in a rat model, diminishing colon and kidney damage and mortality	Silberstein et al. (2011)
O157 Engineered Pyocin	A specific engineered bacteriocin, Avr2-V10.3, prevented or ameliorated EHEC disease in an infant rabbit model when given as late as 3 days after O157 infection. It also reduced bacterial carriage and fecal shedding	Ritchie et al. (2011)
Quorum Sensing Inhibition	LED209, by inhibiting signal binding to the membrane histidine sensor kinase QseC, decreased the expression of Stx genes in response to a stressor or autoinducer	Rasko et al. (2008)
Stx Trafficking Inhibition	A cell-permeable peptide (TVP) that binds to Stx2 reduced disease severity in a baboon model and rescued primates from a lethal dose of toxin given beginning 24 h after Stx2 by preventing toxin from reaching its target	Stearns-Kurosawa et al. (2011) Watanabe-Takahashi et al. (2010)

5.5.2.3 Experimental Methodology

The following clinically available medications (licensed for other purposes) may be useful in the future management of EHEC disease, especially if instituted early in the course of illness. Certain other potential medical interventions are found in Table 5.5.

Manganese: Mukhopadhyay and Linstedt (2012) reported that non-toxic levels of manganese (Mn^{++}), a trace element that can used in intravenous and oral therapy, could protect HeLa cells against Stx-induced cell death. With a protective factor reported to be 3,800, the effect is mediated by degradation of trafficking protein GPP130, preventing Stx from avoiding inactivation. Pretreatment with non-toxic doses of Mn^{++} protected mice from subsequent Stx1-induced toxicity clinically and pathologically. Subsequent reports did not confirm this with either Stx (Gaston et al. 2013).

Plerixafor: As previously noted, Petruzziello-Pelligrini et al. (2012) reported that higher SDF-1 levels could distinguish children who subsequently developed HUS. The authors also showed that Stx enhanced transcription of CXCR4, CXCR7 and SDF-1 RNA and increased association of these RNAs with ribosomes. Inhibition of the CXCR4/SDF-1 receptor/ligand interaction with plerixafor, also known as AMD3100 or Mozobil (Debnath et al. 2013), decreased endothelial activation and organ injury with increased murine survival. CXCR4 (CXC chemokine receptor type 4), also known as fusin or CD (cluster of differentiation) 184, is mainly expressed on hematopoetic, immune and neuronal cells where it serves multiple functions. This receptor was initially discovered as one of the co-receptors for HIV infection. Plerixafor is already US FDA approved for use in autologous stem cell transplants in certain patients.

Eptifibatide: In a microvascular model (Tsai et al. 2012), the glycoprotein IIa/IIIa fibrinogen receptor antagonist eptifibatide that inhibits platelet aggregation was able to attenuate Stx2-induced thrombi formation and microchannel obstruction, especially at conditions of higher shear. Eptifibatide is currently approved by the US FDA as an anticoagulant for treatment of some patients with acute coronary syndrome. It is a cyclic heptapeptide that was synthesized based on barbourin, an anticoagulant in the venom of the pigmy rattlesnake, *Sistrurus miliarius babouri*.

Eculizumab: This humanized hybrid IgG2/IgG4 monoclonal antibody is approved by the American FDA for the treatment of atypical, non-diarrhea-associated HUS and paroxysmal nocturnal hemoglobinuria. The drug (Soliris) is directed against complement component 5 preventing C5 convertase from cleaving C5 into C5a, a potent proinflammatory moiety, and C5b, the principal initiator of the complement membrane attack complex of terminal complement components. Lapeyraque et al. (2011) reported the utility of eculizumab in 3 children with EHEC-associated HUS associated with dramatic neurological manifestations. With the recognition of an increasing role of complement in typical HUS, this biologic could play a role in severe cases. It was used in almost 200 individuals during the 2011 Germany-centered EHEC/EAEC O104:H4 outbreak but did not clearly produce a positive effect (Kielstein et al. 2012).

5.5.3 EHEC Clinical Outcomes

Depending upon individual risk factors and the strain of EHEC involved, 5–10 % of symptomatic infections will progress to HUS. Garg et al. (2003) reviewed the long term renal complications of EHEC-associated HUS in almost 3,500 children up to age 18 from 49 studies in 18 countries. Follow up ranged from a minimum 1 to 22 years. Using pooled averages, 10 % of survivors had high blood pressure, 15 % proteinuria, and 16 % diminished glomerular filtration rate (less than 80 mL/min). 25 % overall had some evidence of residual renal disease. 3 % had end stage kidney disease on dialysis and 9 % had died, usually during the acute illness from non-renal reasons. Oakes et al. (2006) reported 17 deaths among 358 children with HUS (4.7 %) with all but 2 of the deaths occurring during the acute phase of HUS, primarily from severe brain damage. Factors during the acute illness associated with a worse prognosis in these studies included an elevated white blood cell count (greater than 20,000/uL), central nervous system symptoms, ischemic colitis and hypertension. More prominent or prolonged periods of oliguria or anuria and a longer need for dialysis during the acute stage were also associated with worse prognoses. Indeed, in the Garg report, no patient with a dialysis need of more than 4 weeks achieved full renal recovery.

It is important to note that those with less severe HUS are not necessarily spared renal sequelae with evidence of renal dysfunction presenting after apparent kidney recovery. Rosales et al. (2012), in a report from Austria and Germany, found that 18 % of patients who had no sequelae at 1 year subsequently developed proteinuria or hypertension with no difference in sequelae based on EHEC serotype. Pre-HUS antimicrobial use did increase the need for acute phase dialysis but did not affect long-term sequelae. Additionally, the Rosales study reported that neurological sequelae persisted in 4 % of HUS-affected children, 12 % of those with neurological signs during HUS.

In children with self-limited EHEC gastrointestinal infection without HUS, Garg et al. (2006) has reported that no renal sequelae could be found in a follow up to 4 years but a report relating to adults from the same large outbreak in Walkerton, Ontario, Canada (Clark et al. 2010), found an increased risk of hypertension and renal disease. Another follow up study (Hizo-Abes et al. 2013), reported no long term increased risk of cardiovascular events in the same cohort.

5.6 The 2011 *E. coli* O104:H4 Outbreak

In 2011, an almost 4,000 person outbreak presumably caused by an EHEC occurred in 2011 centered in northern Germany. The organism was, however, found to be a chimeric, nonzoonotic *E. coli* that contained genetic material from both an enterohemorrhagic enteropath and an enteroaggregative one. Since that outbreak, very closely related isolates that were not derived from the 2011 strain have been described, sharing many of the same virulence genes and portending potential future outbreaks (Grad et al. 2013). As this organism is not zoonotic, it will not be discussed further.

References

Abu-Ali GS, Ouellette LM, Henderson ST et al (2010) Increased adherence and expression of virulence genes in a lineage of *Escherichia coli* O157:H7 commonly associated with human infection. PLoS One 5(4):e10167. doi:10.1371/journal.pone.0010167

Ake JE, Jelacic S, Ciol MA et al (2005) Relative nephroprotection during *Escherichia coli* O157:H7 infections: association with intravenous volume expansion. Pediatrics 115:e673–e680

Allerberger F, Dierich MP, Gruber-Moesenbacher U et al (2000) Nontoxigenic sorbitol-fermenting *Escherichia coli* O157:H- associated with a family outbreak of diarrhea. Wien Klin Wochenschr 112:845–850

Arthur TM, Keen JE, Bosilevec JM et al (2009) Longitudinal study of *Escherichia coli* O157:H7 in a beef cattle feedlot and role of high-level shedders in hide contamination. Appl Environ Microbiol 75:6515–6523

Arthur TM, Brichta-Harhay DM, Bosilevac JM et al (2010) Supershedding of *Escherichia coli* O157:H7 by cattle and the impact on beef carcass contamination. Meat Sci 86:32–37

Asadulghani M, Ogura Y, Ooka T et al (2009) The defective prophage pool of *Escherichia coli* O157: prophage-prophage interactions potentiate horizontal transfer of virulence determinants. PLoS Pathog 5(5):e1000408. doi:10.1371/journalppat.1000408

Asai Y, Murase T, Osawa R et al (1999) Isolation of Shiga toxin-producing *Escherichia coli* O157: H7 from processed salmon roe associated with the outbreaks in Japan, 1998, and molecular typing of the isolates by pulsed-field gel electrophoresis. Kanenshogaku Zasshi 73:20–24

Bandyopadhyay S, Lodh C, Rahaman H et al (2012) Characterization of Shiga toxin producing (STEC) and enteropathogenic *Escherichia coli* (EPEC) in raw yak (*Poephagus grunniens*) milk and milk products. Res Vet Sci 93:604–610

Barnard PJ, Kibel M (1965) The haemolytic-uraemic syndrome of infancy and childhood: a report of eleven cases. Cent Afr J Med 11(4–11):31–34

Bell BP, Goldoft M, Griffin PM et al (1994) A multistate outbreak of *Escherichia coli* O157:H7-associated bloody diarrhea and hemolytic uremic syndrome from hamburgers: the Washington experience. JAMA 272:1349–1353

Bell BP, Griffin PM, Lozano P et al (1997) Predictors of hemolytic uremic syndrome in children during a large outbreak of *Escherichia coli* O157:H7 infections. Pediatrics 100:e12

Belongia EA, Osterholm MT, Soler JT et al (1993) Transmission of *Escherichia coli* O157:H7 infection in Minnesota child day-care facilities. JAMA 269:883–888

Bergan J, Dyve Lingelem AB, Simm R et al (2012) Shiga toxin. Toxicon 60:1085–1107

Bielaszewska M, Janda J, Blahova K et al (1997) Human *Escherichia coli* O157:H7 infection associated with the consumption of unpasteurized goat's milk. Epidemiol Infect 119:299–305

Bielaszewska M, Köck R, Fredrich AW et al (2007a) Shiga toxin-mediated hemolytic uremic syndrome: time to change the diagnostic paradigm? PLoS One 2(10):e1024. doi:10.1371/journal.pone.0001024

Bielaszewska M, Prager R, Köck R et al (2007b) Shiga toxin loss and transfer in vitro and in vivo during enterohemorrhagic *Escherichia coli* O26 infection in humans. Appl Environ Microbiol 73:3144–3150

Bielaszewska M, Mellman A, Bletz S et al (2013) Enterohemorrhagic *Escherichia coli* O26/H11/H-: a new virulent clone emerges in Europe. Clin Infect Dis 56:1373–1381

Boqvist S, Aspan A, Eriksson E (2009) Prevalence of verotoxigenic *Escherichia coli* O157:H7 in fecal and ear samples from slaughtered cattle in Sweden. J Food Prot 72:1709–1712

Borriello G, Lucibelli MG, De Carlo E et al (2012) Characterization of enterotoxigenic *E. coli* (ETEC), Shiga-toxin producing *E. coli* (STEC) and necrotoxigenic *E. coli* (NTEC) isolated from diarrhoeic Mediterranean water buffalo calves (*Bubalus bubalis*). Res Vet Sci 93:18–22

Brigotti M (2012) The interactions of human neutrophils with Shiga toxins and related plant toxins: danger or safety? Toxins 4:151–190. doi:10.3390/toxins4030157

Bruce MG, Curtis MB, Payne MM et al (2003) Lake-associated outbreak of *Escherichia coli* O157:H7 in Clark County, Washington, August 1999. Arch Pediatr Adolesc Med 157:1016–1021

Bruce-Grey-Owen Sound Health Unit (2000) The investigative report of the walkerton outbreak of waterborne gastroenteritis May-June 2000. http://water.sesep.drexel.edu/outbreaks/WalkertonReportOct2000/REPORT_Oct00.PDF. Accessed 21 Dec 2012

Brunder W, Khan AS, Hacker J, Karch H (2001) Novel type of fimbriae encode by the large plasmid of sorbitol-fermenting enterohemorrhagic *Escherichia coli* O157:H-. Infect Immun 69:4447–4457

Bu F, Borsa N, Gianiluigi A, Smith RJ (2012) Familial atypical hemolytic uremic syndrome: a review of its genetic and clinical aspects. Clin Dev Immunol. doi:10.1155/2012/370426

Buchholz U, Bernard H, Werber D et al (2011) German outbreak of *Escherichia coli* O104:H4 associated with sprouts. N Engl J Med 365:1763–1770

Buvens G, Pierard D, Hachimi-Idrissi S, Lauwers S (2009) First sorbitol-fermenting Verocytotoxin-producing *Escherichia coli* O157:H- isolated in Belgium. Acta Clin Belg 64:59–64

Cai K, Gao X, Li T et al (2011) Enhanced immunogenicity of a novel Stx2Am-Stx1B fusion protein in a mice model of enterohemorrhagic coli O157:H7 infection. Vaccine 29:946–952

Callaway TR, Carr MA, Edrington TS, Anderson RC, Nisbet DJ (2009) Diet, *Escherichia coli* O157:H7, and cattle: a review after 10 years. Curr Issues Mol Biol 11:67–79

Carter AO, Borczyk AA, Carlson JA et al (1987) A severe outbreak of *Escherichia coli* O157:H7-associated hemorrhagic colitis in a nursing home. N Engl J Med 317:1496–1500
Castillo A, Lucia LM, Goodson KJ et al (1998) Comparison of water wash, trimming, and combined hot water and lactic acid treatments for reducing bacteria of fecal origin on beef carcasses. J Food Prot 61:823–828
CDC (1997) Outbreaks of *Escherichia coli* O157:H7 infection and cryptosporidiosis associated with drinking unpasteurized apple cider – Connecticut and New York, October 1996. MMWR 46:4–8
CDC (1999) Outbreak of *E. coli* O157:H7 and Campylobacter – New York, 1999. MMWR 48:803–804
CDC (2000) Outbreak of *Escherichia coli* O157:H7 infection associated with eating fresh cheese curds –- Wisconsin, June 1998. MMWR 49:911–913
CDC (2001) Outbreaks of *Escherichia coli* O157:H7 infection among children associated with farm visits – Pennsylvania and Washington, 2000. MMWR 50:293–297
CDC (2011) Compendium of measures to prevent disease associated with animals in public settings, 2011. MMWR 60(RR-4):1–24
CDC (2012) National enteric disease surveillance: Shiga toxin-producing *Escherichia coli* (STEC) annual summary, 2009. US Department of Health and Human Services, CDC, Atlanta, GA, pp 1–8
Chalmers RM, Salmon RL, Willshaw GA et al (1997) Vero-cytotoxin-producing *Escherichia coli* O157 in a farmer handling horses. Lancet 349:1816
Clark WF, Sontrop JM, Macnab JJ et al (2010) Long term risk of renal impairment and cardiovascular disease after gastroenteritis for drinking water contaminated with *Escherichia coli* O157:H7: a prospective cohort study. BMJ 341:c6020. doi:10.1136/bmj.c6020
Cobbold RN, Hancock DD, Rice DH et al (2007) Rectoanal junction colonization of feedlot cattle by *Escherichia coli* O157:H7 and its association with supershedders and excretion dynamics. Appl Environ Microbiol 69:1563–1568
Cody SH, Glynn MK, Farrar JA et al (1999) An outbreak of *Escherichia coli* O157:H7 infection from unpasteurized commercial apple juice. Ann Intern Med 130:202–209
Colpoys WE, Cochran BH, Carducci TM, Thorpe CM (2005) Shiga toxins activate translational regulation pathways in intestinal epithelial cells. Cell Signal 17:891–899
Crump JA, Sulka AC, Langer AJ et al (2002) An outbreak of *Escherichia coli* O157:H7 infection among visitors to a dairy farm. N Engl J Med 347:555–560
Cull CA, Paddock ZD, Nagaraja TG et al (2012) Efficacy of a vaccine and a direct-fed microbial against fecal shedding of *Escherichia coli* O157:H7 in a randomized pen-level field trial of commercial feedlot cattle. Vaccine 30:6210–6215
D'Amico D (2008) Incidence, ecology, and fate of target foodborne pathogens in the cheesemaking continuum. University of Vermont. http://library.uvm.edu/jspui/bitstream/123456789/165/1/damicofinal.pdf. Accessed 22 Dec 2012
de Schrijver K, Buvens G, Posse G et al (2008) Outbreak of verocytotoxin-producing *E. coli* O145 and O26 infections associated with the consumption of ice cream produced at a farm, Belgium, 2007. Eurosurveillance 13(7):pii = 8041. http://www.eurosurveillance.org/ViewArticle.aspx?ArticleId = 8041. Accessed 22 Dec 2012
Debnath B, Xu S, Grande F et al (2013) Small molecule inhibitors of CXCR4. Theranostics 3:47–75
Deering AJ, Mauer LJ, Pruitt RE (2012) Internalization of *E. coli* O157:H7 and *Salmonella* spp. in plants: a review. Food Res Int 45:567–575
Doane CA, Pangloli P, Richards HA et al (2007) Occurrence of *Escherichia coli* O157:H7 in diverse farm environments. J Food Prot 70:6–10
Donnenberg MS, Whittam TS (2001) Pathogenesis and evolution of virulence in enteropathogenic and enterohemorrhagic *Escherichia coli*. J Clin Invest 107:539–548
Edrington TS, Callaway TR, Ives SE et al (2006) Seasonal shedding of *Escherichia coli* O157:H7 in ruminants: a new hypothesis. Foodborne Pathog Dis 3:413–421

Effler E, Isaacson M, Arntzen L et al (2001) Factors contributing to the emergence of *Escherichia coli* O157 in Africa. Emerg Infect Dis 7:812–819
EFSA (2012) EU summary report, 2009. EFSA J 9:2090
El-Sayed A, Ahmed S, Awad W (2008) Do camels (*Camelus dromedarius*) play an epidemiological role in the spread of Shiga toxin producing *Escherichia coli* (STEC) infection? Trop Anim Health Prod 40:469–473
Espie E, Vaillant V, Mariani-Kurkdijan P et al (2006) *Escherichia coli* O157 outbreak associated with fresh unpasteurized goats' cheese. Epidemiol Infect 134:143–146
Farfan MJ, Torres AG (2012) Molecular mechanisms that mediate colonization of Shiga toxin-producing *Escherichia coli* strains. Infect Immmun 80:903–913
Farrokh C, Jordan K, Avray F et al (2012) Review of Shiga-toxin-producing *Escherichia coli* (STEC) and their significance in dairy production. Int J Food Microbiol. doi:10.1016/j.ijfoodmicro.2012.08.008
FDA 21 CFR Part 133 Cheeses and Related Cheese Products. http://law.justia.com/us/cfr/title21/21cfr133_main_02.html. Accessed 22 Dec 2012
Feng P, Lampel KA, Karch H, Whittam TS (1998) Genotypic and phenotypic changes in the emergence of *Escherichia coli* O157:H7. J Infect Dis 177:1750–1753
Ferens WA, Hovde CJ (2011) *Escherichia coli* O157:H7: animal reservoir and sources of human infection. Foodborne Pathog Dis 8:465–487
Folster JP, Pecie G, Taylor E, Whichard J (2011) Characterization of isolates from an outbreak of multidrug-resistant Shiga toxin-producing *Escherichia coli* O145 in the United States. Antimicrob Agents Chemother 55:5955–5956
Franiek N, Orth D, Grif K et al (2012) ESBL-producing *E. coli* and EHEC in dogs and cats in the Tyrol as possible source of human infection. Berl Munch Tierarztl Wochenschr 125:469–475
Gadea MD, Deza N, Mota MI et al (2012) Two cases of urinary tract infection caused by Shiga toxin-producing *Escherichia coli* O157:H7 strains. Rev Argent Microbiol 44:94–96
Gallegos KM, Conrady DG, Karve SS et al (2012) Shiga toxin binding to glycolipids and glycans. PLoS One 7(2):e30368. doi:10.1371/journal.pone.oo30368
Gamage SD, Strasser JE, Chalk GL, Weiss AA (2003) Nonpathogenic *Escherichia coli* can contribute to the production of Shiga toxin. Infect Immun 71:3107–3115
Garcia A, Bosques CJ, Wishnok JS et al (2006) Renal injury is a constant finding in Dutch Belted rabbits experimentally infected with enterohemorrhagic *Escherichia coli*. J Infect Dis 193:1125–1134
Garg AX, Suri RS, Barrowman N et al (2003) Long term renal prognosis of diarrhea-associated hemolytic uremic syndrome: a systematic review, meta-analysis, and meta-regression. JAMA 290:1360–1370
Garg AX, Clark WF, Salvadori M et al (2006) Absence of renal sequelae after childhood *Escherichia coli* O157:H7 gastroenteritis. Kidney Int 70:807–812
Gaston MA, Pellino CA, Weiss AA (2013) Failure of maganese to protect from Shiga toxin. PLoS One 8(7):e69823
Gautam R, Rani-Yaghoub M, Neill WH et al (2011) Modeling the effect of seasonal variation in ambient temperature on the transmission dynamics of a pathogen with a free-living stage: example of *Escherichia coli* O157:H7 in a dairy herd. Prev Vet Med 102:10–21
Gilbert M, Monk C, Wang H-L, Diplock K (2008) Screening policies for daycare attendees. Can J Pub Health 99:281–285
Goode B, O'Reilly C, Dunn J et al (2009) Outbreak of *Escherichia coli* O157:H7 infections after petting zoo visits, North Carolina State Fair, October-November 2004. Arch Pediatr Adolesc Med 163:42–48
Gould LH (2012) Update: recommendation for diagnosis of Shiga toxin-producing *Escherichia coli* infection by clinical laboratories. Clin Microbiol Newslett 34:75–83
Grad YH, Godfrey P, Cergulera GC et al (2013) Comparative genomics of recent Shiga toxin-producing *Escherichia coli* O104:H4: short-term evolution of an emerging pathogen. mBio 4 (1):e400452-12. doi:10.1128/mBio.00452-12

Greenland K, de Jager C, Heuvelink A et al (2009) Nationwide outbreak of STEC O157 infection in the Netherlands, December 2008–January 2009: continuous risk of consuming raw beef products. Euro Surveill 14(8):pii = 19129. http://www.eurosurveillance.org/ViewArticle.aspx?ArticleId=19129. Accessed 13 Dec 2012

Grif K, Dierich MP, Karch H, Allerberger F (1998) Strain-specific differences in the amount of Shiga toxin released from enterohemorrhagic *Escherichia coli* O157 following exposure to subinhibitory concentrations of antimicrobial agents. Eur J Clin Microbiol Infect Dis 17:761–767

Griffin PM, Bell BP, Cieslak PR et al (1994) Large outbreak of *Escherichia coli* O157:H7 in the western United States: the big picture. In: Karmali MA, Goglio AG (eds) Recent Advances in Verocytotoxin-producing *Escherichia coli* Infections. Elsevier Science, Amsterdam, pp 7–12

Gurtler JB, Douds DD Jr, Dirks BA et al (2013) Survival of *Salmonella* and *E. coli* O157:H7 in soil and translocation into leek (*Allium porrum*) as influenced by arbuscular mycorrhizal fungi (*G. intraradices*). Appl Environ Microbiol 79:1813–1820

Hajian S, Rahimi E, Mommtaz H (2011) A 3-year study of *Escherichia coli* O157:H7 in cattle, camel, sheep, goat, chicken and beef minced meat. Proc Int Proc Chem Biol Environ Eng 9:162–165

Hamano S-I, Nakanishi Y, Nara T et al (1993) Neurologic manifestations of hemorrhagic colitis in the outbreak of *Escherichia coli* O157:H7 infection in Japan. Acta Paediatr 82:454–458

Harris D, Brashears MM, Garmyn AJ et al (2012) Microbiological and organoleptic characteristics of beef trim and ground beef treated with acetic acid, lactic acid, acidifed sodium chlorite, or sterile water in a simulated commercial processing environment to reduce *Escherichia coli* O157:H7 and *Salmonella*. Meat Sci 90:783–788

Headrick ML, Korangy S, Bean NH et al (1998) The epidemiology of raw-milk-associated foodborne disease outbreaks reported in the United States, 1973 through 1992. Am J Public Health 88:1219–1221

Hizo-Abes P, Clark WF, Sontrop JM et al (2013) Cardiovascular disease after *Escherichia coli* O157:H7 gastroenteritis. CMAJ 185:E70–E77

Hoey DE, Sharp L, Currie C et al (2003) Verotoxin 1 binding to intestinal crypt epithelial cells results in localization to lysosomes and abrogation of toxicity. Cell Microbiol 5:85–97

Honish L, Predy G, Hislop N et al (2005) An outbreak of *E. coli* O157:H7 hemorrhagic colitis associated with unpasteurized gouda cheese. Can J Public Health 96:182–184

Horoi M, Takahashi N, Harada T et al (2012) Serotype, Shiga toxin (Stx) type, and antimicrobial resistance of Stx-producing *Escherichia coli* isolates from humans in Shiazuoka Prefecture, Japan (2003–2007). Jpn J Infect Dis 65:198–202

Islam MA, Mondol AS, de Boer E et al (2008) Prevalence and genetic characterization of Shiga toxin-producing *Escherichia coli* isolates from slaughtered animals in Bangladesh. Appl Environ Microbiol 74:5414–5421

Jackson LA, Keene WE, McAnulty JM et al (2000) Where's the beef? The role of cross-contamination in 4 chain restaurant-associated outbreaks of *Escherichia coli* O157:H7 in the Pacific Northwest. Arch Intern Med 160:2380–2385

Johnson EA, Nelson JH, Johnson M (1990) Microbiological safety of cheese made from heat-treated milk, Part 1. Executive summary, introduction and history. J Food Prot 53:441–452

Kang SJ, Ryu SJ, Chae JS et al (2004) Occurrence and characteristics of enterohemorrhagic *Escherichia coli* O157 in calves associated with diarrhea. Vet Microbiol 98:323–328

Kaplan BS, Chesney RW, Drummond KN (1975) Hemolytic uremic syndrome in families. N Engl J Med 298:927–933

Karch H, Bielaszewska M (2001) Sorbitol-fermenting Shiga toxin-producing *Escherichia coli* O157:H7 strains: epidemiology, phenotypic and molecular characteristics, and microbiological diagnosis. J Clin Microbiol 39:2043–2049

Karmali MA, Steele BT, Petric M, Lim C (1983) Sporadic cases of haemolytic-uraemic syndrome associated with faecal cytotoxin and cytotoxin-producing *Escherichia coli* in stools. Lancet 1:619–620

Karmali MA, Mascarenhas M, Shen S et al (2003) Association of genomic O island of *Escherichia coli* EDL 933 with verocytotoxin-producing *Escherichia coli* seropathotypes that are linked to epidemic and/or serious disease. J Clin Microbiol 41:4930–4940
Kataoke Y, Irie Y, Sawada T, Nakazawa M (2010) A 3-year epidemiological surveillance of *Escherichia coli* O157:H7 in dogs and cats in Japan. J Vet Med Sci 72:791–794
Keen JE, Elder RO (2002) Shiga-toxigenic *Escherichia coli* O157 from hide surfaces and the oral cavity of finished beef feedlot cattle. J Am Vet Med 220:756–763
Keene WE, Sazie E, Kok J et al (1997) An outbreak of *Escherichia coli* O157:H7 infections traced to jerky made from deer meat. JAMA 277:1229–1231
Keene W, de Broekert M, Gillette K (2004) A large *Escherichia coli* O157:H7 outbreak at a country farm (Abstract 55:77). In: Program and abstracts of international conference on emerging infectious disease, Atlanta, GA, February 28– 3 March
Kielstein JT, Beutel G, Fleig S et al (2012) Best supportive care and therapeutic plasma exchange with or without eculizumab in Shiga toxin-producing *E. coli* O104:H4 induced haemolytic uraemic syndrome. Nephrol Dial Transplant 27:3807–3815
Konowalchuk J, Speirs JI, Stavric S (1977) Vero response to a cytotoxin of *Escherichia coli*. Infect Immun 18:775–779
Kudva IT, Hatfield PG, Hovde CJ (1996) *Escherichia coli* O157:H7 in microbial flora of sheep. J Clin Microbiol 34:431–433
Kudva IT, Griffin RW, Krastins B et al (2012) Proteins other than the focus of enterocyte effacement-encoded proteins contribute to *Escherichia coli* O157:H7 adherence to bovine rectoanal junction stratified squamous epithelial cells. BMC Microbiol 12:103. doi:10.1186/1471-2180-12-103
La Ragione RM, Best A, Woodward MJ, Wales AD (2009) *Escherichia coli* O157:H7 colonization in small domestic animals. FEMS Microbiol Rev 33:394–410
Lahti E, Ruoho O, Rantala L et al (2003) Longitudinal study of *Escherichia coli* O157 in a cattle finishing unit. Appl Environ Microbiol 69:554–561
Laine ES, Scheftel JM, Boxrud DJ et al (2005) Outbreak of *Escherichia coli* O157:H7 infections associated with nonintact blade-tenderized frozen steaks sold by door-to-door vendors. J Food Prot 68:1198–1202
Lainhart W, Stolfa G, Koudelka GB (2009) Shiga toxin as a bacterial defense against a eukaryotic predator, *Tetrahymena thermophila*. J Bacteriol 191:5116–5122
Lapeyraque A-L, Malina M, Fremeaux-Bacchi V et al (2011) Eculizumab in severe Shiga-toxin-associated HUS. N Engl J Med 364:2561–2563
Lawrence JG, Ochman H (1998) Molecular archeology of the *Escherichia coli* genome. Proc Natl Acad Sci U S A 95:9413–9417
LeJeune JT, Hancock DD, Besser TE (2006) Sensitivity of *Escherichia coli* O157 detection in bovine feces assessed by broth enrichment followed by immunomagnetic separation and direct plating methodology. J Clin Microbiol 44:872–875
Lengacher B, Kline TR, Harpster L et al (2010) Low prevalence of *Escherichia coli* O157:H7 in horses in Ohio, USA. J Food Prot 73:2089–2092
Locking ME, Pollock KGJ, Allison LJ et al (2011) *Escherichia coli* O157 infection and secondary spread, Scotland, 1998–2008. Emerg Infect Dis 17:524–527
Lorusso V, Dambrosio A, Quaglia NC et al (2009) Verocytotoxin-producing *Escherichia coli* O26 in raw water buffalo (*Bubalus bubalis*) milk products in Italy. J Food Prot 72:1705–1708
Lutwick LI (2008) Edible vaccines: to eat, perchance to immunize. Curr Infect Dis Rep 10:439–440
MacDonald DM, Fyfe M, Paccagnella A et al (2004) *Escherichia coli* O157:H7 outbreak linked to salami, British Columbia, Canada, 1999. Epidemiol Infect 132:283–289
Manning SD, Motiwala AS, Springman AC et al (2008) Variation in virulence among clades of *Escherichia coli* O157:H7 associated with disease outbreaks. Proc Natl Acad Sci U S A 105:4868–4873

Matar GM, Rahal E (2003) Inhibition of the transcription of the *Escherichia coli* O157:H7 genes coding for shiga-like toxin and intimin, and its potential use in the treatment of human infection with the bacterium. Ann Trop Med Parasitol 97:281–287

Matsui T, Takita E, Sato T et al (2011) Production of double repeated B subunit of Shiga toxin 2e at high levels in transgenic lettuce plants as vaccine material for porcine edema disease. Transgenic Res 20:735–748

Matz C, Kjelleberg S (2005) Off the hook - how bacteria survive protozoan grazing. Trends Microbiol 13:302–307

McCarthy TA, Barret NL, Hadler JL et al (2001) Hemolytic-uremic syndrome and *Escherichia coli* O121 at a lake in Connecticut, 1999. Pediatrics 108:e59

McCollum JT, Williams NJ, Beam SW et al (2012) Multistate outbreak of *Escherichia coli* O157: H7 infections associated with in-store sampling of an aged raw-milk Gouda cheese, 2010. J Food Prot 75:1759–1765

McGee P, Scott L, Sheridan JI et al (2004) Horizontal transmission of *Escherichia coli* O157:H7 during cattle housing. J Food Prot 67:2651–2656

Michino H, Araki K, Minami S et al (1999) Massive outbreak of *Escherichia coli* O157:H7 infection in schoolchildren in Sakai City, Japan, associated with consumption of white radish sprouts. Am J Epidemiol 150:787–796

Miko A, Preis K, Haby S et al (2009) Assessment of Shiga toxin-producing *Escherichia coli* isolates from wildlife meat as potential pathogens for humans. Appl Environ Microbiol 75:6462–6470

Miller MB, Bassler BL (2001) Quorum sensing in bacteria. Annu Rev Microbiol 55:165–199

Morgan D, Newman CP, Hutchinson DN et al (1993) Verotoxin producing *Escherichia coli* O157 infections associated with the consumption of yoghurt. Epidemiol Infect 111:181–187

Morigi M, Galbusera M, Gastoldi M et al (2011) Alternative pathway of complement by Shiga toxin promotes exuberant C3a formation that triggers microvascular thrombosis. J Immunol 187:172–180

Mukhopadhyay S, Linstedt AD (2012) Manganese blocks intracellular trafficking of Shiga toxin and protects against Shiga toxicosis. Science 335:332–335

Nathanson S, Kwon T, Elmaleh M et al (2010) Acute neurological involvement in diarrhea-associated hemolytic uremic syndrome. Clin J Am Soc Nephrol 5:1218–1228

Neil KP, Biggerstaff G, MacDonald JK et al (2012) A novel vehicle for transmission of *Escherichia coli* O157:H7 to humans: multistate outbreak of *E. coli* O157:H7 infections associated with consumption of ready-to-bake commercial prepackaged cookie dough–United States, 200. Clin Infect Dis 54:511–518

Neupane M, Abu-Ali GS, Mitra A et al (2011) Shiga toxin 2 overexpression in *Escherichia coli* O157:H7 strains associated with severe human disease. Microbiol Pathog 51:466–470

NIID (2010) Infectious agents surveillance report 31:152–153

Nishikawa K (2011) Recent progress of Shiga toxin neutralizer for treatment of infections by Shiga toxin-producing *Escherichia coli*. Arch Immunol Ther Exp 59:239–247

O'Brien AO, Lively TA, Chen ME et al (1983) *Escherichia coli* O157:H7 strains associated with haemorrhagic colitis in the United States produce a *Shigella dysenteriae* 1 (Shiga)-like cytotoxin. Lancet 1:702

Oakes RS, Siegler RL, McReynolds MA, Pysher T, Pavia AT (2006) Predictors of fatality in post diarrheal hemolytic uremic syndrome. Pediatrics 117:1656–1662

Obrig TG, Moran TP, Brown JE (1987) The mode of action of Shiga toxin on peptide elongation of eukaryotic protein synthesis. Biochem J 244:287–294

Obrig TG, Louise CB, Lingwood CA et al (1993) Endothelial heterogeneity in Shiga toxin receptors and response. J Biol Chem 268:15484–15488

Ochoa TJ, Chen J, Walker CM et al (2007) Rifaximin does not induce toxin production or phage-mediated lysis of Shiga toxin-producing *Escherichia coli*. Antimicrob Agent Chemotherap 51:2837–2841

Ohara S, Kawasaki Y, Abe Y et al (2012) Role of vascular endothelial growth factor and angiopoietin 1 in renal injury in hemolytic uremic syndrome. Am J Nephrol 36:516–523

Olsen SJ, Miller G, Breuer T et al (2002) A waterborne outbreak of *Escherichia coli* O157:H7 infections and hemolytic uremic syndrome: implications for rural water systems. Emerg Infect Dis 8:370–375

Omisakin F, MacRae M, Ogden ID, Strachan NJ (2003) Concentration and prevalence of *Escherichia coli* O157 in cattle feces at slaughter. Appl Environ Microbiol 69:2444–2447

Paton JC, Paton AW (1998) Pathogenesis and diagnosis of Shiga toxin-producing *Escherichia coli* infections. Clin Microbiol Rev 11:450–479

Peng S, Tasara T, Hummerjohann J, Stephan R (2011) An overview of molecular stress response mechanisms in *Escherichia coli* contributing to survival of Shiga toxin-producing *Escherichia coli* during raw milk cheese production. J Food Prot 74:849–864

Petruzziello-Pellegrini TR, Marsden PA (2012) Shiga toxin-associated hemolytic uremic syndrome: advances in pathogenesis and therapeutics. Curr Opin Nephrol Hypertens 21:433–440

Petruzziello-Pelligrini TN, Yuen DA, Page AV et al (2012) The CXCR4/CXCR7/SDF-1 pathway contributes to the pathogenesis of Shiga toxin-associated hemolytic uremic syndrome in humans and mice. J Clin Invest 122:759–776

Pittman FE, Pittman JC, Humphrey CD (1974) Colitis following lincomycin therapy. Arch Intern Med 134:368–372

Potter ME, Kaufmann AF, Blake PA, Feldman RA (1984) Unpasteurized milk. The hazards of a health fetish. JAMA 252:2048–2052

Poulton J, Taylor CM, De Giovanni JV (1987) Dilated cardiomyopathy associated with haemolytic uraemic syndrome. Br Heart J 57:181–183

ProMED-mail (2001) *E. coli* O157, dog, human – UK. http://promedmail.org/direct.php?id=20010510.0899. Accessed 22 Dec 2012

ProMED-mail (2007) *E. coli* O157, steak - USA (PA): recall. http://www.promedmail.org/direct.php?id=20070426.1362. Accessed 13 Dec 2012

ProMED-mail (2010) *E. coli* O157 – USA (05): (Colorado, New York), bison meat, alert, recall. http://promedmail.org/direct.php?id=20100709.2286. Accessed 22 Dec 2012

ProMED-mail (2011a) *E. coli* O157 – UK (04): crab consumption. http://promedmail.org/direct.php?id=20111213.3585. Accessed 22 Dec 2012

ProMED-mail (2011b) *E. coli* O157 – USA (06): (Oregon), strawberry, dropping source. http://promedmail.org/direct.php?id=201108192520. Accessed 22 Dec 2012

Rasko DA, Moreira CG, Li DR et al (2008) Targeting QseC signaling and virulence for antibiotic development. Science 321:1078–1080

Reinstein S, Fox JT, Shi X, Alam MJ, Nagaraja TG (2007) Prevalence of *Escherichia coli* O157:H7 in the American bison (*Bison bison*). J Food Prot 70:2555–2560

Reiss G, Kunz P, Koin D, Keefe EB (2006) *Escherichia coli* O157:H7 infection in nursing homes: review of literature and report of a recent outbreak. JAGS 54:680–684

Reitsma CJ (1995) Survival of enterohemorrhagic *Escherichia coli* O157:H7 during the manufacture and curing of cheddar cheese, South Dakota State University, p 60

Reymond D, Johnson RP, Karmali MA et al (1996) Neutralizing antibodies to *Escherichia coli* Vero cytotoxin 1 and O157 lipopolysaccharide in healthy farm family members and urban residents. J Clin Microbiol 34:2053–2057

Rhee MS, Lee SY, Hillers VN et al (2003) Evaluation of consumer-style cooking methods for reduction of *Escherichia coli* O157:H7 in ground beef. J Food Prot 66:1030–1034

Riley LW, Remis RS, Helgerson SD et al (1983) Hemorrhagic colitis associated with a rare *Escherichia coli* serotype. N Engl J Med 308:681–685

Ritchie JM, Greenwich JL, Davis BM et al (2011) An *Escherichia coli* O157-specific engineered pyocin prevents and ameliorates infection by *E. coli* O157:H7 in an animal model of diarrheal disease. Antimicrob Agents Chemotherap 55:5469–5474

Rosales A, Hofer J, Zimmerhackl LB et al (2012) Need for long-term followup in enterohemorrhagic *Escherichia coli*-associated hemolytic uremic syndrome due to late emerging sequelae. Clin Infect Dis 54:1413–1421

Rounds JM, Rigdon CE, Muhl LJ et al (2012) Non-O157 Shiga toxin-producing *Escherichia coli* associated with venison. Emerg Infect Dis 18:279–282

Safdar N, Said A, Gangnon RE, Maki DG (2002) Risk of hemolytic uremic syndrome after antibiotic treatment of *Escherichia coli* O157:H7 enteritis: a meta-analysis. JAMA 288:996–1001

Sakurai Y, Tsuchiya H, Ikegami F et al (1979) Acute right-sided hemorrhagic colitis associated with oral administration of ampicillin. Dig Dis Sci 24:910–915

Salehi TZ, Tonelli A, Mazza A et al (2012) Genetic characterization of *Escherichia coli* O157:H7 strains isolated from the one-humped camel (*Camelus dromedarius*) by using microarray DNA technology. Mol Biotechnol 51:283–288

Scaife HR, Cowan D, Finney J, Kinghorn-Perry SF, Crook B (2006) Wild rabbits (*Oryctolagus cuniculus*) as potential carriers of verocytotoxin-producing *Escherichia coli*. Vet Rec 159:175–178

Scallan E, Hoekstra RM, Angulo FJ et al (2011) Foodborne illness acquired in the United States – major pathogens. Emerg Infect Dis 17:7–15

Schimmer B, Nygard K, Eriksen H-M (2008) Outbreak of haemolytic uremic syndrome in Norway caused by stx(2)-positive *Escherichia coli* O103:H25 traced to cured mutton sausages. BMC Infect Dis 8:41. doi:10.1186/1471-2334-8-41

Schlesser JE, Gerdes R, Ravishankar S et al (2006) Survival of a five-strain cocktail of *Escherichia coli* O157:H7 during the 60-day aging period of cheddar cheese made from unpasteurized milk. J Food Prot 69:990–998

Schoeni JL, Doyle MP (1994) Variable colonization of chickens perorally inoculated with *Escherichia coli* O157:H7 and subsequent contamination of eggs. Appl Environ Microbiol 60:2958–2962

Schüller S (2011) Shiga toxin interaction with human intestinal epithelium. Toxins 3:626–639. doi:10.3390/toxins3060626

Seker E, Yardimci H (2008) First isolation of *Escherichia coli* O157:H7 from faecal and milk specimens from anatolian water buffaloes (*Bubalus bubalus*) in Turkey. J S Afr Vet Assoc 79:167–170

Shere JA, Kaspar CW, Bartlett KJ et al (2002) Shedding of *Escherichia coli* O157:H7 in dairy cattle housed in a confined environment following waterborne inoculation. Appl Environ Microbiol 68:1947–1954

Sheth KJ, Swick HM, Haworth N (1986) Neurological involvement in hemolytic-uremic syndrome. Ann Neurol 19:90–93

Shimizu T, Sato T, Kawakami S et al (2007) Receptor affinity, stability and binding mode of Shiga toxins are determinants of toxicity. Microbial Pathog 43:88–95

Silberstein C, Lucero MS, Zotta E et al (2011) A glucosylceramide synthetase inhibitor protects rats against the cytotoxic effects of Shiga toxin 2. Pediatr Res 69:390–394

Smith KE, Wilker PR, Reiter PL et al (2012) Antibiotic treatment of *Escherichia coli* O157 and the risk of hemolytic uremic syndrome, Minnesota. Pediatr Infect Dis 31:31–41

Snedeker KG, Shaw DJ, Locking ME, Prescott RJ (2009) Primary and secondary cases of *Escherichia coli* O157 outbreaks: a statistical analysis. BMC Infect Dis 9:144, http://www.biomedcentral.com/1471–2334/9/144. Accessed 21 Dec 2012

Snedeker KG, Campbell M, Sargeant JM (2012) A systematic review of vaccinations to reduce the shedding of *Escherichia coli* O157 in the faeces of domestic ruminants. Zoonoses Public Health 59:126–138

Solheim MT, Sekse C, Urdahl AM et al (2013) Biofilm as an environment for dissemination of stx genes by transduction. Appl Environ Microbiol 79:896–900

Spacek LA, Hurley BP, Acheson DWK et al (2004) Shiga toxin-producing *Escherichia coli* as a possible etiological agent in chronic diarrhea. Clin Infect Dis 39:e46–e48

Sperandio V, Torres AG, Giron JA, Kaper JB (2001) Quorum sensing is a global regulatory mechanism in enterohemorrhagic *Escherichia coli* O157:H7. J Bacteriol 183:5187–5197
Spina N, Zansky S, Dumas N, Kondracki S (2005) Four laboratory-associated cases of infection with *Escherichia coli* O157:H7. J Clin Microbiol 43:2938–2939
Stanford K, Croy D, Bach SJ et al (2005) Ecology of *Escherichia coli* O157:H7 in commercial dairies in southern Alberta. J Dairy Sci 88:4441–4451
Starr M, Bennett-Wood V, Bigham AK et al (1998) Hemolytic-uremic syndrome following urinary tract infection with enterohemorrhagic *Escherichia coli*: case report and review. Clin Infect Dis 27:310–315
Stearns-Kurosawa DJ, Collins V, Freeman S et al (2011) Rescue from lethal Shiga toxin 2-induced renal failure with a cell-permeable peptide. Pediatr Nephrol 26:2031–2039
Stephens TP, Loneragan GH, Karunasena E, Brashears MM (2007) Reduction of *Escherichia coli* O157 and *Salmonella* in feces and on hides of feedlot cattle using various doses of a direct-fed microbial. J Food Prot 70:2386–2391
Stolfa G, Koudelka GB (2013) Entry and killing of *Tetrahymena thermophila* by bacterially produced Shiga toxin. mBio 4(1):e00416-12. doi:10.1128/mBio.00416-12
Su C, Brandt LJ, Sigal SH et al (1998) The immunohistological diagnosis of *E. coli* O157:H7 colitis: possible association with colonic ischemia. Am J Gastroenterol 93:1055–1059
Takahashi K, Narita K, Kato Y et al (1997) Low-level release of Shiga-like toxin (Verocytotoxin) and endotoxin from enterohemorrhagic *Escherichia coli* treated with imipenem. Antimicrob Agents Chemotherap 41:2295–2296
Tarr PI, Gordon CA, Chandler WL (2005) Shiga-toxin-producing *Escherichia coli* and haemolytic uraemic syndrome. Lancet 365:1073–1086
Tarr PI, Sadler JE, Chandler WL et al (2012) Immunoadsoption in patients with haemolytic uraemic syndrome. Lancet 379:516
Tauxe RV, Esteban EJ (2007) Advances in food safety to prevent foodborne diseases in the United States. In: Ward JW, Warren C (eds) Silent victories: the history and practice of public health in the twentieth century America. Oxford University Press, Oxford, UK
Toffler RB, Pingoud EG, Burrell MI (1978) Acute colitis related to penicillin and penicillin derivatives. Lancet 2:707–709
Tozzi AE, Niccolini A, Caprioli A et al (1994) A community outbreak of haemolytic-uraemic syndrome in children occurring in a large area of Northern Italy over a period of several months. Epidemiol Infect 113:209–219
Trachtman H, Cnaan A, Christen E et al (2003) Effect of an oral Shiga toxin-binding agent on diarrhea-associated hemolytic uremic children. JAMA 290:1337–1344
Tsai M, Kita A, Leach J et al (2012) In vitro modeling of the microvascular occlusion and thrombosis that occur in hematologic diseases using microfluidic technology. J Clin Invest 122:408–418
Varela NP, Dick P, Wilson J (2012) Assessing the existing information on the efficacy of bovine vaccination against *Escherichia coli* O157:H7 – a systematic review and meta-analysis. Zoonoses Public Health. doi:10.1111/j.1863-2378.2012.01523
Varma JK, Greene KD, Reller ME et al (2003) An outbreak of *Escherichia coli* O157 infection following exposure to a contaminated building. JAMA 290:2709–2712
Verma A, Bolton FJ, Fiefiled D et al (2007) An outbreak of *E. coli* O157 associated with a swimming pool: an unusual vehicle of transmission. Epidemiol Infect 135:989–992
Walle KV, Vanrompay D, Cox E (2012) Bovine innate and adaptive immune responses against *Escherichia coli* O157:H7 and vaccination strategies to reduce faecal shedding in ruminants. Vet Immunol Immunopath. doi:10.101016/j.vetimm.2012.09.028
Walterspiel JN, Ashkenazi S, Morrow AL, Cleary TG (1992) Effect of subinhibitory concentrations of antibiotics on extracellular Shiga-like toxin I. Infection 20:25–29
Watanabe-Takahashi M, Sato T, Dohi T et al (2010) An orally applicable Shiga toxin neutralizer functions in the intestine to inhibit the intracellular transport of the toxin. Infect Immun 78:177–183

Weagant SD, Bryant JL, Bark DH (1994) Survival of *Escherichia coli* O157:H7 in mayonnaise and mayonnaise-based sauces at room and refrigerator temperatures. J Food Prot 57:629–631
Weightman NC, Kirby PJ (2000) Nosocomial *Escherichia coli* O157 infection. J Hosp Infect 44:107–111
Weil E, Felix A (1916) Zur serologischen diagnose des fleckfiebers. Wien Klin Wochenschr 21:33–35
Werber D, Mason BW, Evans MR, Salmon RL (2008) Preventing household transmission of Shiga toxin-producing *Escherichia coli* O157: promptly separating siblings might be the key. Clin Infect Dis 46:1189–1196
Whitman WB, Coleman DC, Wiebe WJ (1998) Prokaryotes: the unseen majority. Proc Natl Acad Sci U S A 95:6578–6583
Whittam TS, Wachsmuth IK, Wilson RA (1988) Genetic evidence of clonal descent of *Escherichia coli* O157:H7 associated with hemorrhagic colitis and hemolytic uremic syndrome. J Infect Dis 157:1124–1133
Whittam TS, Wolfe ML, Wachsmuth IK et al (1993) Clonal relationships among *Escherichia coli* strains that cause hemorrhagic colitis and infantile diarrhea. Infect Immun 61:1619–1629
Wick LM, Qi W, Lacher DW, Whittam TS (2005) Evolution of genomic content in the stepwise emergence of *Escherichia coli* O157:H7. J Bacteriol 187:1783–1791
Williams DL, Gerba CP, Maxwell S, Sinclair RG (2011) Assessment of the potential for cross-contamination of food products by reusable shopping bags. Food Protect Trends 31:508–513
Wilson JB, Clarke RC, Renwick SA et al (1996) Vero cytotoxigenic *Escherichia coli* infection in dairy farm families. J Infect Dis 174:1021–1027
Wolf MJ, Miller MF, Parks AR et al (2012) Validation comparing the effectiveness of a lactic acid dip with a lactic acid spray for reducing *Escherchia coli* O157:H7, *Salmonella*, and non-O157 Shiga toxigenic *Escherichia coli* on beef trim and ground beef. J Food Prot 75:1968–1973
Wong CS, Jelacic S, Habeeb RL et al (2000) The risk of the hemolytic-uremic syndrome after antibiotic treatment of *Escherichia coli* O157:H7 infections. N Engl J Med 342:1930–1936
Zhang X, McDaniel AD, Wolf LE et al (2000) Quinolone antibiotics induce Shiga toxin-encoding bacteriophage, toxin production, and death in mice. J Infect Dis 181:664–670
Zhang XH, He KW, Zhao PD et al (2012) Intranasal immunization with Stx2B-Tir-Stx1B-Zot protein leads to decreased shedding in goats after challenge with *Escherichia coli* O157:H7. Vet Rec 170:178. doi:10.1136/vr.100325
Zhao T, Doyle MP (1994) Fate of enterohemorrhagic *Escherichia coli* O157:H7 in commercial mayonnaise. J Food Prot 57:780–783

Chapter 6
Bartonellosis: A One Health Perspective

Elizabeth L. Pultorak, Ricardo G. Maggi, and Edward B. Breitschwerdt

Abstract Recently, *Bartonella*, a genus comprised of fastidious, Gram-negative, intracellular bacteria, has rapidly expanded from approximately 4–5 species in the 1990s to 30–40 named or proposed species that are currently described in the literature. While Bartonellae are highly adapted to a number of mammalian reservoir hosts, substantial genetic diversity allows for the transmission of *Bartonella* to humans (i.e. a zoonosis) and other non-reservoir hosts. Improvements in the diagnostic sensitivity of laboratory tests have allowed researchers to identify a spectrum of *Bartonella* species in novel host species, including domestic animals and wildlife. Because *Bartonella* can infect a spectrum of accidental hosts, it is imperative that researchers determine the consequences of *Bartonella* spp. infection on new host populations and the impact of this genus during environmental changes, particularly natural disasters. Rapid, accurate diagnostic tests now afford researchers and clinicians the ability to conduct more reliable epidemiologic and treatment related studies that will better characterize the outcomes of *Bartonella* spp. infection in animals and human patients. Prevalence studies have shown that *Bartonella* spp. DNA can be found in numerous arthropods, though additional research is required to identify epidemiologically important competent vectors of *Bartonella* spp. transmission. Future *Bartonella* spp. research initiatives should strive to not only identify the bacterium in arthropod and host populations, but also to combine those data with epidemiologically relevant case data and ecological risk assessment. Combined efforts from professionals in animal and human medicine, vector ecology, pathogen evolution, microbial ecology, and sociology are necessary in order to create the veterinary and medical infrastructure, training, prevention, and surveillance required to thoroughly understand *Bartonella* spp. ecology, dynamics and control. Bartonellosis may prove to be one of the most important One Health emerging zoonotic diseases of the next decade.

Keywords *Bartonella* • Intracellular pathogen • One Health • Vector-borne • Zoonoses

---

E.L. Pultorak • R.G. Maggi • E.B. Breitschwerdt (✉)
Department of Clinical Sciences, Intracellular Pathogens Research Laboratory, Center for Comparative Medicine and Translational Research, College of Veterinary Medicine, North Carolina State University, Raleigh, NC 27607, USA
e-mail: elpultor@ncsu.edu; ed_breitschwerdt@ncsu.edu

A. Yamada et al. (eds.), *Confronting Emerging Zoonoses*,
DOI 10.1007/978-4-431-55120-1_6

6.1 Introduction

In the past several decades, the emergence or re-emergence of infectious diseases and the increasing number of human pathogens that have emerged from animal reservoirs have gained much attention. Zoonotic diseases represent a constant threat to humans, pets and production animals, and wildlife populations. Persistent endemic zoonoses create substantial regional health problems around the world, and account for one billion cases of human illness and millions of deaths each year (Karesh et al. 2012). Transmission of pathogens is determined by interactions among arthropods, wildlife, domestic animals, and the social, economic and cultural practices of humans. Ecological changes that favor the proliferation, persistence, and movement of arthropod species throughout the world will result in the continued emergence of vector borne diseases.

The genus *Bartonella* is comprised of over 30 species of zoonotic, vector-transmitted intracellular bacteria that are able to infect a large number of vertebrate hosts. Because *Bartonella* spp. can infect a range of mammalian hosts and are found in a spectrum of arthropod vectors, understanding the ecology and transmission dynamics of these pathogenic bacteria requires an integrated approach, more rigorous than typically used for other vector-borne diseases. The existence of Bartonellosis in humans and animals dates back hundreds of years. However, due to advances in biotechnology, molecular diagnostic techniques, and subsequent newly discovered syndromes associated with infection, this genus of bacteria has only recently been acknowledged as an emerging pathogen.

Frequent interactions with companion animals that are infested with fleas and ticks place humans in developed, as well as developing countries, at increased risk of infection with *Bartonella* spp. In recent years, a rapid proliferation of case reports and epidemiological surveys indicate that *Bartonella* spp. have a significant impact on human populations in developing countries. Because *Bartonella* spp. can affect the health of populations in both developed and developing countries, collaboration among various types of researchers, animal and human health clinicians, and governments across countries is required to develop effective prevention and intervention strategies.

Within the past 25 years, the number of *Bartonella* species that have been identified in a variety of mammals has dramatically increased. On an evolutionary basis, *Bartonella* species have become highly adapted to one or more mammalian reservoir hosts, and are able to induce a long-lasting intra-erythrocytic or endotheliotropic infections. Species within the *Bartonella* genus are transmitted mainly by arthropod vectors, such as fleas, sand flies, lice, biting flies, potentially ticks and other vectors.

Frequent reports of *Bartonella* infection began in early 1915 during World War I with *Bartonella quintana*, the causative agent of trench fever (Brenner et al. 1993; Kostrzewski 1950; Swift 1920) which affected over one million soldiers in the trenches of the Western front and across Europe and the Middle East during World War I, and a smaller proportion of soldiers during World War II (Relman 1995).

The manifestations of trench fever ranged from a mild influenza-like illness, including fever, malaise, bone pain, and transient macular rash, to debilitating protracted and recurrent disease (Swift 1920; Relman 1995; Vinson et al. 1969). Patterns of fever had been noted in several soldiers with recurrent disease, including a single episode of fever, continuous fever for 5–7 days, and recurrent episodes of fever occurring every 4–5 days (Vinson et al. 1969; McNee and Renshaw 1916). The cyclical nature of febrile illness in soldiers with trench fever caused it to be nicknamed '5-day fever'. Once a soldier became infected, *B. quintana* could circulate in the blood for weeks to months to longer than a year (Vinson et al. 1969).

Soon after the first reports of trench fever, scientists began to emphasize that transmission of *B. quintana* occurred between persons in frequent contact with each other (Byam and Lloyd 1920). Though initially described as a virus, the bacterium was observed in the blood of infected patients and the feces of human body lice (Byam and Lloyd 1920). Early experimental transmission studies proved that lice were the vector of trench fever. Researchers could successfully transmit the disease by allowing lice to first feed on several patients with trench fever before infecting healthy volunteers (Byam and Lloyd 1920). In the 1960s, Vinson successfully fulfilled Koch's postulates by inoculating volunteers with *B. quintana* and observed them develop clinical disease consistent with trench fever (Vinson et al. 1969). The incidence of trench fever increased during cold, winter months, due to people staying indoors in close contact, and the wearing of underclothing for long periods of time (Kostrzewski 1950).

Trench fever had long been regarded as a disease occurring only in wartime, or associated with louse infestation and crowded, unsanitary conditions. Until recently, this Gram-negative bacterium was rarely recognized within industrialized countries, including the United States. In the 1980s, two members of the *Bartonella* genus, identified as *B. quintana* and *B. henselae*, were implicated as agents of severe or fatal disease in individuals infected with Human Immunodeficiency Virus (HIV) (Regnery et al. 1995). In 1990, an organism with DNA sequences closely related to *B. quintana* was identified in an HIV patient with bacillary angiomatosis, a vascular proliferative disorder that occurs in HIV patients (Koehler et al. 1992). In 1986, special blood culture techniques identified a novel *Bartonella* species in patients with HIV, in addition to immunocompetent individuals with relapsing illness (Slater et al. 1990). This novel *Bartonella* species was named *B. henselae* (Regnery et al. 1992; Welch et al. 1992). The spectrum of clinical manifestations in immunocompromised HIV patients, concomitantly infected with *Bartonella*, has since increased to include blood culture-negative endocarditis, vasoproliferative lesions of the skin, liver, lung, bone, and brain and osteolytic or osteoproliferative lesions. Prior to the diagnostic recognition of this genus, undiagnosed *Bartonella* infection in HIV-positive patients was shown to produce fatal consequences (Spach et al. 1993; Koehler and Tappero 1993; Cockerell et al. 1987).

In addition to HIV-positive patient populations, documentation of infection with *Bartonella* spp. began to emerge in other immunocompromised populations, where infection has caused persistent fever in transplant patients and acute graft rejection complications in two renal transplant patients (Lienhardt et al. 2009; Thudi et al. 2007). In liver transplant patients, *B. henselae* can cause hepatic granulomas, hepatic masses, or disseminated bacteremic disease (Humar and Salit 1999; Bonatti et al. 2006). It has been shown that *Bartonella* species can survive throughout the storage period of red blood cell units, and therefore transmission of *Bartonella* from asymptomatic blood donors to blood transfusion recipients may cause additional morbidity and mortality in immunocompromised populations (Magalhães 2008).

In the early 1990s, reports from the United States and France described cases of *B. quintana* in various homeless populations (Spach et al. 1995a, b; Drancourt et al. 1995). In 1995, researchers in Seattle reported ten patients, not infected with HIV, with bacteremia caused by *B. quintana*. These patients reported symptomatic febrile illness that was clinically similar to classical descriptions of trench fever (Spach et al. 1995b). The infections were mostly in homeless individuals suffering from chronic alcoholism—a condition facilitated by close contact with other individuals with poor hygiene,—previously described risk factors for trench fever. Later studies would show that alcoholism is significantly associated with positive antibody titers (>64) to *B. quintana* (Jackson et al. 1996).

A simultaneously published report from France described *B. quintana* in HIV-negative homeless individuals who later developed endocarditis (Drancourt et al. 1995). *B. quintana* was also found to be a significant cause of *Bartonella* blood culture-negative endocarditis, which had a poor prognosis with a 20 % fatality rate, became a frequent diagnosis in HIV-negative homeless individuals, and may have contributed to increases in endocarditis-related deaths among the homeless (Fournier et al. 2001a). It was later reported that approximately 5–10 % of homeless people in Marseilles, France had persistent *B. quintana* bacteremia (Brouqui et al. 1999). Other epidemiological studies, including one conducted in emergency rooms in Marseilles, have shown that 30 % of 71 tested homeless persons had antibody titers against *B. quintana* and that 14 % were bacteremic (Brouqui et al. 1999). A second study from Marseilles demonstrated that 50 (5.4 %) of 930 non-hospitalized, homeless persons tested during 4 years (2000–2003) had *B. quintana* bacteremia (Brouqui et al. 2005).

The presence of *B. quintana* in body lice has recently been reported in Moscow, Russia (12.3 %), rural Andean villages in Peru (1.4 %), and various countries in Africa (2.3–93.9 %) (Fournier et al. 2002; Raoult and Roux 1999). In Mexico, the presence of *B. quintana* in body lice was first reported in 1954 from a homeless population in Mexico City (Varela et al. 1954). Current reports estimate a *B. quintana* infection rate of 28 % in body lice from prisoners and homeless people in Mexico City (Alcantara et al. 2009). In Tokyo, Japan, 57 % of homeless patients had immunoglobulin (Ig) G titers >128 for *B. quintana* (Sasaki et al. 2006). A later study in Japan examined clothing from 12 homeless persons for body lice. These authors found that lice from 2 (16.7 %) of 12 homeless persons were positive for *B. quintana* by PCR (Sasaki et al. 2002). The recent emergence of *B. quintana* among homeless populations throughout the world may be due to increases in poverty and recrudescent human body louse infestations. A higher seroprevalence of *B. quintana* has been reported in injection drug users (IDUs), and high frequency of injection has been associated with a higher prevalence of *Bartonella* antibodies

(Comer et al. 1996). However, it is currently unclear if transmission can occur through needling sharing of IDUs. Additionally, while not previously documented, transmission of *B. quintana* in homeless/IDU populations through direct contact should not be excluded. Several factors in the homeless population, such as crowded and unsanitary living conditions, inability to change clothes regularly, and close exposure to people potentially carrying ectoparasites, predispose homeless individuals to louse infestations.

6.2 Hosts

6.2.1 Humans

In addition to *B. quintana*, humans are also the natural host reservoir for *B. bacilliformis*, the causative agent of Carrion's disease, which characteristically causes massive hemolytic anemia and vasoproliferative skin lesions, and is found primarily in the Andes region of Peru, Columbia, and Ecuador (Ihler 1996). Until 1993, *B. bacilliformis* was the only member of the *Bartonella* genus. At that time, the *Rochalimaea* and *Bartonella* genera were merged to accommodate an expanding list of *Bartonella* species, including *B. henselae*, *B. quintana*, *B. vinsonii*, and *B. elizabethae* (Brenner et al. 1993). *B. bacilliformis* was first identified in erythrocytes in 1905 by Alberto Barton, though archeological evidence suggests its presence in human society for several millennia (Minnick and Battisti 2009).

Oroya fever, or verruga peruana, a sequelae of chronic infection by *B. bacilliformis*, has been documented in anthropomorphic pottery jugs and carvings made by pre-Columbian Indians. In the 1540s, tales of epidemic acute febrile illness followed by 'warts full of blood' were passed down from Spanish conquistadores in South America (Minnick and Battisti 2009). Infection with *B. bacilliformis* can be relatively short lived and potentially life threatening, due to the severe hemolytic anemia (Minnick and Battisti 2009).

In stark contrast to the often acute and fulminant nature of *B. bacilliformis* infections, other major pathogenic *Bartonella*, including *B. henselae*, *B. quintana*, *B. vinsonii* subsp. *berkhoffii*, and *B. koehlerae*, can establish a chronic intracellular parasitism of erythrocytes, endothelial cells and macrophages in humans (Minnick and Battisti 2009). As previously discussed, *Bartonella* was first documented as a cause of disease in immunocompromised individuals. Subsequently, *B. henselae* was identified as the predominant cause of cat scratch disease (CSD). CSD is typically characterized by self-limiting lymphadenopathy and generalized flu-like symptoms in immunocompetent individuals. However, in 10 % to more recently 20 % of human cases, atypical symptoms of CSD can manifest, including encephalopathy, endocarditis, granulomatous hepatitis and nephritis, bacillary angiomatosis and peliosis hepatis, osteomyelitis, arthropathy, myalgia, and neurological issues. (Houpikian and Raoult 2005; Armengol and Hendley 1999; Kahr et al. 2000; Koehler et al. 1997; Maman et al. 2007; Giladi et al. 2005; Breitschwerdt et al. 2008). While immune suppression is a risk factor for more severe clinical manifestations of CSD, it has been recently shown that a subset of immunocompetent individuals can remain chronically infected with *Bartonella* and produce more atypical symptoms, as listed above. Evidence suggests that at least 14 *Bartonella* spp. can cause pathology in humans, with a wide range of symptoms being reported by persistently bacteremic individuals. Pathogenic *Bartonella* spp. species can be found in Table 6.1.

6.2.1.1 Factors Leading to Re-emergence

Bartonella quintana DNA has been identified in the dental pulp in the remains of soldiers in Napoleon's Grand Army, demonstrating its presence in humans for hundreds of years (Raoult et al. 2006). The detection of *Bartonella* spp., a historically old bacterial pathogen, in patients with new disease manifestations led clinicians and researchers to wonder why bacteremic disease caused by *Bartonella* was beginning to re-emerge. Possible explanations include the heightened surveillance of disease combined with improved methods of microbiologic detection that were being developed throughout the 1990s and 2000s. *Bartonella* spp. comprise a fastidious genus of organisms whose growth is not efficiently supported through routine blood-culture methods used in the majority of clinical microbiology laboratories. Isolation of *Bartonella* spp. on blood agar plates requires a lengthy incubation period of 21 days, and even then, is rarely met with success (Maggi and Breitschwerdt 2005). In 1990, researchers found that isolation of the bacterium was enhanced by lengthening the routine maintenance of plates of lysis-centrifugation cultures to at least 2 weeks, (Slater et al. 1990). Shortly after, Drancourt et al. developed a technique that involved the co-cultivation of blood with endothelial cells (Drancourt et al. 1995). In 2005, Maggi et al. reported an insect-based liquid culture medium that supports the growth of at least seven *Bartonella* species (Maggi et al. 2005a). However, limitations of conventional and new culture-based assays include the prolonged incubation period required for isolation of this fastidious pathogen, special growth requirements, and the potential of a relapsing bacteremia.

The use of polymerase chain reaction (PCR) has given researchers and microbiologists the ability to rapidly detect and identify the DNA of *Bartonella* spp. in a variety of blood, effusion, and tissue specimens. Studies have indicated that a combinatorial approach that independently tests blood, serum, and enrichment culture by PCR can improve the diagnostic documentation of *Bartonella* spp. bacteremia (Maggi et al. 2011). Due to the potential relapsing bacteremia associated with *Bartonella* spp. infected patients, serial testing of blood samples from a 3-day collection period may further enhance the diagnostic sensitivity of *Bartonella* spp. detection in bacteremic patients (Pultorak et al. 2013). Rapid changes in test sensitivity for the detection of *Bartonella* spp., in conjunction with enhanced understanding of the microbial ecology of these bacteria, have resulted in the

Table 6.1 List of pathogenic *Bartonella* spp.

Species	Reservoir	Accidental host	Vector	Distribution	Human diseases	References
B. bacilliformis	Human	No	Sandfly (*Lutzomyia verrucarum*)	Andes (Peru, Ecuador, Colombia, Bolivia, Chile, Guatemala)	Carrions disease: Oroya fever and verruga peruana	Ihler (1996), Minnick and Battisti (2009)
B. quintana	Human	No	Body Louse (*Pediculous humanis corporis*)	Worldwide	Trench fever, persistent bacteremia, endocarditis, bacillary angiomatosis	Raoult and Roux (1999), Byam and Lloyd (1920), Relman (1995)
B. clarridgeiae	Cat (*Felis catus*)	Humans, dogs (*Canis lupus familiaris*)	Cat flea (*Ctenocephalides felis*)	Europe and the United States	Cat-scratch disease	Kordick et al. (1997)
B. elizabethae	Rodents (*Rattus norvegicus*)	Humans, dogs (*C. lupus familiaris*)	Fleas	Worldwide	Endocarditis, neuroretinitis	Fenollar et al. (2005), Raoult et al. 2006, Daly et al. (1993)
B. grahamii	Bank voles (*Clethrionomys glareolus*)	Humans	Fleas, Ticks?	Worldwide	Neuroretinitis, uveitis	Serratrice et al. (2003)
B. henselae	Cat (*F. catus*)	Humans, dogs (*C. lupus familiaris*)	Cat flea (*C. felis*), ticks?	Worldwide	Endocarditis, Cat Scratch Disease, Bacillary angiomatosis & peliosis, neuroretinitis, bacteremia, peliosis hepatis, Granulomatous lymphadenitis	Regnery et al. (1995), Koehler et al. (1992), Slater et al. (1990), Welch et al. (1992), Breitschwerdt et al. (2008), Koehler et al. (1997)

(continued)

Table 6.1 (continued)

Species	Reservoir	Accidental host	Vector	Distribution	Human diseases	References
B. koehlerae	Cat (*F.catus*)	Humans, dogs (*C. lupus familiaris*)	Fleas (*C. felis*)	Worldwide	Endocarditis, Neurological disorders, febrile illness	Avidor et al. (2004), Breitschwerdt et al. (2010b)
B. vinsonii subsp. *arupensis*	White-footed mice (*Peromyscus leucopus*)	Humans, dogs (*C. lupus familiaris*)	Fleas? Ticks?	North America, Russia, Thailand, Nepal, France	Endocarditis, febrile illness	Fenollar (2005), Raoult et al. (2006), Daly et al. (1993)
B. vinsonii subsp. *berkhoffii* genotypes I,II,III	Coyotes (*Canis latrans*)	Humans, dogs (*C. lupus familiaris*), cats (*F.catus*)	Ticks?	Worldwide	Endocarditis, persistent bacteremia, neurological manifestations	Breitschwerdt et al. (2005), Breitschwerdt et al. (2007a), Breitschwerdt et al. (2008), Breitschwerdt et al. (2009), Roux et al. (2000)
B. washoensis	Ground squirrel (*Spermophilus beecheyi*)	Humans, dogs (*C. lupus familiaris*)	Fleas? Ticks?	North America	Endocarditis, fever, myocarditis	Kosoy et al. (2003)

B. alsatica	Rabbits (*Oryctolagus cuniculus*)	Humans	Unknown	Europe	Endocarditis, Granulomatous lymphadenitis	Fenollar et al. (2005), Raoult et al. (2006), Daly et al. (1993)
B. tamiae	Rodents (*Rattus* spp.)	Humans	Mites	Asia	Febrile illness	Kosoy et al. (2008)
B. rochalimaea	Foxes (*Vulpes* spp.), raccoons (*Prycon* spp.), coyotes (*Canis* spp.)	Humans	Fleas	Worldwide	Bacteremia, fever	Eremeeva et al. (2007), Henn et al. (2009)
B. melophagi	Sheep (*Ovis aries*)	Humans	Sheep ked (*Melophagus ovinus*)	United States, Europe	Pericarditis, Febrile illness	Maggi et al. (2009)
B. tribocorum	Rodents (*Rattus* spp.)	Humans	Fleas	Worldwide	Febrile illness	Kosoy et al. (2010b), Gundi et al. (2012)
B. rattimassiliensis	Rodents (*Rattus* spp.)	Humans	Fleas	Worldwide	Febrile illness	Kosoy et al. (2010b), Gundi et al. (2004)

expansion of the *Bartonella* genus from approximately 4–5 organisms in the 1990s, to 30–40 *Bartonella* species that are currently reported in the literature (Kosoy et al. 2012).

Additional factors continue to influence the rapid increase in the number of *Bartonella* species that are being reported. A variety of animals have been identified as reservoirs for different *Bartonella* species. These animals and their respective vectors play an important role in the ecological maintenance of these bacteria within various environmental niches and thereby pose a risk for *Bartonella* transmission to humans and other non-reservoir hosts. Multi-host pathogens have been implicated in the emergence of new diseases in humans and animals, and transmission strategies that increase encounters with a new host, such as transmission through an arthropod vector in the case of *Bartonella*, can increase the range of hosts that a pathogen can infect (Woolhouse et al. 2001). For example, recent reports suggest that *B. henselae* may be a multi-host and multi-vector pathogen, rather than a bacterium that is limited to cats and their fleas. Additionally, the substantial antigenic variation among *Bartonella* strains appears to contribute to differences in virulence and may allow the bacterium to easily exploit new hosts and vectors (Pedersen et al. 2005). The host range of *Bartonella* spp. may be expanding into new populations of mammals, such as *B. henselae* in marine mammals in coastal estuaries (Maggi et al. 2005b), resulting in more frequent documentation of this genus of bacteria.

From an evolutionary perspective, *Bartonella* spp. appear to be highly adapted to both an arthropod vector and a reservoir host that have developed an immunologic adaptation to allow for intravascular persistence of the bacterium. This high degree of host and vector specificity would indicate that spill-over events into non reservoir hosts do not happen frequently. However, recent progress in *Bartonella* spp. research has been aimed at identifying various species of this bacterium in novel wildlife, domestic animal, or human hosts. Although molecular techniques afford researchers and clinicians the ability to continue to identify *Bartonella* spp. among novel host populations, we are still unable to identify which *Bartonella* species are most likely to be both a virulent pathogen within a given host population and also capable of spillover or sustained transmission within that population. Understanding the ecology of *Bartonella* spp. requires not only knowledge of the prevalence of infection in reservoir and non-reservoir hosts, but the underlying issues that drive transmission of *Bartonella* within human, wildlife, and domestic animal populations. In the context of One Health, this genus poses many challenges and many opportunities.

6.2.2 Bartonellae of Rodents and Small Mammals

Several *Bartonellae* found in small mammals, such as wild mice, squirrels and rats, have been linked to human disease, including *B. grahamii* with ocular syndromes, *B. elizabethae* with endocarditis, *B. vinsonii* subsp. *arupensis* with fever and bacteremia, and *B. washoensis* with fever and myocarditis (Gil et al. 2010). Recent evidence from southeast Asia suggests that rodent *Bartonella* spp. may play an important and previously unrecognized role as a source of infection in stray dogs and febrile human patients (Kosoy et al. 2010b; Bai et al. 2012). Small mammals and rodents are responsible for maintaining the highest number of *Bartonella* species, including currently unnamed *Bartonella* (Gundi et al. 2009; Inoue et al. 2010). *Bartonella* species are continuing to be discovered in novel potential reservoir species of small mammal and rodent populations around the world, including Europe, Africa, Asia, North and South America (Lin et al. 2012; Kosoy et al. 1997, 2003; Pretorius et al. 2004; Gundi et al. 2004; Jardine et al. 2005; Castle et al. 2004; Winoto et al. 2005; Ying et al. 2002). Based upon several lines of evidence, studies support the hypothesis that *Bartonellae* have very effectively co-evolved with rodent species as reservoir hosts (Telfer et al. 2007b).[1]

First, closely related phylogenetic groups of *Bartonella* are present in populations of related species of rodents separated by considerable geographical distances and land barriers (Castle et al. 2004; Kosoy et al. 1997; Bai et al. 2008). Additionally, experimental infection of some rodents, including cotton rats (*Sigmodon hispidus*) and white-footed mice (*Peromyscus leucopus*), resulted in bacteremia only when variants were originally isolated from the same species or a close taxonomic relative (Kosoy et al. 2000). However, the diversity of *Bartonella* species that can exist within a single rodent species or individual is highly variable. Multiple distinct *Bartonella* species have been documented in a single rodent species (Birtles et al. 2001) or within a rodent population within a single site (Kosoy et al. 1997).

The relative abundance of each *Bartonella* spp. within these rodent populations is thought to be dependent on host density of the rodent species, density of flea species (with peak activity during peak flea season, or seasonal changes in host behavior that result in significant amounts of transmission by means other than fleas), and spill-over infection between each species of rodent (Telfer et al. 2007a). Importantly, transplacental transmission has been documented for some rodent species (Kosoy et al. 1998). It has recently been shown that species of bats in Kenya, southwest England, Taiwan, Central America, and South America can harbor *Bartonella* spp. (Lin et al. 2012; Kosoy et al. 2010a; Concannon et al. 2005; Bai et al. 2011, 2012). Phylogenetically, *Bartonella* strains exhibit high genetic diversity in bats, but isolates cluster in specific host bat species (Kosoy et al. 2010a, b; Bai et al. 2012).

---

[1] *Bartonella* species that are able to infect rodents and small mammals include *B. elizabethae* (Ying et al. 2002), *B. grahamii* (Ellis et al. 1999), *B. vinsonii* subsp. *arupensis* (Welch et al. 1999), *B. doshiae* (Birtles et al. 1995), *B. taylorii* (Birtles 1995), *B. vinsonii* subsp. *vinsonii* (Ellis et al. 1999), *B. tribocorum* (Heller et al. 1998), *B. washoensis* (Kosoy et al. 2003), *B. rattimassiliensis* (Gundi et al. 2004), *B. phoceensis* (Gundi et al. 2004), *B. birtlesii* (Bermond et al. 2000), *B. rochalimae* (Lin et al. 2008), *B. japonica* (Inoue et al. 2010), and *B. silvatica* (Inoue et al. 2010).

6.2.3 Bartonellae of Domestic and Feral Cats

B. henselae was first isolated from cats in the early 1990s (Koehler et al. 1994). Domestic cats are the primary, but seemingly not the sole reservoir for *B. henselae*, the etiologic agent of cat scratch disease (CSD), *B. clarridgeiae*, and *B. koehlerae*, and have been found to be accidental hosts for other species of *Bartonella*, such as *B. quintana* and *B. bovis*, which are less adapted to the cat host and therefore able to produce a variety of severe clinical manifestations (Chomel et al. 2004, 2006; Breitschwerdt and Kordick 2000; Breitschwerdt et al. 2007b). Cats are carriers of *B. henselae*, showing few clinical manifestations of infection. However, experimental studies of cats with *B. henselae* have shown that some cats may experience transient fever, anorexia, and lymphadenopathy following infection (Boulouis et al. 2005; Guptill et al. 1997). Bacteremia in cats may last for months to years, which may allow for *Bartonella* to be more efficiently transmitted among cat populations by fleas, and potentially other vectors (Chomel et al. 2006; Guptill et al. 2004). Mechanisms of transmission cat-to-cat and cat-to-human are provided below.

Seroprevalence studies have indicated that a high proportion of cats worldwide have been exposed to *Bartonella* spp., and can therefore act as a substantial reservoir for infection in humans and other animals. However, recent molecular microbiological evidence indicates that the predominant *B. henselae* strains isolated from cats are not the predominant strains isolated from human patients with CSD (Bouchouicha et al. 2009). Seroprevalence to *B. henselae* antigens in cats tends to vary by geographic region and is highest (40–70 %) in areas with warm or humid climates that are well suited to support flea populations.

In the United States, regional seroprevalence in domestic and feral cats can vary from 13 to 90 %, and a higher seroprevalence correlates to outdoor exposure (Jameson et al. 1995; Guptill et al. 2004). Elsewhere around the world, seroprevalence has ranged from 7 to 48 %.[2] There is often a great discrepancy between antibody and bacteremia prevalence in cats, with no correlation between seropositivity and active infection (McElroy et al. 2010). Studies have shown that cats younger than 1 year of age are more likely to be bacteremic than older cats (Guptill et al. 2004), and that feral cats are more likely to be bacteremic for *Bartonella* spp. than pet cats from within the same geographic region, presumably due to the higher prevalence of parasite infestation within feral cat populations (Breitschwerdt and Kordick 2000).

Historically, it has been thought that *B. henselae* and *B. clarridgeiae* can exist as part of a healthy intravascular microbial flora in cats for months to years

[2] Studies have documented seroprevalences in cats of 47 % in Hawaii, 15 % in Japan, 40.6 % and 41.8 % in pet and feral cats in the United Kingdom, respectively, 54 % in Indonesia, 53 % in France, 40 % in Israel, 11 % in Egypt, 7 % in Portugal, 48 % in Singapore, 21 % in South Africa, and 24 % in Zimbabwe (Kelly et al. 1996; Nasirudeen and Thong 1999; Childs et al. 1995; Maruyama et al. 1996; Demers et al. 1995; Barnes et al. 2000)

(Breitschwerdt and Kordick 2000). While most naturally infected cats are able to tolerate the chronic bacteremia of *Bartonella* spp. infection without obvious clinical abnormalities, epidemiological evidence suggests a link between *Bartonella* infection and various clinical manifestations in cats. Several case reports and studies have suggested a link between *B. henselae* and ocular diseases in cats, such as uveitis and conjunctivitis (Ketring et al. 2004; Lappin and Black 1999; Lappin et al. 2000). *B. henselae* DNA has also been amplified from cats with fatal pyogranulomatous myocarditis and diaphragmatic myositis in the United States (Varanat et al. 2012)

In a non-controlled retrospective study, *B. henselae* antibodies were associated with seizures and other neurological manifestations in cats (Pearce et al. 2006; Leibovitz et al. 2008). In a very recent study, *B. henselae* antibody titer was associated with hyperglobulinemia, which suggests that persistent feline infection with *B. henselae* may be of pathogenic importance, potentially contributing to feline immune complex diseases (Whittemore et al. 2012). However, further studies are needed to fully elucidate any potential role of *Bartonella* spp. in these and other feline diseases.

B. henselae and *B. clarridgeiae* have been shown to be transmitted through inoculation of infected cat blood (Kordick et al. 1999). Therefore, cats should be screened by blood culture and serology, and cats that are seropositive or blood culture positive should not be used as feline blood donors. Prevention of *Bartonella* infection in cats can be accomplished by avoiding exposure to fleas by routine use of arthropod control measures and by limiting exposure to other infected animals. *B. henselae* can be transmitted from cats to humans via a bite or scratch, resulting in clinical illness in the non-reservoir human host. Kim et al. documented *B. henselae* DNA in over 40 % of the saliva, nails, and blood samples of feral cats in Korea (Kim et al. 2009). *Bartonella* DNA has also been detected in the saliva of cats in Iran (Oskouizadeh et al. 2010), in addition to the United States, where either *B. henselae* or *B. clarridgeiae* was amplified from 31 % of skin biopsies, 18 % of both gingival and claw bed swabs, and 57 % of blood samples of feral cats (Lappin and Hawley 2009). It is thought that transmission may occur via bite or scratch if cat blood or flea excrement contaminates the bite site, cat saliva, or nail bed. Despite the risk of transmission via cat scratch, there is no evidence or studies indicating that declawing cats decreases the probability of transmission of *B. henselae* among cats and to non-reservoir species (Guptill 2010). Currently, there is no evidence to support direct cat to cat transmission through bites or scratches that may occur during tomcat fights.

6.2.4 Bartonellae of Canids

Bartonellae have been reported in wild canids, such as coyotes (*Canis latrans*), gray foxes (*Urocyon cinereargenteus*), red foxes (*V. vulpes*), and domestic dogs across North America and Europe, and wild canids have been implicated as reservoir hosts for *B. vinsonii* subsp. *berkhoffii* (Henn et al. 2009; Gabriel et al. 2009; Marquez et al. 2009). Based upon sequence differences in the 16–23 s intergenic spacer region (ITS) and bacteriophage associated heme binding protein Pap31 gene, four genotypes of *B. vinsonii* subsp. *berkhoffii* have been described (Breitschwerdt et al. 2007a; Chang et al. 2000b; Maggi et al. 2006). Studies have reported a *B. vinsonii* subsp. *berkhoffii* seroprevalence of 35–76 % in coyotes in California and a PCR detection rate of 28 % (Chang et al. 1999, 2000b). The seroprevalence of *Bartonella* in gray foxes in the United States is extraordinarily high, ranging from 89 % in Northern California (Sykes et al. 2007), 63 % in Santa Rosa Island, Florida (Schaefer et al. 2011), to 50 % in Texas (Schaefer et al. 2012).

B. vinsonii subsp. *berkhoffii* is able to cause persistent, prolonged bacteremia in domestic dogs that may result in severe pathophysiological and clinical manifestations (Kordick and Breitschwerdt 1998; Breitschwerdt and Kordick 2000). There is evidence for substantial strain variation. The variability among strains may reflect host adaptation, strain variation due to geographic origin, or random events (Cadenas et al. 2008). *B. vinsonii* subsp. *berkhoffii* was first isolated from a domestic dog with endocarditis in 1995 (Breitschwerdt et al. 1999). A later study using molecular methods found a prevalence of *Bartonella* infection in 28 % of dogs with infective endocarditis at the UC-Davis School of Veterinary Medicine in California, United States (MacDonald et al. 2004). Since that time, *Bartonella* infection in dogs has been serologically associated with or detected in dogs with myocarditis, cardiac arrhythmias, anemia/thrombocytopenia, arthropathy, epistaxis, neurological and neurocognitive abnormalities, granulomatous inflammatory diseases, vasoproliferative disorders, such as *peliosis hepatis*, bacillary angiomatosis, and splenic disease, such as hemangiosarcoma (Breitschwerdt et al. 1999, 2004, 2005; Tuttle et al. 2003; MacDonald et al. 2004; Smarick et al. 2004; Pesavento et al. 2005; Cadenas et al. 2008; Diniz et al. 2009; Pappalardo et al. 2000; Kitchell et al. 2000; Yager et al. 2010; Varanat et al. 2011).

In the context of One Health, dogs may serve as an important naturally-occurring model for human disease manifestations and vice versa (Breitschwerdt et al. 2013). Domestic dogs may also serve as environmental sentinels, as dogs can be infected with several *Bartonella* spp. or subspecies.[3] (Breitschwerdt et al. 2010a), and the clinical manifestations of *Bartonella* infection in dogs are similar to those seen in human patients. Early studies documented a seroprevalence of 4 % to *B. vinsonii* subsp. *berkhoffii* antibodies among sick dogs in North Carolina and Virginia (Pappalardo et al. 1997). A more recent study that measured the molecular prevalence of *Bartonella* spp. documented 9 % of dogs with *Bartonella* spp. DNA in their blood (Perez et al. 2011). Seroprevalence studies in other parts of the world have documented higher exposure rates of *Bartonella* spp. in dogs, including 38 % in Bangkok, Thailand, 7 % in Sri Lanka, 0.8–6 % in San Paulo, Brazil, 0 % in rural dogs in Vietnam, and 47 % of dogs in Iraq (Diniz et al. 2007; Suksawat et al. 2001;

[3] Including *B. vinsonii* subsp. berkhoffii, *B. henselae*, *B. vinsonii* subsp. *arupensis*, *B. clarridgeiae*, *B. washoensis*, *B. elizabethae*, *B. quintana*, *B. koehlerae*, *B. bovis*, *B. rochalimae*.

Brenner et al. 2013; Chomel et al. 2012). Dogs have been implicated in recent reports of *Bartonella* transmission to humans. In a prevalence study in 52 dogs, *Bartonella* DNA was found in five of nine oral swabs and five of nine nail clippings (Tsukahara et al. 1998). Studies have also identified *Bartonella* spp. in the saliva of dogs (Duncan et al. 2007). A recent seroprevalence study in China identified an association between having being bitten by a dog and human *Bartonella* seropositivity, suggesting that a large stray dog population may pose a risk for *Bartonella* spp. transmission, as is the case for rabies transmission (Sun et al. 2010).

6.2.5 Bartonellae of Additional Domestic Animals: Cattle and Horses

B. bovis (formerly *B. weissii*), *B. chomelii*, *B. schoenbuchensis*, and *B. henselae* have been identified in cattle (Rolain et al. 2003; Raoult et al. 2005; Cherry et al. 2009), and research over the past decade implicates cattle as the natural reservoir for *B. bovis*. *B. bovis* has been documented in cattle around the world, including North America (Chang et al. 2000a), Europe (Bermond et al. 2002; Rolain et al. 2003), and Africa (Raoult et al. 2005). In the United States, studies have shown a high prevalence of *B. bovis* in cattle; in one study, nearly half of all beef and dairy cattle sampled in Oklahoma and California were positive for *Bartonella* spp. (Chang et al. 2000a), while nearly 90 % of North Carolina beef cattle tested PCR positive in a second study (Cherry et al. 2009). The high prevalence of *B. bovis* in cattle indicates that *B. bovis* may have successfully co-evolved with cattle as its primary host, similar to the evolutionary adaptation of *B. henselae* and cats (Cherry et al. 2009). However, recent reports have described *B. bovis* in a cow that died of endocarditis, indicating *B. bovis* may be able to induce clinical manifestations in some cows (Erol et al. 2013). In addition to cattle, preliminary reports suggest that other ruminants, such as deer and elk may also be possible reservoirs of *Bartonella* spp. infection (Chang et al. 2000a; Chitwood et al. 2013; Dehio et al. 2001, 2004; Lobanov et al. 2012).

Bartonella spp. was first reported in horses in 2008 when *B. henselae* was isolated from a 7 year old mare with presumptive vasculitis and from an 11 year old gelding with chronic arthropathy (Jones et al. 2008). In contrast to cattle, horses are not thought to be a reservoir host for any *Bartonella* species. Various case reports have suggested that infection with *Bartonella* spp. in horses may potentially lead to severe clinical illness, including high grade hemolytic anemia (Cherry et al. 2011), spontaneous abortion (Johnson et al. 2009), and musculoskeletal disease (Cherry et al. 2012), though additional studies are necessary to elucidate the association between *Bartonella* infection and these clinical manifestations in horses.

6.2.6 Bartonellae of Wildlife

Bartonella also have been identified in a variety of other wildlife species, including marine mammals. Recent reports have identified infections in stranded harbor porpoises (*Phocoena phocoena*), harbor seals (*Phoca vitulina*), clinically normal loggerhead sea turtles (*Caretta caretta*), free-ranging wild bottlenose dolphins (*Tursiops truncatus*), and captive and hunter harvested beluga whales (*Delphinapterus leucas*) (Maggi et al. 2005b, 2008; Harms et al. 2008; Turnbull and Cowan 1999). *Bartonella* spp. has been identified in the seal louse (*Echynophtyrus horridus*), which can infest pinneped such as harbor seals (*Phoca vitulina*), though the mechanism of transmission from terrestrial mammals to sea mammals is not readily apparent (Harms et al. 2008). *Bartonella* spp. have been implicated in the development of neurologic disorders in animals and people, and therefore the relationship between *Bartonella* spp. infection and the stranding of marine mammals warrants investigation (Harms et al. 2008).

Bartonella spp. have also been amplified from blood samples taken from feral pigs in the southeastern United States. Feral pigs may represent a transmission risk to domesticated pigs and other livestock, in addition to hunters and butchers that are exposed to large quantities of pig blood (Beard et al. 2011). *Bartonella* spp. DNA and antibodies have been identified in an expanding list of wildlife species, including raccoons (*Procyon lotor*), Florida panthers (*Puma concolor coryi*), mountain lions (*P. concolor stanleyana*) (Rotstein et al. 2000; Henn et al. 2009), and African cheetahs (Kelly et al. 1998).

6.3 Vectors

As scientists, clinicians, and epidemiologists discovered more about the medical importance of *Bartonella* spp. infection, increasing efforts were placed on the study of known and suspected arthropod vectors. Reports have shown that *Bartonella* spp. have been identified in a variety of arthropod vectors (Billeter et al. 2008). Many cases of *Bartonella* spp. detection in arthropods have been determined by culture or by PCR, which documents the presence of the pathogen within the arthropod or the arthropod's feces, but does not provide definitive proof of vector competence. Proof of vector competence of an arthropod requires demonstration of both the susceptibility of the particular arthropod to acquire a pathogen and also the arthropod's ability to transmit the pathogen (Eldridge and Edman 2003). Vector competence must be proven through experimental studies that demonstrate reliable transmission between the vector and the host (Billeter et al. 2008). However, it is not simply enough to document whether an arthropod is a competent vector of a given pathogen; the epidemiological importance of the vector-pathogen relationship must also be investigated at the population level (Eldridge and Edman 2003). A competent vector may have no current epidemiologic relationship to the pathogen, may produce rare or occasional transmission involvement (as a secondary vector), or may provide the primary means of pathogen transmission among a population (primary vector). Medical entomologists have detailed a description of criteria (Barnett 1962) that can be used to incriminate pathogen vectors in humans and animals as follows:

1. Demonstration that members of the suspected arthropod population frequently feed upon vertebrate hosts of the pathogen or make effective contact with the hosts under natural conditions.
2. Demonstration of a biological association between the suspected vectors and clinical or subclinical infections in vertebrate hosts in both time and space.
3. Repeated demonstration that suspected vectors harbor the infective stage of the pathogen under natural conditions.
4. Demonstration of efficient transmission of the identifiable pathogen by the suspected vectors under controlled experimental conditions.

An extraordinarily varied amount of information exists pertaining to the vector competence and vector potential of arthropods for *Bartonella* spp. A majority of the publications are epidemiologically based upon case reports or case-controlled studies. However, several experimental transmission studies have been conducted to examine the vector competence of several arthropod species for *Bartonella* transmission. With the exception of louse and sandfly transmission of *B. quintana* and *B. bacilliformis* respectively, the role of other vectors, such as the common, worldwide distributed cat flea, is not known.

6.3.1 Lice

As previously mentioned, the human body louse (*Pediculus humanus humanus*) has been identified as a vector of *B. quintana* for several decades. Scientists, including Töpfer, Jungmann and Kuczynski, and da Rocha-Lima, described Rickettsia-like organisms within the intestinal mucosa and feces of infected lice (Swift 1920), and experimental studies demonstrated the lack of transovarial transmission of *B. quintana* to offspring of infected lice (Bruce 1921).

A more recent study conducted by Fournier et al. (2001b) demonstrated the ability of body lice to excrete viable *B. quintana* (Fournier et al. 2001b). The observations from these studies indicate that transmission of *B. quintana* by body lice occurs when an adult louse becomes infected, by way of a blood meal, and maintains viable organisms in its intestinal tract. In contrast to other louse-borne pathogens that cause obstruction of the louse intestinal tract, as seen with Rickettsial bacteria and *Yersinia pestis*, *Bartonella* are able to multiply in the gut lumen without interfering with louse viability and are excreted in the feces (Fournier et al. 2001b). Transmission to humans occurs during contamination of the louse bite site or when a wound becomes contaminated with louse feces.

Bartonella spp. have been detected in a number of other louse species, including human head lice. The presence of *B quintana* DNA has been detected in Nepalese children that were heavily infested with the head louse (*P. humanus capitis*) (Sasaki et al. 2006), and in head lice from a homeless population in San Francisco, California (Bonilla et al. 2009). In addition to human lice, *Bartonella* spp. has been detected in various lice infesting rodents and mammals. *Bartonella* spp. DNA, including *B. henselae*, *B. phoceensis*, *B. tribocorum*, and *B. rattimassiliensis*, has been detected in rodent and small mammal infesting lice within the *Neohaemotopinus Hoplopleur*, *Polyplax*, *and Echinophtirius* genera throughout the United States, Egypt, and Taiwan (Durden et al. 2004; Nelder et al. 2009).

6.3.2 Sandflies

While *B. quintana* infection with louse transmission is the most widely known *Bartonella* in many parts of the world, the first *Bartonella* species to be described was *B. bacilliformis*. Arthropod transmission of *B. bacilliformis* was proposed in the early 1900s by Strong et al., and the sandfly (*Lutzomyia verrucarum*) was hypothesized to be the potential vector for *B. bacilliformis* in 1913 (Strong 1915). As previously mentioned, *B. bacilliformis* is the causative agent of Oroya fever and verruga peruana, a clinical manifestation of the bacterium that is restricted to the Andean cordillera in Peru, Ecuador, and Colombia, with sporadic cases in Bolivia, Chile, and possibly Guatemala (Schultz 1968; Gray et al. 1990). In 1913, a British scientist named C.H.T. Townsend observed that the distribution and feeding habits of *Lu. verrucarum* overlapped with cases of Oroya fever in the Peruvian Andes. Townsend's speculation heightened when a British sailor on his ship willingly became exposed to wild sandflies and later developed intermittent fever and papules, presumably Oroya fever, though a diagnosis could never confirm *B. bacilliformis* as the cause of illness (Townsend 1914).

Additional culture experiments detected Gram-negative rod organisms in the proboscis of female sandflies (Hertig 1942). Interestingly, similar organisms, presumably *B. bacilliformis*, have been detected in male sandflies, which do not take a blood meal, and in unfed females. Though this would suggest transovarian transmission, limited studies have been conducted on the replication/survival/transmission of *B. bacilliformis* in the sandflies, and scientists have speculated that transmission of *Bartonella* spp. between sandflies may occur through the commingling of breeding areas, contaminated water supplies and various other locations (Billeter et al. 2008).

6.3.3 Fleas

The cat flea (*Ctenocephalides felis*) has been experimentally and epidemiologically shown to be an important vector of *B. henselae*, the agent of cat scratch disease. Studies have shown that the seroprevalence of *Bartonella* spp. in cats is dependent on average daily temperature and annual precipitation. The seroprevalence is higher in warmer climate areas, where fleas are more common (Jameson et al. 1995). The first epidemiological studies conducted on patients with cat scratch disease identified owning a cat or kitten with fleas as a major risk factor for *B. henselae* infection (Zangwill et al. 1993). Various experimental studies have been conducted to demonstrate the successful acquisition of *B. henselae* by cats or kittens through cat flea transmission (Chomel et al. 1996; Higgins et al. 1996; Foil et al. 1998). Other studies have demonstrated transmission of *Bartonella* spp. through flea feces. In 1998, *B. henselae* was transmitted via intradermal injection of saline containing infected flea feces to five cats (Foil et al. 1998). It has been hypothesized that *Bartonella* spp. can remain reproductively viable in flea feces within the environment, and that transmission of *Bartonella* to humans or other animals can occur via inoculation of contaminated flea feces onto the skin through an animal bite, scratch, or other open wound (Finkelstein et al. 2002). It is also possible that infection might occur by inhalation of infected flea feces, potentially resulting in a pulmonary presentation for bartonellosis (Breitschwerdt et al. 2010b).

It is thought that cat fleas can maintain infection with other *Bartonella* species and transmit the organism among cats, which can subsequently transmit these species to people via bite or scratch (Breitschwerdt et al. 2007a). Cat scratches are able to transmit *Bartonella* through both cat blood and flea feces, in addition to blood that has been ingested by fleas that may contaminate cat claws. *Bartonella* spp. have also been detected in fleas from various other mammals, including human fleas (*Pulex* spp.) infesting people in Peru. The species identified in these fleas was closely related to *B. vinsonii* subsp. *berkhoffii*, which has been associated with endocarditis and myocarditis in humans and dogs (Parola et al. 2002). Human fleas (*P. irritans*) have also been shown to harbor *B. quintana* DNA collected from a pet monkey (*Cercopithecus cephus*) in Gabon, Africa (Rolain et al. 2005), *B. rochalimae* collected from red foxes (*Vulpes vulpes*) in Hungary (Sreter-Lancz et al. 2006) and dogs in Chile (Perez-Martinez et al. 2009), while a similar flea, *P. simulans*, has been shown to harbor *B. vinsonii* subsp. *berkhoffii* and *B. rochalimae* in gray foxes (*Urocyon cineroagentus*) in northern California (Gabriel et al. 2009).

Additional studies have suggested that other species of fleas, including rodent and bat fleas, may be important vectors for transmission of *Bartonella* spp. Experimental studies have indicated that wild-caught *Ctenophthalmus nobilis* are competent vectors for transmission of *B. grahamii* and *B. taylorii* to bank voles (*Myodes glareolus*) (Bown et al. 2004). Many other rodent fleas have been tested for the presence of *Bartonella* DNA. Species of *Bartonella*, *including* included *B. elizabethae*, *B. vinsonii*-like, *B. quintana*, *B. henselae*, *B. clarridgeiae*,

B. schoenbuchensis-like, *B. birtlesii*, *B. koehlerae*, *B. taylorii*, *B. tribocorum*, *B. queenslandensis*, and *B. rochalimae*, have been detected in various flea genera in China, Indonesia, Egypt, Israel, Thailand, Myanmar, Taiwan, including *Xenopsylla*, *Nosopsyllus*, *Sternopsylla*, and *Leptopsylla* (Winoto et al. 2005; Loftis et al. 2005; Li et al. 2007; Reeves et al. 2007; Tsai et al. 2010; Morick et al. 2010). Recently, a proposed new species of *Bartonella* (Candidatus *B. antechini*) was detected in fleas collected on yellow-footed antechinus (*Antechinus flavipes*) in Western Australia (Kaewmongkol et al. 2011). While it is unclear how many of these flea species are competent vectors for *Bartonella* transmission, the increasingly identified number of flea and host species and diversifying geographic range suggest that current literature may only be reporting a fraction of species and ranges that *Bartonella* may be present in.

6.3.4 Ticks

It is currently unclear whether ticks are involved in the transmission of *Bartonella* spp. to humans and animals under natural conditions. However, the high prevalence of *Bartonella* spp. in various tick species highlights the need for increased discussion relative to transmission of *Bartonella* spp. following a tick bite. The first documentation of *Bartonella* spp. detection or isolation from ticks was in the mid-1990s, when a *Bartonella* strain was cultured from a questing *Ixodes ricinus* tick collected from a park in Poland (Kruszewska and Tylewska-Wierzbanowska 1996). Since then, there have been numerous reports of *Bartonella* spp., mostly based on PCR amplification of organism specific DNA sequences, though rarely by culture, from various tick species. In the early 2000s, two studies out of California, United States, provided evidence that two separate tick species might serve as vectors for *Bartonella* transmission. The first study identified several pathogenic *Bartonella*, including *B. henselae*, *B. quintana*, *B. washoensis*, *B. vinsonii* subsp. *berkhoffii*, and *B. bovis* in 19.2 % of questing adult *I. pacificus* ticks that were collected by flagging local vegetation (Chang et al. 2001). The second study identified *Bartonella* spp. in adult and nymph *I. pacificus* ticks, in addition to *Dermacentor variabilis* and *D. occidentalis* ticks (Chang et al. 2002). Near that same time, researchers identified *Bartonella* spp. from *I. scapularis* deer ticks that were collected from the household of two patients co-infected with *B. henselae* and *Borrelia burgdoferi* (the etiologic agent of Lyme disease) detected from their spinal fluid (Eskow et al. 2001), and from 34.5 % of field collected *I. scapularis* in New Jersey (Adelson et al. 2004). *Bartonella* spp. DNA has since been detected in ticks throughout the world, including *Carios kelleyi*, *I. ricinus*, *I. persulcatus*, *Dermacentor reticulates*, *Hyalomma longicornis*, *Rhipicephalus microplus*, *I. tasmani*, *I. sinensis*, *H. rufipes*, and *H. longicornis*.(Loftis et al. 2005)

Additional epidemiological and experimental evidence exists that suggests ticks may be a relevant vector for disease transmission. Ticks have been identified as a risk factor for *B. vinsonii* subsp. *berkhoffii* infection in dogs; a case control study determined that *Bartonella* seropositive dogs are 14 times more likely to have been exposed to previous heavy tick infestations (Pappalardo et al. 1997). In a clinical survey performed in Connecticut, United States, patients were more likely to have been bitten by a tick than controls (Zangwill et al. 1993). Several case reports describe human *Bartonella* infection shortly following a tick bite (Lucey et al. 1992) (Angelakis et al. 2010) (Eskow et al. 2001), and co-infections with *Bartonella* and *Borrelia burgdoferi*, the etiologic agent of Lyme disease, have been frequently reported (Halperin and Wormser 2001). A recent report detected a high prevalence of bacteremia with and antibodies against *Bartonella* spp. in patients diagnosed with Lyme disease or chronic arthralgia and myalgia, and statistically associated *Bartonella* spp. bacteremia with neurologic symptoms (Maggi et al. 2011). While some researchers may argue that cases of Bartonellosis may be potentially misdiagnosed as Lyme disease, rigorous scientific studies will clarify the role of *Bartonella* infection with rheumatic and neurologic symptoms.

Experimental transmission of *Bartonella* by ticks has also been demonstrated. In one study, *I. ricinus* ticks were infected with *B. henselae* in artificially infected ovine blood using an artificial feeding system. The ticks maintained infection throughout the molt, and were able to transmit *B. henselae* during a subsequent blood meal through an artificial feeding system (Cotte et al. 2008). A more recent study demonstrated vector competence of *I. ricinus* through transmission of *Bartonella* by ticks from an infected mouse and subsequent transmission to a naïve mouse, in addition to localization of *Bartonella* within the adult ticks (Reis et al. 2011). Also, Billeter et al. (2009) demonstrated invasion and replication of seven *Bartonella* species in an *Amblyomma americanum* tick cell line in 2009, visualized *Bartonella* within the cells and quantified bacterial growth through qPCR amplification (Billeter et al. 2009). Recently, Billeter was able to detect *B. vinsonii* subsp. *berkhoffii* DNA in *R. sanguineus* ticks post-capillary tube feeding, as well as in tick feces (Billeter et al. 2012). Similar to other vectors for *Bartonella* that may transmit the bacterium through contamination of arthropod feces, it has been hypothesized that *R. sanguineus* may transmit *Bartonella* through inoculation of tick feces through broken skin. Despite the evidence that *Bartonella* spp. might be transmitted through a tick vector, many scientists argue that transmission of *Bartonella* spp. to animals or humans through ticks is unlikely. While no well-documented case of transmission by ticks to humans or animals has been reported, additional studies will clarify the role of ticks in the transmission of this pathogen.

6.3.5 Biting Flies and Mites

A potential role in the transmission of *Bartonella* spp. has been suggested for a variety of other insects and arthropods. A study conducted in France revealed that 86 % of 83 sample *Hippoboscidae* flies, including 94 % of deer keds (*Lipoptena cervi*), 71 % of louse flies (*H. equina*), and 100 % of sheep keds (*Melophagus ovinus*) harbored *Bartonella* DNA, notably *B. schoenbuchensis* or *B. chomelii* (Halos et al. 2004). Dehio et al. (2004) isolated *B. schoenbuchensis* from deer keds, which are flies that lose their wings after finding a host animal, collected from roe deer and red deer in Germany and demonstrated the presence of large bacterial aggregates in the midgut of the flies, suggesting that the arthropod may be a viable vector for *Bartonella* transmission (Dehio et al. 2004). *B. schoenbuchensis* has been detected in deer keds in Georgia and South Carolina, United States (Reeves et al. 2006), and it has been hypothesized that *B. schoenbuchensis* and deer keds may be widely distributed by elk or caribou across continents (Matsumoto et al. 2008).

Bartonella has also been detected in pools of *Mesostigmatid* mites collected from wild rodents and insectivores in Korea, in addition to house dust mites (*Dermatophagoides farina* and *D. pteronyssinus*) in the United States (Kim et al. 2005; Valerio et al. 2005). *B. tamiae* has been detected in chigger mite pools (*Leptotrombidium*, *Schoengastia*, and *Blankarrtia*) that had been collected on rodents in Thailand (Kabeya et al. 2010). The bat fly (*Trichobius major Coquillett*) and the Eastern bat bed bug (*Cimex adjunctus Barber*) are ectoparasites of bats. A bat fly collected from Florida caverns and a Eastern bat bed bug collected from the Santee Caves in South Carolina, USA, were shown to harbor a unique *Bartonella* species (Reeves et al. 2005). *Bartonella* has also been detected in honey bees (*Apis mellifera capensis Eschscholtz*). It has been suggested that the organism was ingested through contact with the environment and therefore it does not appear as though *Bartonella* can commonly be found in honey bees (Jeyaprakash et al. 2003). Recently, *Bartonella* spp. has also been detected in woodlouse hunter spiders and their associated prey, the woodlouse, though vector competence has yet to be investigated in these arthropods (Mascarelli et al. 2013).

6.4 *Bartonella* and One Health

Prior to 1990, *Bartonella* infection was not reported in human or animal species in North America. The AIDs epidemic and subsequent recognition of *Bartonella*-induced vasoproliferative disorders in immunocompromised individuals has helped to unveil *Bartonella* as a complex, stealth pathogen of substantial medical importance. The difficulties in studying *Bartonella* spp. risk factors, pathology, and potential treatment options lie in the bacterium's generalist and versatile host affiliation and its ability to induce a spectrum of disease expression in a variety of species that ranges from self-limiting and benign to chronic and complex. *Bartonella's* strain variability in combination with the nuances of a specific host response (influenced by various nutritional, genetic and other medically relevant factors) creates an unpredictable pattern of disease expression within or across infected species. As such, it remains difficult to reproduce and study an accurate model of *Bartonella* infection in a scientific setting. Furthermore, as a fastidious organism, it remains technically difficult to detect and diagnose cases of naturally occurring Bartonellosis within non-reservoir populations; misclassification of true cases of disease as false negatives limits the ability of researchers to conduct controlled and insightful epidemiologic studies that will better elucidate risk factors for infection or subtle manifestations of disease. To combat the difficulties in studying this complex pathogen, a One Health approach that combines clinical, microbiological, and pathological observations across animal genera and species has proved necessary to further elucidate the ability of *Bartonella* spp. to induce disease processes in humans and animals. A broad scale analysis of *Bartonella* literature has observed that similar pathologies following *Bartonella* infection can be seen across multiple mammalian species. The DNA of *Bartonella* species has been detected in both humans and dogs with clinically similar and histolopathologically confirmed vasoproliferative lesions; granulomatous inflammation has occurred in naturally infected cats, dogs, and humans. And most notably, *Bartonella* induced culture-negative endocarditis has been observed across various mammalian species, including humans, dogs, cats, and cattle (Breitschwerdt et al. 2013). These observations have proved crucially insightful to researchers and medical professionals that have previously struggled to attribute disease causation to *Bartonella* spp. The importance of a comparative examination of the biological and pathological behavior of *Bartonella* and other stealth pathogens across different animal/arthropod species cannot be understated. As humans, pets, domestic animals, and wildlife are expanding their number of overlapping habitats, a close integration of veterinary, medical, and public health professionals is a necessary framework of disease investigation for pathogens shared between humans and animals, with *Bartonella* spp. as only an example.

6.5 Conclusions

Zoonotic diseases present a constant threat to humans, domestic animals, and wildlife. Transmission of pathogens into human populations and non-reservoir hosts from other animal species occurs from changes in the interactions between host, vector, and environment. The emergence and re-emergence of *Bartonella* spp. in humans and a variety of mammalian hosts and arthropod vectors draws attention to the need for enhanced diagnostic techniques for documenting infection in hosts, effective vector control, and treatment of individuals with associated diseases. Certainly, enhanced pathogen detection and diagnostic techniques have enabled the discovery of a wide spectrum of *Bartonella* species, hosts and vectors outlined within this chapter. However, substantial gaps in the knowledge of *Bartonella* disease ecology and transmission diagnostics have prevented researchers from developing effective strategies for disease prevention and control. Key challenges in combating zoonotic infections, such as *Bartonella*, have to do with the non-specificity and similarity of symptoms caused by a variety of infections. This highlights the need for continued development of high-quality laboratory based diagnostics so as to facilitate the rapid identification of infections for proper surveillance and appropriate treatment. Increased use of rapid, accurate diagnostic tests will afford researchers and clinicians with the ability to conduct more reliable epidemiologic and treatment-related studies and will more accurately characterize the outcomes of *Bartonella* spp. infection.

Environmental changes that affect vector prevalence and distributions and also impact wildlife reservoir hosts, in addition to social changes that affect human exposure to vectors may contribute to the local emergence and increases in human incidence in *Bartonella* spp. infection, and should not be overlooked. Though the catalogue of animal hosts and insect vectors that harbor *Bartonella* spp. continues to expand, researchers are still unable to accurately quantify the epidemiological and medical significance of these bacteria or the effects of *Bartonella* in animal and insect vector populations. As presented earlier, reports of severe clinical manifestations in domestic animals and livestock, such as cattle and horses, infected with *Bartonella* spp. are continuing to emerge. The distribution of reservoir host populations and vector communities is easily altered through land modification activities such as increased farming, cyclical deforestation/reforestation, and habitat fragmentation. Modifying the distribution of vectors and reservoir hosts, such as rodents and small mammals, in a given area will increase the exposure of domestic animals and livestock to *Bartonella* spp. infection and could easily result in substantial morbidity and mortality within the domestic animal production industry. The spill-over effects of *Bartonella* spp. into non-reservoir domestic animal hosts remains an understudied aspect of current research.

Introductions and increases in the incidence of vector borne pathogens clearly call attention to the need for effective vector control measures. Increased outdoor activity, poverty, and low income may allow for humans and other non-reservoir populations, such as dogs and cats, to come into increased contact with infected arthropod vectors (Boulouis et al. 2005). Control of arthropod transmitted *Bartonella* requires combination efforts by clinicians, researchers, and public health officials to educate and promote behavior that will minimize the risk of infection and reduce sociologic and human-driven environmental factors that may facilitate the transmission of *Bartonella*. But before this can be done, additional research is required to identify epidemiologically important competent vectors of *Bartonella* spp. transmission. It is currently unclear which arthropod vectors are able to amplify the bacterium and successfully transmit it to new host populations. Studies that aim to not only identify *Bartonella* spp. in arthropod populations but also combine epidemiologically relevant case data with ecological risk assessment should be encouraged. While *Bartonella* spp. appear to have distinct host-vector specificity, epidemiological studies are needed to identify the effects of transmission into new hosts and the ability of *Bartonella* spp. to adapt and become a successful part of a new host-vector-pathogen lifecycle.

Changes in the abundance of available animal hosts caused by events such as habitat fragmentation, new animal production systems, and global travel may increase the risk of spill-over or zoonotic transmission of *Bartonella* into novel non-reservoir populations. With the ability of *Bartonella* to jump to accidental hosts, it is imperative that we determine the consequences of *Bartonella* spp.

infection on new host populations and environments. Co-factors that exacerbate pathologic manifestations, such as co-infections and immunosuppression, need to be investigated. An aging immunosenescent human population, behavioral and societal changes that bring animals and humans into increasingly close contact, and alterations in vector biology and reservoir host density all contribute to the community interactions among *Bartonella* species, host, and vectors (Boulouis et al. 2005). A One Health approach incorporating the combined efforts of professionals in human and animal medicine, vector ecology, pathogen evolution, microbial ecology, and sociology are necessary in order to create the veterinary and medical infrastructure, training, prevention, and surveillance required to thoroughly understand *Bartonella* spp. ecology, dynamics and control.

References

Adelson ME, Rao RV, Tilton RC, Cabets K, Eskow E, Fein L, Occi JL, Mordechai E (2004) Prevalence of *Borrelia burgdorferi*, *Bartonella* spp., *Babesia microti*, and *Anaplasma phagocytophila* in Ixodes scapularis ticks collected in Northern New Jersey. J Clin Microbiol 42(6):2799–2801. doi:10.1128/JCM.42.6.2799-2801.2004

Alcantara V, Rolain JM, Eduardo AG, Raul MJ, Raoult D (2009) Molecular detection of *Bartonella quintana* in human body lice from Mexico City. Clin Microbiol Infect 15(Suppl 2):93–94. doi:10.1111/j.1469-0691.2008.02176.x

Angelakis E, Pulcini C, Waton J, Imbert P, Socolovschi C, Edouard S, Dellamonica P, Raoult D (2010) Scalp eschar and neck lymphadenopathy caused by *Bartonella henselae* after Tick Bite. Clin Infect Dis 50(4):549–551. doi:10.1086/650172

Armengol CE, Hendley JO (1999) Cat-scratch disease encephalopathy: a cause of status epilepticus in school-aged children. J Pediatr 134(5):635–638

Avidor B, Graidy M, Efrat G, Leibowitz C, Shapira G, Schattner A, Zimhony O, Giladi M (2004) Bartonella koehlerae, a new cat-associated agent of culture-negative human endocarditis. J Clin Microbiol 42(8):3462–3468. doi:10.1128/JCM.42.8.3462-3468.2004

Bai Y, Kosoy MY, Ray C, Brinkerhoff RJ, Collinge SK (2008) Temporal and spatial patterns of *Bartonella* infection in black-tailed prairie dogs (*Cynomys ludovicianus*). Microb Ecol 56 (2):373–382. doi:10.1007/s00248-007-9355-6

Bai Y, Kosoy M, Recuenco S, Alvarez D, Moran D, Turmelle A, Ellison J, Garcia DL, Estevez A, Lindblade K, Rupprecht C (2011) *Bartonella* spp. in bats, Guatemala. Emerg Infect Dis 17 (7):1269–1272. doi:10.3201/eid1707.101867

Bai Y, Recuenco S, Gilbert AT, Osikowicz LM, Gomez J, Rupprecht C, Kosoy MY (2012) Prevalence and diversity of *Bartonella* spp. in bats in Peru. Am J Trop Med Hyg 87(3):518–523. doi:10.4269/ajtmh.2012.12-0097

Barnes A, Bell SC, Isherwood DR, Bennett M, Carter SD (2000) Evidence of *Bartonella henselae* infection in cats and dogs in the United Kingdom. Vet Rec 147(24):673–677

Barnett H (1962) The incrimination of arthropods as vectors of disease. In: Proceedings of the 11th congress on entomology. Vienna, Austria

Beard AW, Maggi RG, Kennedy-Stoskopf S, Cherry NA, Sandfoss MR, DePerno CS, Breitschwerdt EB (2011) *Bartonella* spp. in feral pigs, southeastern United States. Emerg Infect Dis 17(5):893–895. doi:10.3201/eid1705.100141

Bermond D, Heller R, Barrat F, Delacour G, Dehio C, Alliot A, Monteil H, Chomel B, Boulouis HJ, Piemont Y (2000) *Bartonella birtlesii* sp. nov., isolated from small mammals (*Apodemus* spp.). Int J Syst Evol Microbiol 50(Pt 6):1973–1979

Bermond D, Boulouis HJ, Heller R, Van Laere G, Monteil H, Chomel BB, Sander A, Dehio C, Piemont Y (2002) *Bartonella bovis* Bermond et al. sp. nov. and *Bartonella capreoli* sp. nov., isolated from European ruminants. Int J Syst Evol Microbiol 52(Pt 2):383–390

Billeter SA, Levy MG, Chomel BB, Breitschwerdt EB (2008) Vector transmission of *Bartonella* species with emphasis on the potential for tick transmission. Med Vet Entomol 22(1):1–15. doi:10.1111/j.1365-2915.2008.00713.x

Billeter SA, Diniz PP, Battisti JM, Munderloh UG, Breitschwerdt EB, Levy MG (2009) Infection and replication of *Bartonella* species within a tick cell line. Exp Appl Acarol 49(3):193–208. doi:10.1007/s10493-009-9255-1

Billeter SA, Kasten RW, Killmaster LF, Breitschwerdt EB, Levin ML, Levy MG, Kosoy MY, Chomel BB (2012) Experimental infection by capillary tube feeding of *Rhipicephalus sanguineus* with *Bartonella vinsonii* subspecies *berkhoffii*. Comp Immunol Microbiol Infect Dis 35(1):9–15. doi:10.1016/j.cimid.2011.09.004

Birtles RJ (1995) Differentiation of *Bartonella* species using restriction endonuclease analysis of PCR-amplified 16S rRNA genes. FEMS Microbiol Lett 129(2–3):261–265

Birtles RJ, Harrison TG, Saunders NA, Molyneux DH (1995) Proposals to unify the genera *Grahamella* and *Bartonella*, with descriptions of *Bartonella talpae* comb. nov., *Bartonella peromysci* comb. nov., and three new species, *Bartonella grahamii* sp. nov., *Bartonella taylorii* sp. nov., and *Bartonella doshiae* sp. nov. Int J Syst Bacteriol 45(1):1–8

Birtles RJ, Hazel SM, Bennett M, Bown K, Raoult D, Begon M (2001) Longitudinal monitoring of the dynamics of infections due to *Bartonella* species in UK woodland rodents. Epidemiol Infect 126(2):323–329

Bonatti H, Mendez J, Guerrero I, Krishna M, Ananda-Michel J, Yao J, Steers JL, Hellinger W, Dickson RC, Alvarez S (2006) Disseminated *Bartonella* infection following liver transplantation. Transpl Int 19(8):683–687. doi:10.1111/j.1432-2277.2006.00336.x

Bonilla DL, Kabeya H, Henn J, Kramer VL, Kosoy MY (2009) *Bartonella quintana* in body lice and head lice from homeless persons, San Francisco, California, USA. Emerg Infect Dis 15 (6):912–915. doi:10.3201/eid1506.090054

Bouchouicha R, Durand B, Monteil M, Chomel BB, Berrich M, Arvand M, Birtles RJ, Breitschwerdt EB, Koehler JE, Maggi R, Maruyama S, Kasten R, Petit E, Boulouis HJ, Haddad N (2009) Molecular epidemiology of feline and human *Bartonella henselae* isolates. Emerg Infect Dis 15(5):813–816. doi:10.3201/eid1505.080995

Boulouis HJ, Chang CC, Henn JB, Kasten RW, Chomel BB (2005) Factors associated with the rapid emergence of zoonotic *Bartonella* infections. Vet Res 36(3):383–410. doi:10.1051/vetres:2005009

Bown KJ, Bennet M, Begon M (2004) Flea-borne *Bartonella grahamii* and *Bartonella taylorii* in bank voles. Emerg Infect Dis 10(4):684–687. doi:10.3201/eid1004.030455

Breitschwerdt EB, Kordick DL (2000) *Bartonella* infection in animals: carriership, reservoir potential, pathogenicity, and zoonotic potential for human infection. Clin Microbiol Rev 13 (3):428–438

Breitschwerdt EB, Atkins CE, Brown TT, Kordick DL, Snyder PS (1999) *Bartonella vinsonii* subsp. berkhoffii and related members of the alpha subdivision of the Proteobacteria in dogs with cardiac arrhythmias, endocarditis, or myocarditis. J Clin Microbiol 37(11):3618–3626

Breitschwerdt EB, Blann KR, Stebbins ME, Munana KR, Davidson MG, Jackson HA, Willard MD (2004) Clinicopathological abnormalities and treatment response in 24 dogs seroreactive to *Bartonella vinsonii* (berkhoffii) antigens. J Am Anim Hosp Assoc 40(2):92–101

Breitschwerdt EB, Hegarty BC, Maggi R, Hawkins E, Dyer P (2005) Bartonella species as a potential cause of epistaxis in dogs. J Clin Microbiol 43(5):2529–2533. doi:10.1128/jcm.43.5.2529-2533.2005

Breitschwerdt EB, Maggi RG, Duncan AW, Nicholson WL, Hegarty BC, Woods CW (2007a) *Bartonella* species in blood of immunocompetent persons with animal and arthropod contact. Emerg Infect Dis 13(6):938–941

Breitschwerdt EB, Maggi RG, Sigmon B, Nicholson WL (2007b) Isolation of *Bartonella quintana* from a woman and a cat following putative bite transmission. J Clin Microbiol 45(1):270–272. doi:10.1128/jcm.01451-06

Breitschwerdt EB, Maggi RG, Nicholson WL, Cherry NA, Woods CW (2008) *Bartonella* sp. bacteremia in patients with neurological and neurocognitive dysfunction. J Clin Microbiol 46(9):2856–2861. doi:10.1128/JCM.00832-08

Breitschwerdt EB, Maggi RG, Varanat M, Linder KE, Weinberg G (2009) Isolation of *Bartonella vinsonii* subsp. *berkhoffii* genotype II from a boy with epithelioid hemangioendothelioma and a dog with hemangiopericytoma. J Clin Microbiol 47(6):1957–1960. doi:10.1128/JCM.00069-09

Breitschwerdt EB, Maggi RG, Chomel BB, Lappin MR (2010a) Bartonellosis: an emerging infectious disease of zoonotic importance to animals and human beings. J Vet Emerg Crit Care 20(1):8–30. doi:10.1111/j.1476-4431.2009.00496.x

Breitschwerdt EB, Maggi RG, Robert Mozayeni B, Hegarty BC, Bradley JM, Mascarelli PE (2010b) PCR amplification of *Bartonella koehlerae* from human blood and enrichment blood cultures. Parasit Vectors 3:76. doi:10.1186/1756-3305-3-76

Breitschwerdt EB, Linder KL, Day MJ, Maggi RG, Chomel BB, Kempf VA (2013) Koch's postulates and the pathogenesis of comparative infectious disease causation associated with *Bartonella* species. J Comp Pathol 148(2–3):115–125. doi:10.1016/j.jcpa.2012.12.003

Brenner DJ, O'Connor SP, Winkler HH, Steigerwalt AG (1993) Proposals to unify the genera *Bartonella* and *Rochalimaea*, with descriptions of *Bartonella quintana* comb. nov., *Bartonella vinsonii* comb. nov., *Bartonella henselae* comb. nov., and *Bartonella elizabethae* comb. nov., and to remove the family Bartonellaceae from the order Rickettsiales. Int J Syst Bacteriol 43 (4):777–786

Brenner EC, Chomel BB, Singhasivanon OU, Namekata DY, Kasten RW, Kass PH, Cortes-Vecino JA, Gennari SM, Rajapakse RP, Huong LT, Dubey JP (2013) *Bartonella* infection in urban and rural dogs from the tropics: Brazil, Colombia, Sri Lanka and Vietnam. Epidemiol Infect 141(1):54–61. doi:10.1017/S0950268812000519

Brouqui P, Lascola B, Roux V, Raoult D (1999) Chronic *Bartonella quintana* bacteremia in homeless patients. N Engl J Med 340(3):184–189. doi:10.1056/NEJM199901213400303

Brouqui P, Stein A, Dupont HT, Gallian P, Badiaga S, Rolain JM, Mege JL, La Scola B, Berbis P, Raoult D (2005) Ectoparasitism and vector-borne diseases in 930 homeless people from Marseilles. Medicine (Baltimore) 84(1):61–68

Bruce D (1921) Trench fever: final report of the war office trench fever investigation committee. J Hyg 20(3):258–288

Byam W, Lloyd L (1920) Trench fever: its epidemiology and endemiology. Proc R Soc Med 13 (Sect Epidemiol State Med):1–27

Cadenas MB, Bradley J, Maggi RG, Takara M, Hegarty BC, Breitschwerdt EB (2008) Molecular characterization of *Bartonella vinsonii* subsp. *berkhoffii* genotype III. J Clin Microbiol 46 (5):1858–1860. doi:10.1128/jcm.02456-07

Castle KT, Kosoy M, Lerdthusnee K, Phelan L, Bai Y, Gage KL, Leepitakrat W, Monkanna T, Khlaimanee N, Chandranoi K, Jones JW, Coleman RE (2004) Prevalence and diversity of *Bartonella* in rodents of northern Thailand: a comparison with *Bartonella* in rodents from southern China. Am J Trop Med Hyg 70(4):429–433

Chang C, Yamamoto K, Chomel BB, Kasten RW, Simpson DC, Smith CR, Kramer VL (1999) Seroepidemiology of *Bartonella vinsonii* subsp. *berkhoffii* infection in California coyotes, 1994–1998. Emerg Infect Dis 5(5):711–715. doi:10.3201/eid0505.990514

Chang CC, Chomel BB, Kasten RW, Heller RM, Ueno H, Yamamoto K, Bleich VC, Pierce BM, Gonzales BJ, Swift PK, Boyce WM, Jang SS, Boulouis HJ, Piemont Y, Rossolini GM, Riccio ML, Cornaglia G, Pagani L, Lagatolla C, Selan L, Fontana R (2000a) *Bartonella* spp. isolated from wild and domestic ruminants in North America. Emerg Infect Dis 6(3):306–311

Chang CC, Kasten RW, Chomel BB, Simpson DC, Hew CM, Kordick DL, Heller R, Piemont Y, Breitschwerdt EB (2000b) Coyotes (*Canis latrans*) as the reservoir for a human pathogenic

Bartonella sp.: molecular epidemiology of *Bartonella vinsonii* subsp. *berkhoffii* infection in coyotes from central coastal California. J Clin Microbiol 38(11):4193–4200
Chang CC, Chomel BB, Kasten RW, Romano V, Tietze N (2001) Molecular evidence of *Bartonella* spp. in questing adult *Ixodes pacificus* ticks in California. J Clin Microbiol 39 (4):1221–1226. doi:10.1128/JCM.39.4.1221-1226.2001
Chang CC, Hayashidani H, Pusterla N, Kasten RW, Madigan JE, Chomel BB (2002) Investigation of *Bartonella* infection in ixodid ticks from California. Comp Immunol Microbiol Infect Dis 25 (4):229–236
Cherry NA, Maggi RG, Cannedy AL, Breitschwerdt EB (2009) PCR detection of *Bartonella bovis* and *Bartonella henselae* in the blood of beef cattle. Vet Microbiol 135(3–4):308–312. doi:10.1016/j.vetmic.2008.09.063
Cherry NA, Maggi RG, Rossmeisl JH, Hegarty BC, Breitschwerdt EB (2011) Ecological diversity of *Bartonella* species infection among dogs and their owner in Virginia. Vector Borne Zoonotic Dis 11(11):1425–1432. doi:10.1089/vbz.2010.0201
Cherry NA, Jones SL, Maggi RG, Davis JL, Breitschwerdt EB (2012) *Bartonella* spp. infection in healthy and sick horses and foals from the southeastern United States. J Vet Intern Med 26 (6):1408–1412. doi:10.1111/j.1939-1676.2012.00998.x
Childs JE, Olson JG, Wolf A, Cohen N, Fakile Y, Rooney JA, Bacellar F, Regnery RL (1995) Prevalence of antibodies to *Rochalimaea* species (cat-scratch disease agent) in cats. Vet Rec 136(20):519–520
Chitwood MC, Maggi RG, Kennedy-Stoskopf S, Toliver M, Deperno CS (2013) *Bartonella vinsonii* subsp. *berkhoffii* in free-ranging white-tailed deer (*Odocoileus virginianus*). J Wildl Dis 49(2):468–470. doi:10.7589/2012-11-286
Chomel BB, Kasten RW, Floyd-Hawkins K, Chi B, Yamamoto K, Roberts-Wilson J, Gurfield AN, Abbott RC, Pedersen NC, Koehler JE (1996) Experimental transmission of *Bartonella henselae* by the cat flea. J Clin Microbiol 34(8):1952–1956
Chomel BB, Boulouis HJ, Breitschwerdt EB (2004) Cat scratch disease and other zoonotic *Bartonella* infections. J Am Vet Med Assoc 224(8):1270–1279
Chomel BB, Kasten RW, Henn JB, Molia S (2006) Bartonella infection in domestic cats and wild felids. Ann N Y Acad Sci 1078:410–415. doi:10.1196/annals.1374.080
Chomel BB, McMillan-Cole AC, Kasten RW, Stuckey MJ, Sato S, Maruyama S, Diniz PP, Breitschwerdt EB (2012) *Candidatus Bartonella merieuxii*, a potential new zoonotic *Bartonella* species in canids from Iraq. PLoS Negl Trop Dis 6(9):e1843. doi:10.1371/journal.pntd.0001843
Cockerell CJ, Whitlow MA, Webster GF, Friedman-Kien AE (1987) Epithelioid angiomatosis: a distinct vascular disorder in patients with the acquired immunodeficiency syndrome or AIDS-related complex. Lancet 2(8560):654–656
Comer JA, Flynn C, Regnery RL, Vlahov D, Childs JE (1996) Antibodies to *Bartonella* species in inner-city intravenous drug users in Baltimore, MD. Arch Intern Med 156(21):2491–2495
Concannon R, Wynn-Owen K, Simpson VR, Birtles RJ (2005) Molecular characterization of haemoparasites infecting bats (Microchiroptera) in Cornwall, UK. Parasitology 131(Pt 4):489–496. doi:10.1017/S0031182005008097
Cotte V, Bonnet S, Le Rhun D, Le Naour E, Chauvin A, Boulouis HJ, Lecuelle B, Lilin T, Vayssier-Taussat M (2008) Transmission of *Bartonella henselae* by *Ixodes ricinus*. Emerg Infect Dis 14(7):1074–1080. doi:10.3201/eid1407.071110
Daly JS, Worthington MG, Brenner DJ, Moss CW, Hollis DG, Weyant RS, Steigerwalt AG, Weaver RE, Daneshvar MI, O'Connor SP (1993) *Rochalimaea elizabethae* sp. nov. isolated from a patient with endocarditis. J Clin Microbiol 31(4):872–881
Dehio C, Lanz C, Pohl R, Behrens P, Bermond D, Piemont Y, Pelz K, Sander A (2001) *Bartonella schoenbuchii* sp. nov., isolated from the blood of wild roe deer. Int J Syst Evol Microbiol 51 (Pt 4):1557–1565

Dehio C, Sauder U, Hiestand R (2004) Isolation of *Bartonella schoenbuchensis* from *Lipoptena cervi*, a blood-sucking arthropod causing deer ked dermatitis. J Clin Microbiol 42(11):5320–5323. doi:10.1128/JCM.42.11.5320-5323.2004

Demers DM, Bass JW, Vincent JM, Person DA, Noyes DK, Staege CM, Samlaska CP, Lockwood NH, Regnery RL, Anderson BE (1995) Cat-scratch disease in Hawaii: etiology and seroepidemiology. J Pediatr 127(1):23–26

Diniz PP, Maggi RG, Schwartz DS, Cadenas MB, Bradley JM, Hegarty B, Breitschwerdt EB (2007) *Canine bartonellosis*: serological and molecular prevalence in Brazil and evidence of co-infection with *Bartonella henselae* and *Bartonella vinsonii* subsp. *berkhoffii*. Vet Res 38(5):697–710. doi:10.1051/vetres:2007023

Diniz PP, Billeter SA, Otranto D, De Caprariis D, Petanides T, Mylonakis ME, Koutinas AF, Breitschwerdt EB (2009) Molecular documentation of *Bartonella* infection in dogs in Greece and Italy. J Clin Microbiol 47(5):1565–1567. doi:10.1128/jcm.00082-09

Drancourt M, Mainardi JL, Brouqui P, Vandenesch F, Carta A, Lehnert F, Etienne J, Goldstein F, Acar J, Raoult D (1995) *Bartonella* (Rochalimaea) *quintana* endocarditis in three homeless men. N Engl J Med 332(7):419–423. doi:10.1056/NEJM199502163320702

Duncan AW, Maggi RG, Breitschwerdt EB (2007) *Bartonella* DNA in dog saliva. Emerg Infect Dis 13(12):1948–1950

Durden LA, Ellis BA, Banks CW, Crowe JD, Oliver JH Jr (2004) Ectoparasites of gray squirrels in two different habitats and screening of selected ectoparasites for bartonellae. J Parasitol 90(3):485–489. doi:10.1645/GE-3299

Eldridge BF, Edman J (2003) Medical entomology: a textbook on public health and veterinary problems caused by arthropods, 2nd edn. Springer, New York

Ellis BA, Regnery RL, Beati L, Bacellar F, Rood M, Glass GG, Marston E, Ksiazek TG, Jones D, Childs JE (1999) Rats of the genus Rattus are reservoir hosts for pathogenic *Bartonella* species: an old world origin for a new world disease? J Infect Dis 180(1):220–224. doi:10.1086/314824

Eremeeva ME, Gerns HL, Lydy SL, Goo JS, Ryan ET, Mathew SS, Ferraro MJ, Holden JM, Nicholson WL, Dasch GA, Koehler JE (2007) Bacteremia, fever, and splenomegaly caused by a newly recognized *Bartonella* species. N Engl J Med 356(23):2381–2387

Erol E, Jackson C, Bai Y, Sells S, Locke S, Kosoy M (2013) *Bartonella bovis* isolated from a cow with endocarditis. J Vet Diagn Invest 25(2):288–290. doi:10.1177/1040638713477408

Eskow E, Rao RV, Mordechai E (2001) Concurrent infection of the central nervous system by *Borrelia burgdorferi* and *Bartonella henselae*: evidence for a novel tick-borne disease complex. Arch Neurol 58(9):1357–1363

Fenollar F, Sire S, Raoult D (2005) *Bartonella vinsonii* subsp. arupensis as an agent of blood culture-negative endocarditis in a human. J Clin Microbiol 43(2):945–947

Finkelstein JL, Brown T, O'Reilly KL, Wedincamp J Jr, Foil LD (2002) Studies on the growth of *Bartonella henselae* in the cat flea (Siphonaptera: Pulicidae). J Med Entomol 39(6):915–919

Foil L, Andress E, Freeland RL, Roy AF, Rutledge R, Triche PC, O'Reilly KL (1998) Experimental infection of domestic cats with *Bartonella henselae* by inoculation of *Ctenocephalides felis* (Siphonaptera: Pulicidae) feces. J Med Entomol 35(5):625–628

Fournier PE, Lelievre H, Eykyn SJ, Mainardi JL, Marrie TJ, Bruneel F, Roure C, Nash J, Clave D, James E, Benoit-Lemercier C, Deforges L, Tissot-Dupont H, Raoult D (2001a) Epidemiologic and clinical characteristics of *Bartonella quintana* and *Bartonella henselae* endocarditis: a study of 48 patients. Medicine (Baltimore) 80(4):245–251

Fournier PE, Minnick MF, Lepidi H, Salvo E, Raoult D (2001b) Experimental model of human body louse infection using green fluorescent protein-expressing *Bartonella quintana*. Infect Immun 69(3):1876–1879. doi:10.1128/iai.69.3.1876-1879.2001

Fournier PE, Ndihokubwayo JB, Guidran J, Kelly PJ, Raoult D (2002) Human pathogens in body and head lice. Emerg Infect Dis 8(12):1515–1518. doi:10.3201/eid0812.020111

Gabriel MW, Henn J, Foley JE, Brown RN, Kasten RW, Foley P, Chomel BB (2009) Zoonotic *Bartonella* species in fleas collected on gray foxes (*Urocyon cinereoargenteus*). Vector Borne Zoonotic Dis 9(6):597–602. doi:10.1089/vbz.2008.0134

Gil H, Garcia-Esteban C, Barandika JF, Peig J, Toledo A, Escudero R, Jado I, Rodriguez-Vargas-M, Garcia-Amil C, Lobo B, Roales P, Rodriguez-Moreno I, Olmeda AS, Garcia-Perez AL, Anda P (2010) Variability of *Bartonella* genotypes among small mammals in Spain. Appl Environ Microbiol 76(24):8062–8070. doi:10.1128/AEM.01963-10

Giladi M, Maman E, Paran D, Bickels J, Comaneshter D, Avidor B, Varon-Graidy M, Ephros M, Wientroub S (2005) Cat-scratch disease-associated arthropathy. Arthritis Rheum 52(11):3611–3617. doi:10.1002/art.21411

Gray GC, Johnson AA, Thornton SA, Smith WA, Knobloch J, Kelley PW, Obregon Escudero L, Arones Huayda M, Wignall FS (1990) An epidemic of Oroya fever in the Peruvian Andes. Am J Trop Med Hyg 42(3):215–221

Gundi VA, Davoust B, Khamis A, Boni M, Raoult D, La Scola B (2004) Isolation of *Bartonella rattimassiliensis* sp. nov. and *Bartonella phoceensis* sp. nov. from European *Rattus norvegicus*. J Clin Microbiol 42(8):3816–3818. doi:10.1128/JCM.42.8.3816-3818.2004

Gundi VA, Taylor C, Raoult D, La Scola B (2009) Bartonella rattaustraliani sp. nov., Bartonella queenslandensis sp. nov. and Bartonella coopersplainsensis sp. nov., identified in Australian rats. Int J Syst Evol Microbiol 59(Pt 12):2956–2961. doi:10.1099/ijs.0.002865-0

Gundi VA, Billeter SA, Rood MP, Kosoy MY (2012) *Bartonella* spp. in rats and zoonoses, Los Angeles, California, USA. Emerg Infect Dis 18(4):631–633. doi:10.3201/eid1804.110816

Guptill L (2010) Feline bartonellosis. Vet Clin North Am Small Anim Pract 40(6):1073–1090. doi:10.1016/j.cvsm.2010.07.009

Guptill L, Slater L, Wu CC, Lin TL, Glickman LT, Welch DF, HogenEsch H (1997) Experimental infection of young specific pathogen-free cats with *Bartonella henselae*. J Infect Dis 176 (1):206–216

Guptill L, Wu CC, HogenEsch H, Slater LN, Glickman N, Dunham A, Syme H, Glickman L (2004) Prevalence, risk factors, and genetic diversity of *Bartonella henselae* infections in pet cats in four regions of the United States. J Clin Microbiol 42(2):652–659

Halos L, Jamal T, Maillard R, Girard B, Guillot J, Chomel B, Vayssier-Taussat M, Boulouis HJ (2004) Role of Hippoboscidae flies as potential vectors of *Bartonella* spp. infecting wild and domestic ruminants. Appl Environ Microbiol 70(10):6302–6305. doi:10.1128/aem.70.10.6302-6305.2004

Halperin JJ, Wormser GP (2001) Of fleas and ticks on cats and mice. Arch Neurol 58(9):1345–1347

Harms CA, Maggi RG, Breitschwerdt EB, Clemons-Chevis CL, Solangi M, Rotstein DS, Fair PA, Hansen LJ, Hohn AA, Lovewell GN, McLellan WA, Pabst DA, Rowles TK, Schwacke LH, Townsend FI, Wells RS (2008) *Bartonella* species detection in captive, stranded and free-ranging cetaceans. Vet Res 39(6):59. doi:10.1051/vetres:2008036

Heller R, Riegel P, Hansmann Y, Delacour G, Bermond D, Dehio C, Lamarque F, Monteil H, Chomel B, Piemont Y (1998) *Bartonella tribocorum* sp. nov., a new *Bartonella* species isolated from the blood of wild rats. Int J Syst Bacteriol 48(Pt 4):1333–1339

Henn JB, Chomel BB, Boulouis HJ, Kasten RW, Murray WJ, Bar-Gal GK, King R, Courreau JF, Baneth G (2009) *Bartonella rochalimae* in raccoons, coyotes, and red foxes. Emerg Infect Dis 15(12):1984–1987. doi:10.3201/eid1512.081692

Hertig M (1942) Phlebotomus and Carrion's disease. Am J Trop Med Hygiene 22(Suppl.):1–80

Higgins JA, Radulovic S, Jaworski DC, Azad AF (1996) Acquisition of the cat scratch disease agent *Bartonella henselae* by cat fleas (Siphonaptera:Pulicidae). J Med Entomol 33(3):490–495

Houpikian P, Raoult D (2005) Blood culture-negative endocarditis in a reference center: etiologic diagnosis of 348 cases. Medicine 84(3):162–173

Humar A, Salit I (1999) Disseminated *Bartonella* infection with granulomatous hepatitis in a liver transplant recipient. Liver Transpl Surg 5(3):249–251. doi:10.1002/lt.500050308

Ihler GM (1996) *Bartonella bacilliformis*: dangerous pathogen slowly emerging from deep background. FEMS Microbiol Lett 144(1):1–11

Inoue K, Kabeya H, Shiratori H, Ueda K, Kosoy MY, Chomel BB, Boulouis HJ, Maruyama S (2010) *Bartonella japonica* sp. nov. and *Bartonella silvatica* sp. nov., isolated from *Apodemus* mice. Int J Syst Evol Microbiol 60(Pt 4):759–763. doi:10.1099/ijs.0.011528-0

Jackson LA, Spach DH, Kippen DA, Sugg NK, Regnery RL, Sayers MH, Stamm WE (1996) Seroprevalence to *Bartonella quintana* among patients at a community clinic in downtown Seattle. J Infect Dis 173(4):1023–1026

Jameson P, Greene C, Regnery R, Dryden M, Marks A, Brown J, Cooper J, Glaus B, Greene R (1995) Prevalence of *Bartonella henselae* antibodies in pet cats throughout regions of North America. J Infect Dis 172(4):1145–1149

Jardine C, Appleyard G, Kosoy MY, McColl D, Chirino-Trejo M, Wobeser G, Leighton FA (2005) Rodent-associated *Bartonella* in Saskatchewan, Canada. Vector Borne Zoonotic Dis 5(4):402–409. doi:10.1089/vbz.2005.5.402

Jeyaprakash A, Hoy MA, Allsopp MH (2003) Bacterial diversity in worker adults of *Apis mellifera capensis* and *Apis mellifera scutellata* (Insecta: Hymenoptera) assessed using 16S rRNA sequences. J Invertebr Pathol 84(2):96–103

Johnson R, Ramos-Vara J, Vemulapalli R (2009) Identification of *Bartonella henselae* in an aborted equine fetus. Vet Pathol 46(2):277–281. doi:10.1354/vp.46-2-277

Jones SL, Maggi R, Shuler J, Alward A, Breitschwerdt EB (2008) Detection of *Bartonella henselae* in the blood of 2 adult horses. J Vet Intern Med 22(2):495–498. doi:10.1111/j.1939-1676.2008.0043.x

Kabeya H, Colborn JM, Bai Y, Lerdthusnee K, Richardson JH, Maruyama S, Kosoy MY (2010) Detection of *Bartonella tamiae* DNA in ectoparasites from rodents in Thailand and their sequence similarity with bacterial cultures from Thai patients. Vector Borne Zoonotic Dis 10 (5):429–434. doi:10.1089/vbz.2009.0124

Kaewmongkol G, Kaewmongkol S, Owen H, Fleming PA, Adams PJ, Ryan U, Irwin PJ, Fenwick SG (2011) *Candidatus Bartonella antechini:* a novel *Bartonella* species detected in fleas and ticks from the yellow-footed antechinus (*Antechinus flavipes*), an Australian marsupial. Vet Microbiol 149(3–4):517–521. doi:10.1016/j.vetmic.2010.12.003

Kahr A, Kerbl R, Gschwandtner K, Heinzl B, Lackner H, Schwinger W, Stunzner D, Lindbichler F, Urban EC (2000) Visceral manifestation of cat scratch disease in children. A consequence of altered immunological state? Infection 28(2):116–118

Karesh WB, Dobson A, Lloyd-Smith JO, Lubroth J, Dixon MA, Bennett M, Aldrich S, Harrington T, Formenty P, Loh EH, Machalaba CC, Thomas MJ, Heymann DL (2012) Ecology of zoonoses: natural and unnatural histories. Lancet 380(9857):1936–1945. doi:10.1016/S0140-6736(12)61678-X

Kelly PJ, Matthewman LA, Hayter D, Downey S, Wray K, Bryson NR, Raoult D (1996) *Bartonella* (*Rochalimaea*) *henselae* in southern Africa–evidence for infections in domestic cats and implications for veterinarians. J S Afr Vet Assoc 67(4):182–187

Kelly PJ, Rooney JJ, Marston EL, Jones DC, Regnery RL (1998) *Bartonella henselae* isolated from cats in Zimbabwe. Lancet 351(9117):1706

Ketring KL, Zuckerman EE, Hardy WD Jr (2004) *Bartonella*: a new etiological agent of feline ocular disease. J Am Anim Hosp Assoc 40(1):6–12

Kim CM, Kim JY, Yi YH, Lee MJ, Cho MR, Shah DH, Klein TA, Kim HC, Song JW, Chong ST, O'Guinn ML, Lee JS, Lee IY, Park JH, Chae JS (2005) Detection of *Bartonella* species from ticks, mites and small mammals in Korea. J Vet Sci 6(4):327–334

Kim YS, Seo KW, Lee JH, Choi EW, Lee HW, Hwang CY, Shin NS, Youn HJ, Youn HY (2009) Prevalence of *Bartonella henselae* and *Bartonella clarridgeiae* in cats and dogs in Korea. J Vet Sci 10(1):85–87

Kitchell BE, Fan TM, Kordick D, Breitschwerdt EB, Wollenberg G, Lichtensteiger CA (2000) *Peliosis hepatis* in a dog infected with *Bartonella henselae*. J Am Vet Med Assoc 216(4):519–523

Koehler JE, Tappero JW (1993) Bacillary angiomatosis and bacillary peliosis in patients infected with human immunodeficiency virus. Clin Infect Dis 17(4):612–624

Koehler JE, Quinn FD, Berger TG, LeBoit PE, Tappero JW (1992) Isolation of *Rochalimaea* species from cutaneous and osseous lesions of bacillary angiomatosis. N Engl J Med 327 (23):1625–1631. doi:10.1056/NEJM199212033272303

Koehler JE, Glaser CA, Tappero JW (1994) *Rochalimaea henselae* infection. A new zoonosis with the domestic cat as reservoir. JAMA 271(7):531–535

Koehler JE, Sanchez MA, Garrido CS, Whitfeld MJ, Chen FM, Berger TG, Rodriguez-Barradas MC, LeBoit PE, Tappero JW (1997) Molecular epidemiology of bartonella infections in patients with bacillary angiomatosis-peliosis. N Engl J Med 337(26):1876–1883. doi:10.1056/NEJM199712253372603

Kordick DL, Breitschwerdt EB (1998) Persistent infection of pets within a household with three *Bartonella* species. Emerg Infect Dis 4(2):325–328. doi:10.3201/eid0402.980225

Kordick DL, Hilyard EJ, Hadfield TL, Wilson KH, Steigerwalt AG, Brenner DJ, Breitschwerdt EB (1997) Bartonella clarridgeiae, a newly recognized zoonotic pathogen causing inoculation papules, fever, and lymphadenopathy (cat scratch disease). J Clin Microbiol 35(7):1813–1818

Kordick DL, Brown TT, Shin K, Breitschwerdt EB (1999) Clinical and pathologic evaluation of chronic *Bartonella henselae* or *Bartonella clarridgeiae* infection in cats. J Clin Microbiol 37 (5):1536–1547

Kosoy MY, Regnery RL, Tzianabos T, Marston EL, Jones DC, Green D, Maupin GO, Olson JG, Childs JE (1997) Distribution, diversity, and host specificity of *Bartonella* in rodents from the Southeastern United States. Am J Trop Med Hyg 57(5):578–588

Kosoy MY, Regnery RL, Kosaya OI, Jones DC, Marston EL, Childs JE (1998) Isolation of *Bartonella* spp. from embryos and neonates of naturally infected rodents. J Wildl Dis 34 (2):305–309

Kosoy MY, Saito EK, Green D, Marston EL, Jones DC, Childs JE (2000) Experimental evidence of host specificity of *Bartonella* infection in rodents. Comp Immunol Microbiol Infect Dis 23 (4):221–238

Kosoy M, Murray M, Gilmore RD Jr, Bai Y, Gage KL (2003) *Bartonella* strains from ground squirrels are identical to *Bartonella washoensis* isolated from a human patient. J Clin Microbiol 41(2):645–650

Kosoy M, Bai Y, Lynch T, Kuzmin IV, Niezgoda M, Franka R, Agwanda B, Breiman RF, Rupprecht CE (2010a) *Bartonella* spp. in bats, Kenya. Emerg Infect Dis 16(12):1875–1881. doi:10.3201/eid1612.100601

Kosoy M, Bai Y, Sheff K, Morway C, Baggett H, Maloney SA, Boonmar S, Bhengsri S, Dowell SF, Sitdhirasdr A, Lerdthusnee K, Richardson J, Peruski LF (2010b) Identification of *Bartonella* infections in febrile human patients from Thailand and their potential animal reservoirs. Am J Trop Med Hyg 82(6):1140–1145. doi:10.4269/ajtmh.2010.09-0778

Kosoy M, Hayman DT, Chan KS (2012) *Bartonella* bacteria in nature: where does population variability end and a species start? Infect Genet Evol 12(5):894–904. doi:10.1016/j.meegid.2012.03.005

Kosoy M, Morway C, Sheff KW, Bai Y, Colborn J, Chalcraft L, Dowell SF, Peruski LF, Maloney SA, Baggett H, Sutthirattana S, Sidhirat A, Maruyama S, Kabeya H, Chomel BB, Kasten R, Popov V, Robinson J, Kruglov A, Petersen LR (2008) *Bartonella tamiae* sp. nov., a newly recognized pathogen isolated from three human patients from Thailand. J Clin Microbiol 46 (2):772–775

Kostrzewski J (1950) Epidemiology of trench fever. Med Dosw Mikrobiol 2(1):19–51

Kruszewska D, Tylewska-Wierzbanowska S (1996) Unknown species of rickettsiae isolated from *Ixodes ricinus* tick in Walcz. Rocz Akad Med Bialymst 41(1):129–135

Lappin MR, Black JC (1999) *Bartonella* spp infection as a possible cause of uveitis in a cat. J Am Vet Med Assoc 214(8):1205–1207

Lappin MR, Hawley J (2009) Presence of *Bartonella* species and *Rickettsia* species DNA in the blood, oral cavity, skin and claw beds of cats in the United States. Vet Dermatol 20(5–6):509–514. doi:10.1111/j.1365-3164.2009.00800.x

Lappin MR, Kordick DL, Breitschwerdt EB (2000) *Bartonella* spp antibodies and DNA in aqueous humour of cats. J Feline Med Surg 2(1):61–68. doi:10.1053/jfms.2000.0067

Leibovitz K, Pearce L, Brewer M, Lappin MR (2008) *Bartonella* species antibodies and DNA in cerebral spinal fluid of cats with central nervous system disease. J Feline Med Surg 10(4):332–337. doi:10.1016/j.jfms.2007.12.011

Li DM, Liu QY, Yu DZ, Zhang JZ, Gong ZD, Song XP (2007) Phylogenetic analysis of *Bartonella* detected in rodent fleas in Yunnan, China. J Wildl Dis 43(4):609–617

Lienhardt B, Irani S, Gaspert A, Weishaupt D, Boehler A (2009) Disseminated infection with *Bartonella henselae* in a lung transplant recipient. J Heart Lung Transplant 28(7):736–739. doi:10.1016/j.healun.2009.03.016

Lin JW, Chen CY, Chen WC, Chomel BB, Chang CC (2008) Isolation of *Bartonella* species from rodents in Taiwan including a strain closely related to 'Bartonella rochalimae' from *Rattus norvegicus*. J Med Microbiol 57(Pt 12):1496–1501. doi:10.1099/jmm.0.2008/004671-0

Lin JW, Hsu YM, Chomel BB, Lin LK, Pei JC, Wu SH, Chang CC (2012) Identification of novel *Bartonella* spp. in bats and evidence of Asian gray shrew as a new potential reservoir of *Bartonella*. Vet Microbiol 156(1–2):119–126. doi:10.1016/j.vetmic.2011.09.031

Lobanov VA, Gajadhar AA, Al-Adhami B, Schwantje HM (2012) Molecular study of free-ranging mule deer and white-tailed deer from British Columbia, Canada, for evidence of *Anaplasma* spp. and *Ehrlichia* spp. Transbound Emerg Dis 59(3):233–243. doi:10.1111/j.1865-1682.2011.01259.x

Loftis AD, Gill JS, Schriefer ME, Levin ML, Eremeeva ME, Gilchrist MJ, Dasch GA (2005) Detection of Rickettsia, Borrelia, and Bartonella *in Carios kelleyi* (Acari: Argasidae). J Med Entomol 42(3):473–480

Lucey D, Dolan MJ, Moss CW, Garcia M, Hollis DG, Wegner S, Morgan G, Almeida R, Leong D, Greisen KS et al (1992) Relapsing illness due to *Rochalimaea henselae* in immunocompetent hosts: implication for therapy and new epidemiological associations. Clin Infect Dis 14(3):683–688

MacDonald KA, Chomel BB, Kittleson MD, Kasten RW, Thomas WP, Pesavento P (2004) A prospective study of canine infective endocarditis in northern California (1999–2001): emergence of *Bartonella* as a prevalent etiologic agent. J Vet Intern Med 18(1):56–64

Magalhães RF, Pitassi LH, Salvadego M, de Moraes AM, Barjas-Castro ML, Velho PE (2008) *Bartonella henselae* survives after the storage period of red blood cell units: is it transmissible by transfusion? Transfus Med 18(5):287–291

Maggi RG, Breitschwerdt EB (2005) Potential limitations of the 16S-23S rRNA intergenic region for molecular detection of *Bartonella* species. J Clin Microbiol 43(3):1171–1176. doi:10.1128/jcm.43.3.1171-1176.2005

Maggi RG, Duncan AW, Breitschwerdt EB (2005a) Novel chemically modified liquid medium that will support the growth of seven *Bartonella* species. J Clin Microbiol 43(6):2651–2655. doi:10.1128/jcm.43.6.2651-2655.2005

Maggi RG, Harms CA, Hohn AA, Pabst DA, McLellan WA, Walton WJ, Rotstein DS, Breitschwerdt EB (2005b) *Bartonella henselae* in porpoise blood. Emerg Infect Dis 11(12):1894–1898

Maggi RG, Chomel B, Hegarty BC, Henn J, Breitschwerdt EB (2006) A *Bartonella vinsonii berkhoffii* typing scheme based upon 16S-23S ITS and Pap31 sequences from dog, coyote, gray fox, and human isolates. Mol Cell Probes 20(2):128–134. doi:10.1016/j.mcp.2005.11.002

Maggi RG, Raverty SA, Lester SJ, Huff DG, Haulena M, Ford SL, Nielsen O, Robinson JH, Breitschwerdt EB (2008) Bartonella henselae in captive and hunter-harvested beluga (*Delphinapterus leucas*). J Wildl Dis 44(4):871–877

Maggi RG, Kosoy M, Mintzer M, Breitschwerdt EB (2009) Isolation of *Candidatus Bartonella melophagi* from human blood. Emerg Infect Dis 15(1):66–68. doi:10.3201/eid1501.081080

Maggi RG, Mascarelli PE, Pultorak EL, Hegarty BC, Bradley JM, Mozayeni BR, Breitschwerdt EB (2011) *Bartonella* spp. bacteremia in high-risk immunocompetent patients. Diagn Microbiol Infect Dis 71(4):430–437. doi:10.1016/j.diagmicrobio.2011.09.001

Maman E, Bickels J, Ephros M, Paran D, Comaneshter D, Metzkor-Cotter E, Avidor B, Varon-Graidy M, Wientroub S, Giladi M (2007) Musculoskeletal manifestations of cat scratch disease. Clin Infect Dis 45(12):1535–1540. doi:10.1086/523587

Marquez FJ, Millan J, Rodriguez-Liebana JJ, Garcia-Egea I, Muniain MA (2009) Detection and identification of Bartonella sp. in fleas from carnivorous mammals in Andalusia, Spain. Med Vet Entomol 23(4):393–398. doi:10.1111/j.1365-2915.2009.00830.x

Maruyama S, Nogami S, Inoue I, Namba S, Asanome K, Katsube Y (1996) Isolation of Bartonella henselae from domestic cats in Japan. J Vet Med Sci 58(1):81–83

Mascarelli PE, Maggi RG, Hopkins S, Mozayeni BR, Trull CL, Bradley JM, Hegarty BC, Breitschwerdt EB (2013) Bartonella henselae infection in a family experiencing neurological and neurocognitive abnormalities after woodlouse hunter spider bites. Parasit Vectors 6(1):98. doi:10.1186/1756-3305-6-98

Matsumoto K, Berrada ZL, Klinger E, Goethert HK, Telford SR 3rd (2008) Molecular detection of *Bartonella schoenbuchensis* from ectoparasites of deer in Massachusetts. Vector Borne Zoonotic Dis 8(4):549–554. doi:10.1089/vbz.2007.0244

McElroy KM, Blagburn BL, Breitschwerdt EB, Mead PS, McQuiston JH (2010) Flea-associated zoonotic diseases of cats in the USA: bartonellosis, flea-borne rickettsioses, and plague. Trends Parasitol 26(4):197–204. doi:10.1016/j.pt.2010.01.001

McNee JW, Renshaw A (1916) "Trench fever": a relapsing fever occurring with the British forces in France. Br Med J 1(2876):225–234

Minnick MF, Battisti JM (2009) Pestilence, persistence and pathogenicity: infection strategies of *Bartonella*. Future Microbiol 4(6):743–758. doi:10.2217/fmb.09.41

Morick D, Krasnov BR, Khokhlova IS, Shenbrot GI, Kosoy MY, Harrus S (2010) *Bartonella* genotypes in fleas (insecta: siphonaptera) collected from rodents in the Negev Desert, Israel. Appl Environ Microbiol 76(20):6864–6869

Nasirudeen AM, Thong ML (1999) Prevalence of *Bartonella henselae* immunoglobulin G antibodies in Singaporean cats. Pediatr Infect Dis J 18(3):276–278

Nelder MP, Reeves WK, Adler PH, Wozniak A, Wills W (2009) Ectoparasites and associated pathogens of free-roaming and captive animals in zoos of South Carolina. Vector Borne Zoonotic Dis 9(5):469–477. doi:10.1089/vbz.2008.0008

Oskouizadeh K, Zahraei-Salehi T, Aledavood S (2010) Detection of *Bartonella henselae* in domestic cats' saliva. Iran J Microbiol 2(2):80–84

Pappalardo BL, Correa MT, York CC, Peat CY, Breitschwerdt EB (1997) Epidemiologic evaluation of the risk factors associated with exposure and seroreactivity to *Bartonella vinsonii* in dogs. Am J Vet Res 58(5):467–471

Pappalardo BL, Brown T, Gookin JL, Morrill CL, Breitschwerdt EB (2000) Granulomatous disease associated with *Bartonella* infection in 2 dogs. J Vet Intern Med 14(1):37–42

Parola P, Shpynov S, Montoya M, Lopez M, Houpikian P, Zeaiter Z, Guerra H, Raoult D (2002) First molecular evidence of new *Bartonella* spp. in fleas and a tick from Peru. Am J Trop Med Hyg 67(2):135–136

Pearce LK, Radecki SV, Brewer M, Lappin MR (2006) Prevalence of *Bartonella henselae* antibodies in serum of cats with and without clinical signs of central nervous system disease. J Feline Med Surg 8(5):315–320. doi:10.1016/j.jfms.2006.04.001

Pedersen AB, Altizer S, Poss M, Cunningham AA, Nunn CL (2005) Patterns of host specificity and transmission among parasites of wild primates. Int J Parasitol 35(6):647–657. doi:10.1016/j.ijpara.2005.01.005

Perez C, Maggi RG, Diniz PP, Breitschwerdt EB (2011) Molecular and serological diagnosis of *Bartonella* infection in 61 dogs from the United States. J Vet Intern Med 25(4):805–810. doi:10.1111/j.1939-1676.2011.0736.x

Perez-Martinez L, Venzal JM, Gonzalez-Acuna D, Portillo A, Blanco JR, Oteo JA (2009) *Bartonella rochalimae* and other *Bartonella* spp. in fleas, Chile. Emerg Infect Dis 15 (7):1150–1152. doi:10.3201/eid1507.081570

Pesavento PA, Chomel BB, Kasten RW, McDonald KA, Mohr FC (2005) Pathology of *Bartonella endocarditis* in six dogs. Vet Pathol 42(3):370–373. doi:10.1354/vp.42-3-370

Pretorius AM, Beati L, Birtles RJ (2004) Diversity of bartonellae associated with small mammals inhabiting Free State province, South Africa. Int J Syst Evol Microbiol 54(Pt 6):1959–1967. doi:10.1099/ijs.0.03033-0

Pultorak EL, Maggi RG, Mascarelli PE, Breitschwerdt EB (2013) Serial testing from a three-day collection period using the BAPGM platform may enhance the sensitivity of *Bartonella* spp. detection in bacteremic human patients. J Clin Microbiol. doi:10.1128/JCM.00123-13

Raoult D, Roux V (1999) The body louse as a vector of reemerging human diseases. Clin Infect Dis 29(4):888–911. doi:10.1086/520454

Raoult D, La Scola B, Kelly PJ, Davoust B, Gomez J (2005) *Bartonella bovis* in cattle in Africa. Vet Microbiol 105(2):155–156. doi:10.1016/j.vetmic.2004.10.013

Raoult D, Dutour O, Houhamdi L, Jankauskas R, Fournier PE, Ardagna Y, Drancourt M, Signoli M, La VD, Macia Y, Aboudharam G (2006) Evidence for louse-transmitted diseases in soldiers of Napoleon's Grand Army in Vilnius. J Infect Dis 193(1):112–120. doi:10.1086/498534

Reeves WK, Loftis AD, Gore JA, Dasch GA (2005) Molecular evidence for novel bartonella species in *Trichobius major* (Diptera: Streblidae) and *Cimex adjunctus* (Hemiptera: Cimicidae) from two southeastern bat caves, U.S.A. J Vector Ecol 30(2):339–341

Reeves WK, Nelder MP, Cobb KD, Dasch GA (2006) *Bartonella* spp. in deer keds, *Lipoptena mazamae* (Diptera: Hippoboscidae), from Georgia and South Carolina, USA. J Wildl Dis 42 (2):391–396

Reeves WK, Loftis AD, Szumlas DE, Abbassy MM, Helmy IM, Hanafi HA, Dasch GA (2007) Rickettsial pathogens in the tropical rat mite *Ornithonyssus bacoti* (Acari:Macronyssidae) from Egyptian rats (*Rattus* spp.). Exp Appl Acarol 41(1–2):101–107

Regnery RL, Anderson BE, Clarridge JE 3rd, Rodriguez-Barradas MC, Jones DC, Carr JH (1992) Characterization of a novel *Rochalimaea* species, *R. henselae* sp. nov., isolated from blood of a febrile, human immunodeficiency virus-positive patient. J Clin Microbiol 30(2):265–274

Regnery RL, Childs JE, Koehler JE (1995) Infections associated with *Bartonella* species in persons infected with human immunodeficiency virus. Clin Infect Dis 21(Suppl 1):S94–S98

Reis C, Cote M, Le Rhun D, Lecuelle B, Levin ML, Vayssier-Taussat M, Bonnet SI (2011) Vector competence of the tick Ixodes ricinus for transmission of *Bartonella birtlesii*. PLoS Negl Trop Dis 5(5):e1186. doi:10.1371/journal.pntd.0001186

Relman DA (1995) Has trench fever returned? N Engl J Med 332(7):463–464. doi:10.1056/NEJM199502163320710

Rolain JM, Rousset E, La Scola B, Duquesnel R, Raoult D (2003) *Bartonella schoenbuchensis* isolated from the blood of a French cow. Ann N Y Acad Sci 990:236–238

Rolain JM, Bourry O, Davoust B, Raoult D (2005) *Bartonella quintana* and *Rickettsia felis* in Gabon. Emerg Infect Dis 11(11):1742–1744. doi:10.3201/eid1111.050861

Rotstein DS, Taylor SK, Bradley J, Rieitschwerdt EB (2000) Prevalence of *Bartonella henselae* antibody in Florida panthers. J Wildl Dis 36(1):157–160

Roux V, Eykyn SJ, Wyllie S, Raoult D (2000) *Bartonella vinsonii* subsp. *berkhoffii* as an agent of afebrile blood culture-negative endocarditis in a human. J Clin Microbiol 38(4):1698–1700

Sasaki T, Kobayashi M, Agui N (2002) Detection of *Bartonella quintana* from body lice (Anoplura: Pediculidae) infesting homeless people in Tokyo by molecular technique. J Med Entomol 39(3):427–429

Sasaki T, Poudel SK, Isawa H, Hayashi T, Seki N, Tomita T, Sawabe K, Kobayashi M (2006) First molecular evidence of *Bartonella quintana* in *Pediculus humanus capitis* (Phthiraptera: Pediculidae), collected from Nepalese children. J Med Entomol 43(1):110–112

Schaefer JD, Kasten RW, Coonan TJ, Clifford DL, Chomel BB (2011) Isolation or detection of *Bartonella vinsonii* subspecies *berkhoffii* and *Bartonella rochalimae* in the endangered island foxes (Urocyon littoralis). Vet Microbiol 154(1–2):135–139. doi:10.1016/j.vetmic.2011.06.031

Schaefer JD, Moore GM, Namekata MS, Kasten RW, Chomel BB (2012) Seroepidemiology of *Bartonella* infection in gray foxes from Texas. Vector Borne Zoonotic Dis 12(5):428–430. doi:10.1089/vbz.2011.0805

Schultz MG (1968) A history of bartonellosis (Carrion's disease). Am J Trop Med Hyg 17(4):503–515

Serratrice J, Rolain JM, Granel B, Ene N, Conrath J, Avierinos JF, Disdier P, Raoult D, Weiller PJ (2003) Bilateral retinal artery branch occlusions revealing *Bartonella grahamii* infection. Rev Med Interne 24(9):629–630

Slater LN, Welch DF, Hensel D, Coody DW (1990) A newly recognized fastidious gram-negative pathogen as a cause of fever and bacteremia. N Engl J Med 323(23):1587–1593. doi:10.1056/NEJM199012063232303

Smarick SD, Jandery KE, Chomel BB, Thomas WP, Aldrich J (2004) Aortic valvular endocarditis caused by *Bartonella vinsonii* subsp. *berkhoffii* in 2 dogs presenting for fulminant pulmonary edema. J Vet Emerg Crit Care 14:42–51

Spach DH, Callis KP, Paauw DS, Houze YB, Schoenknecht FD, Welch DF, Rosen H, Brenner DJ (1993) Endocarditis caused by *Rochalimaea quintana* in a patient infected with human immunodeficiency virus. J Clin Microbiol 31(3):692–694

Spach DH, Kanter AS, Daniels NA, Nowowiejski DJ, Larson AM, Schmidt RA, Swaminathan B, Brenner DJ (1995a) *Bartonella* (*Rochalimaea*) species as a cause of apparent "culture-negative" endocarditis. Clin Infect Dis 20(4):1044–1047

Spach DH, Kanter AS, Dougherty MJ, Larson AM, Coyle MB, Brenner DJ, Swaminathan B, Matar GM, Welch DF, Root RK et al (1995b) *Bartonella* (*Rochalimaea*) *quintana* bacteremia in inner-city patients with chronic alcoholism. N Engl J Med 332(7):424–428. doi:10.1056/nejm199502163320703

Sreter-Lancz Z, Tornyai K, Szell Z, Sreter T, Marialigeti K (2006) *Bartonella* infections in fleas (Siphonaptera: Pulicidae) and lack of bartonellae in ticks (Acari: Ixodidae) from Hungary. Folia Parasitol 53(4):313–316

Strong RP, Tyzzer E, Brues CT, Sellards AW, Gastiaburu JC (1915) Report of the first expedition to South America. Havard University Press, Cambridge

Suksawat J, Pitulle C, Arraga-Alvarado C, Madrigal K, Hancock SI, Breitschwerdt EB (2001) Coinfection with three *Ehrlichia* species in dogs from Thailand and Venezuela with emphasis on consideration of 16S ribosomal DNA secondary structure. J Clin Microbiol 39(1):90–93. doi:10.1128/JCM.39.1.90-93.2001

Sun J, Fu G, Lin J, Song X, Lu L, Liu Q (2010) Seroprevalence of *Bartonella* in Eastern China and analysis of risk factors. BMC Infect Dis 10:121. doi:10.1186/1471-2334-10-121

Swift HF (1920) Trench fever. Arch Intern Med 26(1):76–98

Sykes JE, Henn JB, Kasten RW, Allen C, Chomel BB (2007) *Bartonella henselae* infection in splenectomized domestic cats previously infected with hemotropic *Mycoplasma* species. Vet Immunol Immunopathol 116(1–2):104–108. doi:10.1016/j.vetimm.2006.12.004

Telfer S, Begon M, Bennett M, Bown KJ, Burthe S, Lambin X, Telford G, Birtles R (2007a) Contrasting dynamics of *Bartonella* spp. in cyclic field vole populations: the impact of vector and host dynamics. Parasitology 134(Pt 3):413–425. doi:10.1017/S0031182006001624

Telfer S, Clough HE, Birtles LR, Bennett M, Carslake D, Helyar S, Begon M (2007b) Ecological differences and coexistence in a guild of microparasites: *Bartonella* in wild rodents. Ecology 88(7):1841–1849

Thudi KR, Kreikemeier JT, Phillips NJ, Salvalaggio PR, Kennedy DJ, Hayashi PH (2007) Cat scratch disease causing hepatic masses after liver transplant. Liver Int 27(1):145–148. doi:10.1111/j.1478-3231.2006.01386.x

Townsend CH (1914) On the identity of Verruga and Carrion's fever. Science 39(994):99–100. doi:10.1126/science.39.994.99

Tsukahara M, Tsuneoka H, Iino H, Ohno K, Murano I (1998) *Bartonella henselae* infection from a dog. Lancet 352(9141):1682

Tsai YL, Chuang ST, Chang CC, Kass PH, Chomel BB (2010) *Bartonella* species in small mammals and their ectoparasites in Taiwan. Am J Trop Med Hyg 83(4):917–923

Turnbull BS, Cowan DF (1999) Angiomatosis, a newly recognized disease in Atlantic bottlenose dolphins (*Tursiops truncatus*) from the Gulf of Mexico. Vet Pathol 36(1):28–34

Tuttle AD, Birkenheuer AJ, Juopperi T, Levy MG, Breitschwerdt EB (2003) Concurrent bartonellosis and babesiosis in a dog with persistent thrombocytopenia. J Am Vet Med Assoc 223(9):1306–1310, 1280–1301

Valerio CR, Murray P, Arlian LG, Slater JE (2005) Bacterial 16S ribosomal DNA in house dust mite cultures. J Allergy Clin Immunol 116(6):1296–1300. doi:10.1016/j.jaci.2005.09.046

Varanat M, Maggi RG, Linder KE, Breitschwerdt EB (2011) Molecular prevalence of *Bartonella*, *Babesia*, and hemotropic *Mycoplasma* sp. in dogs with splenic disease. J Vet Intern Med 25 (6):1284–1291. doi:10.1111/j.1939-1676.2011.00811.x

Varanat M, Broadhurst J, Linder KE, Maggi RG, Breitschwerdt EB (2012) Identification of *Bartonella henselae* in 2 cats with pyogranulomatous myocarditis and diaphragmatic myositis. Vet Pathol 49(4):608–611. doi:10.1177/0300985811404709

Varela G, Fournier R, Mooser H (1954) Presence of *Rickettsia quintana* in *Pediculus humanus* from Mexico City; experimental inoculation. Rev Inst Salubr Enferm Trop 14(1):39–42

Vinson JW, Varela G, Molina-Pasquel C (1969) Trench fever. 3. Induction of clinical disease in volunteers inoculated with *Rickettsia quintana* propagated on blood agar. Am J Trop Med Hyg 18(5):713–722

Welch DF, Pickett DA, Slater LN, Steigerwalt AG, Brenner DJ (1992) *Rochalimaea henselae* sp. nov., a cause of septicemia, bacillary angiomatosis, and parenchymal bacillary peliosis. J Clin Microbiol 30(2):275–280

Welch DF, Carroll KC, Hofmeister EK, Persing DH, Robison DA, Steigerwalt AG, Brenner DJ (1999) Isolation of a new subspecies, *Bartonella vinsonii* subsp. *arupensis*, from a cattle rancher: identity with isolates found in conjunction with *Borrelia burgdorferi* and *Babesia microti* among naturally infected mice. J Clin Microbiol 37(8):2598–2601

Whittemore JC, Hawley JR, Radecki SV, Steinberg JD, Lappin MR (2012) *Bartonella* species antibodies and hyperglobulinemia in privately owned cats. J Vet Intern Med 26(3):639–644. doi:10.1111/j.1939-1676.2012.00925.x

Winoto IL, Goethert H, Ibrahim IN, Yuniherlina I, Stoops C, Susanti I, Kania W, Maguire JD, Bangs MJ, Telford SR 3rd, Wongsrichanalai C (2005) *Bartonella* species in rodents and shrews in the greater Jakarta area. Southeast Asian J Trop Med Public Health 36(6):1523–1529

Woolhouse ME, Taylor LH, Haydon DT (2001) Population biology of multihost pathogens. Science 292(5519):1109–1112

Yager JA, Best SJ, Maggi RG, Varanat M, Znajda N, Breitschwerdt EB (2010) Bacillary angiomatosis in an immunosuppressed dog. Vet Dermatol 21(4):420–428. doi:10.1111/j.1365-3164.2010.00879.x

Ying B, Kosoy MY, Maupin GO, Tsuchiya KR, Gage KL (2002) Genetic and ecologic characteristics of *Bartonella* communities in rodents in southern China. Am J Trop Med Hyg 66 (5):622–627

Zangwill KM, Hamilton DH, Perkins BA, Regnery RL, Plikaytis BD, Hadler JL, Cartter ML, Wenger JD (1993) Cat scratch disease in Connecticut. Epidemiology, risk factors, and evaluation of a new diagnostic test. N Engl J Med 329(1):8–13

Chapter 7
A One Health Approach to Influenza Pandemics

Robert E. Kahn and Jürgen A. Richt

Abstract Controlling influenza pandemics requires extensive scientific research, both in the laboratory and in the field. Focusing on the experience with influenza A H1N1, H5N1 and H7N9, this chapter sets out a historical framework for implementing the One Health paradigm within the context of interspecies medicine. Effective control and eradication of influenza viruses requires extensive surveillance, improved biosecurity, improved detection devices, new vaccines and monitoring.

Keywords Influenza • Interspecies medicine • One Health • Pandemic

7.1 A One Health Perspective

Writing about 400 BCE, in his book *On the Nature of Man*, the Greek physician Hippocrates sought to achieve an unified approach to health by balancing four humors (blood, phlegm, yellow bile and black bile). His humoral theory influenced the treatment of disease for more than 2,000 years, with each of the humors linked to a particular characteristic—courageousness, calm, anger, or depression. Today we are still engaged in that search for unity in understanding, our tools are no longer the discredited four humors, but rather, the triad of the person, the animal, and the environment. These three factors are interdependent. The One Health concept recognizes these three interdependent factors because human health is inextricably linked with animal and environmental health. The One Health concept offers a viable and informed approach to controlling pandemics.

At least three major influenza outbreaks have occurred in the sixteenth, eighteenth, nineteen and twentieth centuries, as well as regional outbreaks in the

R.E. Kahn
College of Veterinary Medicine, Kansas State University, Manhattan, KS, USA
e-mail: rs_kahn@hotmail.com

J.A. Richt (✉)
Center of Excellence for Emerging and Zoonotic Animal Diseases (CEEZAD),
College of Veterinary Medicine, Diagnostic Medicine/Pathobiology, Kansas State University, Manhattan, KS, USA
e-mail: jricht@vet.k-state.edu

A. Yamada et al. (eds.), *Confronting Emerging Zoonoses*,
DOI 10.1007/978-4-431-55120-1_7

seventeenth century (Crosby 1993). The most famous and most lethal of these pandemics began in 1918, afflicting approximately one-quarter of the world's population and killing some 50 million people with a 2.5 % case fatality rate (Steel and Palese 2008). The origins of the pandemic were either in Europe or the United States (or both), but the experience became known as "The Spanish Flu" because the Spaniards, neutral in World War I, were not censoring their press, unlike all the combatant nations (Barry 2006; Oxford et al. 2005). Initially, a bacterium was suspected of being the cause of this emerging zoonotic disease which infected birds, pigs and people, while moving probably solely by inter-human transmission to kill more than 50 million people.

In 1931, Richard Shope isolated a virus that caused swine influenza (Crosby 2003). But it wasn't until 1944, after years of investigation, that Oswald Avery with the support of others, identified that viruses contain nucleic acids that carry genetic information (Barry 2006). One of the reasons why it took so long to identify the cause of influenza was that the highly respected bacteriologist, Richard F. J. Pfeiffer, thought that the secondary bacterial infections were its primary cause (Crosby 1993). However, even more significant was the common thought that the influenza virus was primarily a human pathogen, thereby limiting a fuller epidemiological understanding of the sources and transmission events of the disease. With the rise of interspecies medicine, it has become increasingly clear that scientific investigations of one species are seldom sufficient to uncover the mysteries of a zoonotic disease.

Pandemics are often evaluated in the context of their impact on human health. This is understandable but consistently obscures the origins of a pandemic, its original host species and reservoir, its means of transmission, its ecology, and how it can best be controlled. The problems and solutions for human health require an awareness of all forms of life—humans, animals and plants—each in their own environment, yet also connected to other forms of life. Without such an interdependent awareness within the One Health paradigm, no future pandemic will likely ever be swiftly and effectively brought under control.

Influenza has been a global threat to human health for many centuries. While acute viral and bacterial infections of the human *upper* respiratory tract are generally mild, infections of human *lower* respiratory tract are often severe, leading to increased morbidity and mortality, and killing a significant number of children under age five every year—the vast majority in low-income countries (Reingold and Phares 2005). It has only been within the last 100 years that zoonotic influenza viruses, especially in chickens and pigs have been identified as a cause of human influenza.

Of the five major routes by which zoonotic pathogens are transmitted to humans, it is the first four that are pertinent to the interspecies transmission of influenza: (1) the inhalation of infectious materials; (2) the ingestion of contaminated materials; (3) direct or indirect contact of the human skin with the hides, hair, excreta, blood or carcasses of infected animals; (4) animal bites and scratches; and (5) biting arthropods that seek out animals and people on which to feed (Reingold and Phares 2005). As zoonotic agents, influenza viruses move back and forth between avian species and mammals, including humans. The original viruses change and reassort with other influenza viruses, creating new pathogens whose biological characteristics are initially unknown. Therefore, every new influenza pandemic is disturbingly unique; and its impact on any species will only be discovered after careful laboratory and epidemiological investigation.

7.2 The H7N9 Influenza Virus

Between 1996 and 2012, sporadic human outbreaks with H7 influenza viruses (H7N2, H7N3 and H7N7) had been reported in the United Kingdom, the United States of America, Italy, the Netherlands, Canada and Mexico. These human outbreaks were usually linked to outbreaks in poultry. With the exception of the death of one veterinarian in the Netherlands in 2003, human infections have usually been minor, mainly conjunctivitis and upper respiratory symptoms (WHO 2013a). However, the H7N9 outbreak that began in southern China in March 2013 and spread in less than 2 months throughout various parts of the country was of a different order. The latest information on H7N9 infections and deaths is available on the web from the World Health Organization (WHO) at *Disease Outbreak News* at: www.who.int/csr/don/en/index.html.

The H7N9 virus probably originated from the reassortment of at least three virus strains that infected only birds, with the six genes coding for the internal proteins similar to recent H9N2 viruses, the gene for the NA protein similar to avian N9, and the gene for the HA protein similar to Eurasian H7 viruses (WHO European Office 2013; Liu et al. 2013; Van Ranst and Lemey 2013). However, the unusual behavior of the new virus made tracing its origins difficult, because this H7N9 virus has low pathogenicity in birds and poultry, but high pathogenicity in humans. Furthermore, as a WHO Risk Assessment warned: "Some genetic changes, including amino acid substitutions associated with increased affinity to alpha 2–6 receptors, suggest that H7N9 may have greater ability to infect mammals, including humans, than other avian influenza viruses" (WHO 2013b; WHO 2013c; Holmes 2013).

In less scientific language, this newly emerged virus is dangerous, both because it is asymptomatic within China's poultry and wildfowl, and because it is a zoonotic agent that might be more likely to make the leap to human beings than the deadlier H5N1 virus. However, research published after this WHO Risk Assessment suggests that although the human H7N9 virus has already acquired certain receptor-binding characteristics that are typical of pandemic viruses, its further evolution toward a virus that could bind to cellular receptors in the upper human respiratory tract might be prevented by the preference of human H7N9 viruses to bind mainly to avian influenza receptors (Xiong et al. 2013).

The task of finding the reservoir(s) of emerging influenza viruses in China is daunting. The statistical background is important: 1.36 billion people—19 % of the world's population in 2013—live in the midst of immense livestock populations—an estimated five billion chickens (bred for consumption and egg production, often in small farms or in backyards) along with 716 million pigs slaughtered in 2013 (National Bureau of Statistics of China 2014; The Economist 2011). Furthermore, although China's population is now nearly stable with an annual increase of less than 0.5 % in 2013, the annual output of pork, beef, mutton and poultry increased by 1.8 % from 2012 to 2013, with the number of slaughtered pigs increasing by 2.5 % (National Bureau of Statistics of China 2014). With the Gross Domestic Product of China having risen by a further 7.7 % in 2013 more and more people have sufficient income to add meat to their diets, often purchased live or recently slaughtered from live animal markets (National Bureau of Statistics of China 2014; The Economist 2011).

Neither the precise reservoirs nor host species of Influenza A H7N9 has been definitively determined. Chickens appear to be the primary reservoir, but wildfowl, ducks and pigeons have also been implicated as host species (WHO 2013c; Kahn and Richt 2013). Furthermore, it has now been shown through laboratory testing that the H7N9 virus can be transmitted to both pigs and ferrets as intermediate hosts (Zhu et al 2013). In effect, the initial "reservoir" may well be the live animal markets themselves, because of the mixing of species in confined spaces with poor hygienic conditions.

Temporarily closing many live animal markets in urban areas, the Chinese health authorities appear to have considerably reduced the presence of the H7N9 virus, at least as measured in transmission to humans. Happily, despite extensive follow-up of human patients, in the summer and fall of 2013, there was no evidence of sustained human-to-human transmission. However, in a broad One Health perspective, for a number of reasons this reduction in human illness linked to the H7N9 virus may be misleading because: (1) influenza viruses are generally less prevalent in the summer months than in the winter months; (2) closing live animal markets in rural areas is not a sustainable policy; and (3) as noted above, the genetic characteristics of this already mutated H7N9 virus suggest that it is close to being able to sustain mammalian transmission; and the reality of efficient replication in both ferrets and swine has been demonstrated under laboratory conditions (Zhu et al. 2013).

Of even greater significance for transmission is the undisputed evidence that H7N9 has now been found in poultry in animal markets in all 31 provinces of the Chinese mainland (Holmes 2013; Chen et al. 2005). With such extensive presence in chickens in mainland China, as well as a potential intermediate host such as pigs, the number of human exposure events to this new virus will most likely increase significantly. However, whether occasional cross-species transmission of the H7N9 virus will lead to a human pandemic is unknown (Morens et al. 2013).

Given the billions of people and animals interacting within China, the continuing emergence of influenza viruses, is inevitable. Both scientists and the public need to know what is happening in the midst of a fast moving scene in which new data is constantly emerging and being evaluated about how the novel influenza virus is being transmitted, as well as possible difficulties in making an efficacious vaccine (Bao et al. 2013; Koopmans and de Jong 2013; Branswell 2013).

The study of evolutionary genetics is essential because although environmental changes often trigger viral emergence, some viruses are better equipped than others to jump species barriers (Holmes and Rambaut 2005). To ensure the future control of H7N9, an intensive combination of laboratory investigation, epidemiological study, continuing surveillance of both human and animal outbreaks, and the preparation and manufacture of appropriate antiviral drugs and vaccines is necessary for disease control. Such a One Health-oriented strategy is essential for reducing the risk of sporadic disease transmissions from chickens and other animals to humans (Schenk et al. 2013).

7.3 The Ubiquity of Influenza A Viruses

The identification and characterization the influenza H5N1 virus has required considerable sustained scientific research. Most avian influenza A viruses are low pathogenic avian influenza (LPAI) viruses and have been isolated in many wild and domestic birds, with a multitude of serotypes (H1 to H16, N1 to N9) having been found. Moreover, there is now a scientific consensus that wild aquatic birds provide the natural hosts of all LPAI viruses, thereby forming a large genetic pool that serves as a source for highly pathogenic avian influenza (HPAI) viruses of the H5 and H7 subtype (Klenk et al. 2008). The mixing of numerous influenza viruses in many species is an ever-present reality that requires considerable surveillance and at times medical intervention.

7.4 Reservoir and Host Species of Influenza

Although it is clear that wild birds provide both a reservoir and a host species for the majority of the influenza A viruses, other reservoirs and host species are also significant, especially chickens, quail, geese, ducks and pigs. The interaction of these multiple hosts and multiple reservoirs creates a formidable challenge to scientists in the laboratory, farmers in the field, public health officials and doctors, veterinarians and hospital workers. Furthermore, in the US, the classical swine H1N1 has been replaced in the past 10–15 years by reassortment viruses of both H1 and H3 subtypes that contain genes from classical swine, human and avian viruses (Zhou et al. 1999; Webby et al. 2000; Scholtissek 2008; Khiabanian et al. 2009). Prior to the pandemic H1N1/2009 virus outbreak in Mexico, both Scholtissek (2008) and Ma et al. (2009) and their colleagues pointed out the danger that reassorted pig viruses from North America, rather than a H5N1 virus from Southeast Asia, might start a future pandemic. This was indeed what happened, but in this instance the reassortment virus was not as deadly as initially feared (York and Donis 2013). However, direct infections from avian species to humans are possible and could generate new pandemics without pigs or other animals serving as intermediary hosts.

7.5 Mode of Transmission

For any influenza virus present in nonhuman species to adapt to human species and become pandemic, considerable changes are necessary during interspecies transmission (Matrosovich et al. 2008). These changes may be by either adaptation of the virus to a new host or reassortment of the virus, causing changes in the specificity of the viral receptor binding sites of the novel virus (Matrosovich et al. 2008). The complexity of receptor specificity is so great that in trying to trace the origins of a specific virus such as the 1918 pandemic influenza virus, definitive answers are often not possible without sequencing strains of earlier viruses (Steel and Palese 2008).

There is an emerging consensus that a key variable in whether a pandemic viral strain emerges is the receptor specificity of the HA protein. However, in the context of H5N1 and other avian viruses "it is believed that a poor fit of avian viruses to receptors in humans limits the emergence of new pandemic strains" (Matrosovich et al. 2008). Unfortunately, it is not clear which hosts can transmit viruses to humans. Therefore, for all of the influenza viruses, "the challenge for the future is to determine what chain of events makes it possible for a virus to be successfully transmitted *to*, and more importantly *within*, an alternative host population" (Webby et al. 2007). Much further research, both in the field and in the laboratory, is essential in order to understand how and why transmission of influenza viruses occurs.

7.6 Ecology and Control

On a global scale, the interaction between species has increased vastly in the last 40 years, as the number of chickens grown for human consumption has increased from 5.2 billion in 1970 to 19.4 billion in 2010, while during the same time, pigs have increased from 347 million to 965 million, ducks from 256 million to 1.187 billion, and cattle from 1.081 billion to 1.428 billion (Worldwatch Institute 2012). The human population of the world during these 40 years has also increased from 3.71 billion to 6.86 billion (United States Census Bureau 2013); and this has led David Quammen (2012, pp. 496–497, 503) to question whether from an ecological perspective *homo sapiens* itself should be characterized as "an outbreak".

In considering this question with the University of Chicago ecologist, Greg Dwyer, the two men agreed that clearly both human and animal behavior have to be considered; and that the answer "it depends" indicates that there is no one definitive answer to the question of whether the growth in population of any one species will lead to its ultimate destruction (Quammen 2012, pp. 518–519). What is striking is that simply posing this ecological quandary led Quammen and

Dwyer to a consideration of influenza, because the influenza virus itself in all its forms poses a great threat to many species, including people, wild birds, chickens and pigs. Whether influenza A viruses or any other virus should be viewed as a living organism is much debated among scientists. However, as the virologist Nathan Wolfe rightly insists "viruses are part of the living systems of our planet . . . and should be interpreted in that way" (Wolfe 2011; cf. Miller and Spoolman 2011).

The possibility of exercising meaningful control or even modest mitigation of any virus outbreak must begin with effective surveillance of both HP and LP influenza viruses in both animals and humans. Such surveillance is essential to identify the emerging virus and provide the time necessary to establish viable virus repositories with appropriate seed viruses from which to formulate and manufacture both animal and human vaccines (Osterhaus et al. 2008). The genetic diversity of the many different influenza viruses has clearly established multiple genetic clades not only in the traditional influenza epicenter of Southeastern Asia, but throughout the whole of Asia and many other continents (Yen et al. 2008). The scale of the problem is indicated by the fact that in China alone during the years from 2004 to 2006 the attempt to minimize the impact of the H5N1 virus led to the death of around 35 million poultry either by infection or "stamping out" (i.e. killing) as well as the use of 20 billion doses of vaccines in order to control H5N1 outbreaks (Chen et al. 2008).

Effective control and eradication of influenza viruses will require an integrated global program with four distinct policy arms: (1) extensive surveillance, especially of multiple viruses in pigs with their potential for interspecies transmission (Brown 2008); (2) improved biosecurity; (3) the administration of inactivated, recombinant or live virus-vectored vaccines (Werner et al. 2008) or the use of reverse genetics-based vaccines (Neumann et al. 2008); and (4) monitoring (Furuse et al. 2011). The complexity of this challenge is indicated by the fact that for monitoring alone, it is necessary not only to isolate as many as possible virus strains of different subtypes, but also to develop full phenotypic and genotypic characterizations, hopefully with considerable awareness of the genetic relationship of emerging virus to earlier influenza A virus strains and subtypes (Lvov and Kavein 2008). Furthermore, such a four-pronged attack on influenza viruses in both animals and people will need to be ongoing throughout the twenty-first century with a preventive orientation, not a sporadic, isolated response to emerging outbreaks. Clearly, extensive translational scientific research, improved manufacturing capacity of vaccines, national legislation and proactive international guidelines will be essential to formulate and implement an effective program for controlling influenza (cf. Capua et al. 2008; Miller et al. 2009).

In conclusion, it is only with a One Health approach that sufficient scientific knowledge can be gained from many disciplines to make possible the control of an influenza pandemic. Interspecies medicine, with all its uncertainties, is only possible with a One Health perspective that is open to new ideas and new links among people, animals and the environment. Furthermore, controlling pandemics and emerging infections requires not only extensive scientific knowledge and an awareness of the importance of interspecies medicine, but also effective collaboration among scientists, doctors and public health professionals, all of whom must learn how to communicate the validity of the One Health approach to the public.

Acknowledgements This material is based upon work supported by the U. S. Department of Homeland Security under Grant Award Number 2010-ST-AG0001 and the Kansas Bioscience Authority. The views and conclusions contained in this chapter are those of the authors and should not be interpreted as necessarily representing the official policies, either explicit or implicit, of the U. S. Department of Homeland Security or the Kansas Bioscience Authority.

References

Bao C-J, Cui L-B, Zhou M-H et al (2013) Live-animal markets and Influenza A (H7N9) virus infection. N Engl J Med. www.nejm.org/doi/full/10.1056/NEJMc1306100. Accessed 16 July 2013

Barry JM (2006) The great influenza: the epic story of the deadliest plague in history, 2nd edn. Penguin, New York

Branswell H (2013) Making vaccine for new flu H7N9 flu virus could be a challenge, experts say. www.huffingtonpost.ca/2013/04/11/making-vaccine-for-new-fl_n_3064478.html. Accessed 31 May 2013

Brown I (2008) The role of pigs in interspecies transmission. In: Klenk HD, Matrosovich MN, Stech J (eds) Avian influenza, vol 27, Monographs in virology. Karger, Basel, pp 88–100

Capua I, Cattoli G, Terregino C, Marangon S (2008) Avian influenza in Italy 19972006. In: Klenk HD, Matrosovich MN, Stech J (eds) Avian influenza, vol 27, Monographs in virology. Karger, Basel, pp 59–70

Chen H, Smith GJD, Zhang SY et al (2005) H5N1 virus outbreak in migratory waterfowl. Nature 436:191–192. doi:10.1038/nature03974

Chen H, Bu Z, Wang J (2008) Epidemiology and control of H5N1Avian influenza in China. In: Klenk HD, Matrosovich MN, Stech J (eds) Avian influenza, vol 27, Monographs in virology. Karger, Basel, pp 7–8

Crosby AW (1993) Influenza. In: Kipple KF (ed) The Cambridge world history of human disease. Cambridge University Press, Cambridge, pp 807–811

Crosby AW (2003) America's forgotten pandemic: the influenza of 1918, 2nd edn. Cambridge University Press, Cambridge

Furuse Y, Suzuki A, Kamigaki T et al (2011) Viruses that cross borders: factors responsible for global dissemination of viral infections. Intervirology 54:246–252. doi:10.1159/000320967

Holmes D (2013) The world waits for H7N9 to yield up its secrets. Lancet Infect Dis 13:477–478

Holmes EC, Rambaut A (2005) Evolutional genetics and the emergence of SARS coronavirus. In: McLean AR, May RM, Pattison J, Weiss RA (eds) SARS: a case study in emerging infections. Oxford University Press, Oxford, pp 16–23

Kahn RE, Richt JA (2013) The novel H7N9 influenza A virus: its present impact and indeterminate future. www.onehealthinitiative.com/publications/H7N9%20OHI%20comment%20_FINAL_.pdf. Accessed 17 July 2013

Khiabanian H, Trifonov V, Rabadan R (2009) Reassortment patterns in swine influenza viruses. PLoS One 4(10):e7366. doi:10.1371/journal.pone.007366

Klenk HD, Matrosovich MN, Stech J (2008) Foreword. In: Klenk HD, Matrosovich MN, Stech J (eds) Avian influenza, vol 27, Monographs in virology. Karger, Basel, pp 7–8

Koopmans M, de Jong M (2013) Avian influenza A H7N9 in Zhejiang, China. Lancet 381:1882–1883. doi:10.1016/S0140-6736(13)60936-8

Liu D, Shi W, Shi Y et al (2013) Origin and diversity of novel avian influenza A H7N9 virus causing human infection: phylogenetic, structural, and coalescent analyses. Lancet. doi:10.1016/S0140-6736(13)60938-1

Lvov DK, Kavein NV (2008) Avian influenza in northern Eurasia. In: Klenk HD, Matrosovich MN, Stech J (eds) Avian influenza, vol 27, Monographs in virology. Karger, Basel, pp 41–58

Ma W, Kahn RE, Richt JA (2009) The pig as a mixing vessel for influenza viruses: human and veterinary implications. J Mol Genet Med 3(1):158–166

Matrosovich MN, Gambaryan AS, Klenk H-D (2005) Receptor specificity of influenza viruses and its alteration during interspecies transmission. In: Klenk HD, Matrosovich MN, Stech J (eds) Avian influenza, vol 27, Monographs in virology. Karger, Basel, pp 134–155

Miller GT, Spoolman S (2011) Living in the environment: principles, connections, and solutions, 17th edn. Brooks/Cole, Pacific Gove, CA, International edition

Miller MA, Uiboud C, Balinska M et al (2009) The significant features of influenza pandemics—implications for policy. N Engl J Med 360:2295–22298. doi:10.1056/NEJMp0903906

Morens DM, Taubenberger JK, Fauci AG (2013) Pandemic influenza viruses—hoping for the road not taken. N Engl J Med 368:2345–2348. doi:10.1056/NEJMp1307009

National Bureau of Statistics of China (2014) China's economy showed good momentum of steady growth in the year of 2013. www.stats.gov.cn/english/pressrelease/index_6.html. Accessed 20 May 2014

Neumann G, Horimoto T, Kawaoka Y (2008) Reverse genetics of influenza viruses—applications in research and vaccine design. In: Klenk HD, Matrosovich MN, Stech J (eds) Avian influenza, vol 27, Monographs in virology. Karger, Basel, pp 118–133

Osterhaus ADME, Munster VJ, Fouchier RAM (2008) Epidemiology of Avian Influenza. In: Klenk HD, Matrosovich MN, Stech J (eds) Avian influenza, vol 27, Monographs in virology. Karger, Basel, pp 1–26

Oxford JS, Lambkin R, Sefton A et al (2005) A hypothesis: the conjunction of soldiers, gas, pigs, ducks, geese and horses in northern France during the Great War provided the conditions for the emergence of the "Spanish" influenza pandemic of 1918–1919. Vaccine 23(7):940–945. doi:10.1016/j.vaccine.2004.06.035. Accessed 16 July 2013

Quammen D (2012) Spillover: animal infections and the next human pandemic. The Bodley Head, London

Reingold AL, Phares CR (2005) Infectious diseases. In: Merson MH, Black RE, Mills AJ (eds) International public health: diseases, programs, systems, and policies. Jones and Bartlett, Sudbury, MA

Schenk C, Plachouras D, Danioelsson N et al (2013) Outbreak with a novel avian influenza A (H7N9) virus in China—scenarios and triggers for assessing risks and planning responses in the European Union. Euro Surveill 18:20. www.eurosurveillance.org/ViewArticle.aspx?ArticleId=20482. Accessed 31 May 2013

Scholtissek C (2008) History of research on avian influenza. In: Klenk HD, Matrosovich MN, Stech J (eds) Avian influenza, vol 27, Monographs in virology. Karger, Basel, pp 101–117

Steel J, Palese P (2008) The 1918 influenza pandemic: lessons from the past raise questions for the future. In: Klenk HD, Matrosovich MN, Stech J (eds) Avian influenza, vol 27, Monographs in virology. Karger, Basel, pp 272–286

The Economist (2011) Global livestock counts: counting chickens: where the world's livestock lives. www.economist.com/blogs/dailychart/2011/07/global-livestock-counts. Accessed 18 July 2013

United States Census Bureau (2013) International programs, world population. www.census.gov/population/international/data/worldpop/table_population.php. Accessed 16 July 2013

Van Ranst M, Lemey P (2013) Genesis of avian-origin H7N9 influenza A viruses. Lancet 381:1883–1885. doi:10.1016/SO140-6736)13)60959-9

Webby RJ, Swenson SL, Krauss SL et al (2000) Evolution of swine H3N2 influenza viruses in the United States. J Virol 74:8243–8251

Webby RJ, Webster RG, Richt JA (2007) Influenza viruses in animal wildlife populations. In: Childs JE, Mackenzie JS, Richt JA (eds) Wildlife and emerging zoonotic diseases: the biology, circumstances and consequences of cross-species transmission. Springer, Berlin, pp 67–83
Werner O, Harder T, Veits J et al (2008) Avian influenza outbreaks in Germany—development of new avian vaccines. In: Klenk HD, Matrosovich MN, Stech J (eds) Avian influenza, vol 27, Monographs in virology. Karger, Basel, pp 71–87
WHO (2013a) Frequently asked questions on human infection caused by the avian influenza A (H7N) virus. www.who.int/influenza/human_animal_interface/faq_H7N9/en/index.html. Accessed 30 May 2013
WHO (2013b) Risk assessment: human infections with avian influenza A (H7N9) virus. www.who.int/influenza/human_animal_interface/influenza_h7n9/RiskAssessment_H7N9_10May13.pdf. Accessed 30 May 2013
WHO (2013c) China-WHO joint mission on human infection with Avian Influenza A (H7N9) virus. 18–24 April 2013. Mission report. www.who.int/influenza/human_animal_interface/influenza_h7n9/en/index.html. Accessed 30 May 2013
WHO Regional Office for Europe (2013) Public health relevant virological features of Influenza A (H7N9) causing human infection in China. www.euro.who.int/en/what-we-do/health-topics/communicable-diseases/influenza/publications/2013/public-health-relevant-virological-features-of-influenza-ah7n9-causing-human-infection-in-china. Accessed 30 May 2013
Wolfe N (2011) The viral storm: the dawn of a new pandemic age. Allen Lane/Penguin, London
Worldwatch Institute (2012) Vital signs: farm animal populations continue to grow. http://vitalsigns.worldwatch.org/vs-trend/farm-animal-populations-continue-grow. Accessed 3 June 2013
Xiong X, Martin SR, Haire LF et al (2013) Receptor binding by an H7N9 influenza virus from humans. Nature. doi:10.1038/nature12372
Yen HL, Guan Y, Peiris M, Webster RG (2008) H5N1 in Asia. In: Klenk HD, Matrosovich MN, Stech J (eds) Avian influenza, vol 27, Monographs in virology. Karger, Basel, pp 11–26
York I, Donis RO (2013) The 2009 pandemic influenza virus: where did it come from, where is it now and where is it going? Curr Top Microbiol Immunol 370:241–257
Zhou NN, Senne DA, Landgraf JS et al (1999) Genetic reassortment of avian, swine, and human influenza A viruses in American pigs. J Virol 73:8851–8856
Zhu H, Wang D, Kelvin DJ et al (2013) Infectivity, transmission, and pathology of human-isolated H7N9 influenza virus in ferrets and pigs. Science 341(6142):183–186. doi:10.1126/science.1239844

Part III
The Successes and Challenges of Implementing One Health

Chapter 8
One Health: From Concept to Practice

John S. Mackenzie, Moira McKinnon, and Martyn Jeggo

Abstract One Health (OH) is an approach, focusing on emergent infectious diseases, which looks at health in the context of human, animal and environment relationships. Governments worldwide through the International Ministerial Conferences on Avian and Pandemic Influenza (IMCAPI) meetings have made a commitment to OH. There is unanimous agreement from the international organizations to the community level that this is a necessary approach in an increasingly populous world. It is a world, however, in which professions have moved to specialization and expertise within their own realm rather than in collaboration and cross discipline. This chapter reports on the operationalization of OH to date and consideration of the forward path in examination of the key areas of: leadership, relationships, infrastructure, skills and capacity, communication and technology, and resources.

Keywords Cross-species transmission • Food and Agriculture Organization • Food safety and security • Human-animal-ecosystem interfaces • One Health • One Health Global Network • One Health Initiative • World Health Organization • World Organization of Animal Health • Zoonoses

J.S. Mackenzie (✉)
Faculty of Health Sciences, Curtin University, Perth, Australia
e-mail: J.Mackenzie@curtin.edu.au

M. McKinnon
Medical Practitioner and Consultant in Public Health and Health Systems, Canberra, Australia

M. Jeggo
Centre for Emerging Infectious Diseases, C/O Deakin University Medical School, Geelong, Australia

A. Yamada et al. (eds.), *Confronting Emerging Zoonoses*,
DOI 10.1007/978-4-431-55120-1_8

8.1 Introducing and Defining the Scope and Concept of One Health (OH)

8.1.1 Development of the OH Paradigm: The Decade Since SARS

The past two decades have been a momentous period for extending our knowledge of the origin and emergence of novel diseases, and in particular, our recognition of the role of wildlife reservoirs as the source of potential pathogens. It has clearly delineated the importance of the interface between humans and domestic and wild animals in cross-species transmission of zoonotic diseases.

It is now 10 years since the world was faced with the first severe and readily transmissible new disease to emerge in the twenty-first century, Severe Acute Respiratory Syndrome (SARS). Caused by a previously unrecognized coronavirus associated with the animal markets of southern China, the disease spread across the world via major air routes, reaching 29 countries on five continents (Guan et al. 2003). Later evidence strongly implicated insectivorous bats as the probable natural reservoir of the virus (Lau et al. 2005; Li et al. 2005; Ge et al. 2013). This discovery stimulated further investigations into a variety of wildlife species which revealed a plethora of new viruses carried by fruit and insectivorous bats, rodents and other species of wildlife from around the globe (Mackenzie and Jeggo 2013). Thus the SARS outbreak led to a number of important observations. It demonstrated that:

- a previously unknown pathogen could emerge from a wildlife source at any time and in any place and, without warning, threaten the health, well-being and economies of all societies;
- there was a clear need for countries to have the capability and capacity to maintain an effective alert and response system to detect and quickly react to outbreaks of international concern, and to share information about such outbreaks rapidly and transparently; and
- responding to pandemic threats requires global cooperation and global participation.

International concerns about novel zoonotic diseases were subsequently heightened by the spread of virulence of highly pathogenic avian influenza (HPAI) H5N1 and its potential to become the next worldwide pandemic. The H5N1 avian influenza threat persists (World Health Organization 2014a), but the more immediate concerns are the possible global spread of two novel viruses, avian influenza H7N9 and the Middle East Respiratory Syndrome coronavirus (MERS-CoV). At the time of writing, there have been 432 human cases of avian influenza H7N9, most of which have occurred in mainland China with a few cases from China Taiwan and Hong Kong SAR (ProMED 2014; World Health Organization 2014b; Centre for Health Protection 2014). Surveillance at the animal-human interface is proving difficult because of the avirulence of the virus in domestic poultry (Chen et al. 2013). With MERS-CoV, 253 laboratory-confirmed cases have been reported at the time of writing, most of which have occurred in the Middle East (particularly in Saudi Arabia, but also including Jordan, Kuwait, Oman, Qatar, and United Arab Emirates), but with sporadic cases in travellers from the Middle East to Europe (France, Germany, Greece, Italy and the United Kingdom), North Africa (Tunisia) and Asia (Malaysia and Philippines) (World Health Organization 2014c). Recent data has suggested that dromedary camels are the spill-over hosts of MERS-CoV and the source of most human infections (Alagaili et al. 2014; Meyer et al. 2014), although bats have been suggested as a possible reservoir (Memish et al. 2013).

Many emergent pathogens are not only linked to increasing contact between humans and animals, both domestic and wild, but to intensification and integration of food production, to the need for clean drinking water, to climate, and to the expansion of international travel and trade. The role of the wildlife-livestock-human-ecosystem interfaces has been fundamental to the development of the OH paradigm over the past decade, a concept that recognizes that the health of humans, animals, and ecosystems are interconnected, and that to better understand and respond to zoonotic diseases requires coordinated, collaborative, multidisciplinary and cross-sectoral approaches.

The concept of OH is not new; the recognition that the health of people, animals and the ecosystems of which we are part, are inextricably woven together and is as old as human culture (Veterinarians Without Borders 2010). Over 2,500 years ago Hippocrates urged physicians that all aspects of their patients' lives need to be considered including their environment. However, the concept has been more commonly associated with the nineteenth century physician Rudolf Virchow, who acknowledged the similarities between human and animal medicine, and who first used the term 'zoonosis' for infections acquired from animals. Much more recently in the mid-1960s, the eminent American veterinary epidemiologist, Calvin Schwabe, also recognized that the health of humans, animals and ecosystems are interconnected, which he referred to as "One Medicine" in his textbook "Veterinary Medicine and Human Health" (Schwabe 1984).

The significance of zoonoses in the emergence of human infectious diseases was also recognized in the 1992 Institute of Medicine Report 'Emerging Infections: Microbial Threats to Health in the United States (IOM 1992), and the subsequent 2003 Report "Microbial Threats to Health: Emergence, Detection, and Response (IOM 2003). These and a number of subsequent meetings and events have been instrumental in further developing and defining the OH paradigm. They have been listed in Table 8.1 as a series of milestones.

The current momentum associated with the OH concept during the past decade was driven initially by a series of strategic goals, known as the 'Manhattan Principles' which were derived during a 'One World, One Health™' meeting of the Wildlife Conservation Society entitled *Building Interdisciplinary Bridges to Health in a "Globalized World"* held in September 2004 in New York (Wildlife Conservation Society 2004). These goals, developed for combating threats to life on Earth, clearly recognized the link between human health and animal health, and the threats that diseases pose to food supplies and economies, and called for the

Table 8.1 A timeline of major milestones in the development and implementation of the One Health concept

Time	Event
Approximately 400 BCE	Hippocrates urged physicians that all aspects of their patient's lives need to be considered including their environment
1855	Rudolph Virchow first used the term 'zoonosis' for infections acquired from animals
1971	An international group, Wildlife Trust, formed by naturalist Gerald Durrell. The Wildlife Trust is now known as EcoHealth Alliance
1984	Calvin Schwabe, veterinary epidemiologist coined the term 'One Medicine'
1992	Institute Of Medicine release the report, 'Emerging infections: Microbial Threats to Health in the United States.'
1993	Federation of American Scientists and WHO sponsored a meeting in Geneva to discuss the potential threat of biological warfare
1994	Launch of the Program for Monitoring Emerging Diseases (ProMED) (Morse et al. 1996)
2000	The Global Outbreak Alert and Response Network (GOARN) conceived and development begun
2003	'Microbial Threats to Health: Emergence, Detection and Response' Report released by IOM
2004	Human Animal Infections and Risk Surveillance (HAIRS) begins operation in the UK
2004 September	Manhattan Principles are defined in a meeting on 'One World One Health' convened by the Wildlife Conservation Society
2005	International Health Regulations agreed by the World Health Assembly (WHO), and implemented in 2007
2005 September	United Nations Systems for Influenza Coordination formed (UNSIC)
2005	The Asia Pacific Strategy for Emerging Diseases (APSED) implemented
2006 January	First International Ministerial Conference on Avian Influenza and Pandemic Influenza (IMCAPI) Beijng, PR China
2006	The Global Early Warning System for major Animal Diseases (GLEWS) developed and implemented by FAO, OIE and WHO
2006 December	Second IMCAPI Bamako, Mali
2007	American Veterinary Medicine Association (AVMA) was instrumental in forming the One Health Initiative Taskforce (OHITF) (King et al. 2008)
2007December	Third IMCAPI New Delhi, India
2008 January	The South Africa Centre for Infectious Disease Surveillance (SACIDS) established as a One Health Virtual Centre linking research institutions in Tanzania, DRC, Mozambique, Zambia and South Africa
2008	The One Health Initiative (OHI) www.onehealthinitiative.com, a major internet based communications resource launched
2008 October	Fourth IMCAPI Sharm el-Sheikh, Egypt
2008 October	FAO, OIE, WHO, UNICEF, World Bank and UNSIC produce collaborative document 'Contributing to One World, One Health. A Strategic Framework for Reducing Risks of Infectious Disease at the Animal–Human-Ecosystem Interface' endorsed by IMCAPI at Sharm el-Sheikh

(continued)

Table 8.1 (continued)

Time	Event
2009 March	Expert international consultation held in Winnipeg 'One World One Health: from Ideas to Action'
2009	A national (USA) One Health Commission (http://www.onehealthcommission.org) formed following recommendations of the OHITF, with AVMA, the American Medical Association, the American Public Health Association, the Association of American Medical Colleges, the Association of American Veterinary Medical Colleges, the American Society for Microbiology, and the Infectious Diseases Society of America as partners, and funded by the Rockefeller Foundation
2009	Afrique One, an African Research Consortium on Ecosystem and Population Health, launched with members from West and East Africa with three European partners
2009	One Health Alliance of South Asia (OHASA) formed as a network of scientists in Bangladesh, India, Nepal, and Pakistan
2010 April	Fifth IMCAPI Hanoi
2010 April	Tripartite Concept Strategy—An FAO-OIE-WHO collaboration to address health risks at the animal-human-ecosystems interface' is endorsed
2010 May	Centers for Disease Control (CDC) Stone Mountain meeting initiates seven international working groups to progress aspects of One Health
2010	World Bank publishes 'People, Pathogens and Our Planet: Vol 1. Toward a One Health Approach for Controlling Zoonotic Disease.'
2010	'Connecting Organizations for Regional Disease Surveillance (CORDS)', a non-government platform connects global regional surveillance systems
2011 February	First International One Health Congress Melbourne, Australia
2011	One Health Central and Eastern Africa (OHCEA) formed as a network of 14 public health and veterinary institutions in Ethiopia, Uganda, Kenya, Tanzania, DRCongo and Rwanda
2012	One Health Global Network, an information clearing house, is launched
2012	World Bank publishes 'People, Pathogens and Our Planet: Vol 2. The Economics of One Health'
2012 February	Global Risk Forum One Health Summit 'One Health, One Planet, One Future' held in Davos, Switzerland
2013 January	Second International One Health Congress Bangkok, Thailand
2013	The Gates Foundation calls for One Health research through the Grand Challenge program
2014	One Health Summer Schools available in Denmark, England and Australia. Masters in One Health offered in USA and UK, and a doctorate in the USA

inclusion of wildlife health as an essential component of global disease prevention, surveillance, monitoring, control and mitigation. The goals also recognized the need to devise holistic and forward-looking approaches for responding to emerging and resurging diseases.

Support for OH continued to grow at different levels through workshops, meetings and discussion groups, and with government agencies advancing the agenda, and the paradigm broadened as it became clearer that OH extends beyond an examination of the human–animal health interface to encompass the health and sustainability of the world's ecosystems This has led to the recognition that the environment profoundly influences both human and animal health, especially through effects on the water and food supply and through global climate and air quality (Shomaker et al. 2013).

8.1.2 Definitions and Scope

There is no single, accepted definition of OH, but the most frequently used definition is that developed by the AVMA:

> The integrative effort of multiple disciplines working locally, nationally, and globally to attain optimal health for people, animals and the environment. (King et al. 2008).

The breadth and scope of OH, however, makes it difficult to find a definition that covers all aspects, and indeed the lack of consensus in definition gives the OH approach much greater flexibility. Increasingly OH is considered to be systemic-based with its scope of practice broad and growing, ranging from a specific zoonosis focus to a wide multidisciplinary ecosystem approach with a very broad canvas. However, the price for flexibility even in the field of emerging infectious disease (EID) has been diffusion, fragmentation and duplication of effort, and there is a strong argument put forward that it is timely for a unified international voice and structure to counteract these impacts (Lee and Brumme 2012).

There is a general agreement, especially in national OH programs, of the need to focus on the animal–human infectious disease interface as 'doable' whilst extending activities to develop infrastructure and capacity to a broader approach. Many of the important zoonoses relate to animals in the food production chain. Food, therefore, becomes an important vehicle for a substantial number of these zoonotic pathogens (Wielinga and Schlundt 2013), and food safety and food security are major OH issues, as clearly envisaged by the Food and Agriculture Organization (FAO) (Lubroth 2013; FAO 2011), the United Nations (Nabarro 2012) and WHO. Indeed, within WHO, OH is located in the Food Safety and Food Security Division of the Health, Security and Environment cluster. Globalization of the food supply has increased the risks for outbreaks and spread of food-borne pathogens, and also for the emergence and spread of novel antibiotic resistance genes (IOM 2012; Wielinga and Schlundt 2013). The International Food Safety Authorities Network (INFOSAN), a joint program of FAO and WHO, with the Secretariat in WHO, links national authorities responsible for managing food safety emergencies (INFOSAN 2013).

Food security is potentially one of the biggest global issues facing the planet. Kelly et al. (2013) argue that increasing livestock numbers without increasing productivity is a recipe for ecosystem disaster and eventual protein scarcity. The authors call on OH practitioners to "develop a vision for veterinary education that sees livestock disease control and productivity as a continuum, going beyond disease control to include production efficiencies, economically based agricultural development, the potential impact on the food system of changes in technology, ecosystem health, urbanization, public health and the multidisciplinary collaborations necessary to generate food policies". Sadly there has not been a similar passionate cry expressed from the medical world.

8.1.3 Recent Drivers of the One Health Paradigm

8.1.3.1 Concerns About Pandemic Risks from Highly Pathogenic Avian Influenza

After the SARS epidemic, international attention and concern focused on the potential of the HPAI H5N1 virus to mutate and become transmissible from person-to-person. This was the catalyst that led to President George W Bush to announce the International Partnership on Avian and Pandemic Influenza at a high-level Plenary Meeting of the United Nations General Assembly in September 2005. The goals of the partnership were to improve global readiness by elevating the issue on national agendas, leveraging resources, and building capacity to identify, contain and respond to pandemic influenza. The Secretary-General of the United Nations, Kofi Annan, subsequently announced that the UN family would do all it could to ensure all countries—rich and poor—were protected and prepared for a pandemic caused by avian influenza, and appointed a UN Systems Coordinator for Avian and Human Influenza (UNSIC). This proved to be a major leap forward in advancing One Health through bringing about collaboration between international organizations, regions and nations to discuss the ongoing pandemic threat posed by HPAI H5N1, and provided an excellent example of what can be done when specifically targeting collaborations between leaders in order to bring about change and action. This collaboration brought together international organizations, including FAO, WHO, the World Organization for Animal Health (OIE), the United Nations Children's Fund (UNICEF), the World Bank, and the EU, with various country representatives, and was the force behind the development of the International Ministerial Conferences on Avian and Pandemic Influenza (IMCAPI).

IMCAPI have clearly been major drivers of OH over the past decade, and may provide a blueprint for future directions and leadership of OH. The early conferences were concerned with pledging of funds and resources for an international preparedness and response effort, but subsequent IMCAPI have considered the wider impact and cause of EIDs. In 2008, the four international organizations (FAO, OIE, WHO and UNICEF), together with the World Bank and UNSIC, collaborated on developing strategies primarily focused on influenza yet clearly intended to extend to the wider context of disease emergence. The strategies were outlined in a document entitled '*Contributing to One World, One Health. A Strategic Framework for Reducing Risks of Infectious Diseases at the*

Animal-Human-Ecosystems Interface' which focused on how best to diminish the risk and minimize the global impact of epidemics and pandemics due to EID by enhancing disease intelligence, surveillance and emergency response systems at national, regional and international levels, and by supporting them through strong and stable public and animal health services and effective national communication strategies (UNSIC 2008).

The IMCAPI meeting in Hanoi in 2010 concluded with the Hanoi Declaration which proposed a multi-sector array of national measures to detect new diseases that may cross from animals to humans. There was an agreement to promote international surveillance, diagnosis and rapid response, and a clearly stated recognition of the need for community participation, increased capacity particularly in least developed countries, and consideration of societal and cultural elements. It also noted that country strategies should be aligned nationally and regionally to address "One Health" challenges (UNSIC 2010). In addition, FAO, OIE and WHO distributed their Strategy Framework to work more closely together to address the animal-human-ecosystem interfaces. This Tripartite Concept Note, 'The FAO-OIE-WHO Collaboration—sharing responsibilities and coordinating global activities to address health risks at the animal–human-ecosystems interfaces' aligned strategies and resource streamlining and commitment (World Health Organization 2010). Several areas have been advanced through the Tripartite Concept including identification of elements for joint plans, development and implementation of integrated risk assessment projects at country level in Egypt and Vietnam, OH assessment missions in Costa Rica and Kenya, table top exercises, collaboration in surveillance systems, strengthening awareness in food safety, jointly agreed food standards and the development of a tripartite road map to tackle neglected zoonotic diseases (Landford and Nunn 2012).

Early detection and response to outbreaks of emerging infectious diseases were greatly assisted by the introduction of the International Health Regulations (2005) (IHR) which provided a policy infrastructure for OH in keeping with the awareness, prevention, shared data and response. The IHR, developed and implemented by WHO, is a global instrument agreed to by all 194 member States to plan for and respond to public health threats to the international community. The IHR set out the basic public health core capacities that States must develop in order to detect, report and respond to public health risks and public health emergencies of international concern (World Health Organization 2008a). The new regulations offer a new global health governance framework that promotes collective interests above national interests, and is radically transforming the international law applicable to identifying and responding to the international spread of diseases.

8.1.3.2 Developing Plans to Put One Health Concepts into Practice

Two seminal meetings were held in 2009 and 2010 that were directed at developing ways to put the concepts of OH into practice. The first of these was an expert consultation entitled "One World, One Health: From Ideas to Action", sponsored by the Public Health Agency of Canada and bringing together scientific experts, international organizations and government officials to seek ways to advance the concept and build a framework for future activities. The meeting made a number of recommendations including to: (a) foster political will; (b) support partnerships and collaborations: (c) encourage data sharing and integration; (d) build capacity; (e) develop communication strategies (include the media); (f) provide incentives for reporting adverse events; (g) encourage stakeholder and community engagement; and (h) develop 'supra country' (trans-border) approaches (Public Health Agency of Canada 2009).

This meeting was followed a year later in 2010 by a second strategy meeting convened by the US Centers for Disease Control and Prevention (CDC) and the major international organizations at Stone Mountain, entitled 'Operationalizing One Health: a Policy Perspective—Taking Stock and Shaping an Implementation Roadmap' (CDC 2010). The specific aim of the meeting was to identify clear and concrete actions to move the concept of OH from vision to implementation. Seven working groups were identified as being crucial steps in attaining the 3–5 year vision. These were: (1) 'Training' to develop and build skills, expertise and competencies through a OH curriculum; (2) a 'One Health Global Network' as a means of gaining international support and as a vehicle to stimulate further global collaboration; (3) an 'Information Clearing House' to promote advocacy through providing information on success stories and lessons learned; (4) 'Needs Assessment', by developing country-level self-assessment methods to identify activities which could benefit from a OH approach; (5) 'Capacity Building', by identifying ways to leverage existing programs and capacity-building efforts to have a major impact at minimal cost; (6) 'Proof of Concept', through demonstrating a retrospective and prospective evidence base that the use of OH interventions and/or concepts leads to better cross-species health outcomes; and (7) a 'Business Plan', by articulating the concept of and rationale for OH more clearly and presenting this information to policymakers and donors worldwide (Centers for Disease Control and Prevention 2010). Outcomes from the working groups are, with some exceptions, still progressing (Stone Mountain Workgroups 2011; 2012; Rubin 2013). The workgroup to develop an 'Information Clearing House' was subsequently combined with the 'One Health Global Network workgroup and, led by the group chair Dr Vandersmissen, the One Health Global Network (www.onehealthglobal.net) was successfully launched in 2012. The Proof of Concept working group has also reviewed and assembled evidence to demonstrate that proof of concept for a OH approach to EID threats is feasible through cross-sectoral integration between human, animal and environmental health sectors. It concluded that bigger and more controlled comparative studies of OH disease prediction and control strategies are needed, such as larger implementation of surveillance systems integrating human, animal and environmental data, and larger controlled intervention trials (Rabinowitz et al. 2013).

8.2 Implementation of One Health Activities

UNSIC clearly demonstrated the significant impact that an organization dedicated to forming effective international relationships can achieve in the pursuit of determined goals. UNSIC's role and authority have diminished since the influenza H1N1 pandemic and it has not, as many wished, evolved into a collaborating organization to promote a global OH approach. The effectiveness of UNSIC might also suggest that the Tripartite Collaboration of FAO, OIE and WHO should take the lead in implementing OH internationally. While the argument that a global governance approach is needed to oversee implementation of an effective and strong OH agenda, there are currently many important activities happening internationally, regionally, nationally and even locally.

8.2.1 The Involvement of the International Organizations in One Health

Each of the three major international organizations has instituted OH approaches or units within their organizations. The different scope, priorities and resources of the three international organizations however, have proved a challenge for ongoing collaboration and maximizing synergies. The OIE, which has a narrower remit and far less resources than FAO and WHO, has moved to ensure implementation of OH approaches in global animal health policies, reflecting also the activity of veterinary services regionally. FAO's approach aims to strengthen veterinary-public health systems to more clearly encompass disease prevention, with greater emphasis on food production, distribution and marketing practices, and adoption of sustainable animal agriculture and natural resource management. It established the Emergency Center for Transboundary Animal Diseases to assist member countries in responding to threats from transboundary animal diseases, initially HPAI, but subsequently for all transboundary diseases with probable consequences for human and animal health (FAO 2014). WHO has aligned One Health activities to food safety and food security in the Department of Food Safety and Zoonoses in the Health Security and Environment cluster, and despite its close involvement with OIE and FAO in the Tripartite Concept Note for sharing responsibilities in addressing health risks at the animal–human-ecosystems interfaces, makes no mention of the concept of OH at any place on its website. Joint OH programs and policies are limited between the three organizations, although they have made considerable effort to develop collaboration in surveillance and laboratory networks, and FAO, WHO and OIE are strong promoters of OH forums internationally and regionally. These collaborations are described in more detail below.

The World Bank has grasped the value of OH and taken a leadership role, especially with respect to economic evaluation. It has produced two volumes in the series 'People, Pathogens and Our Planet', the first volume entitled 'Toward a

One Health Approach for Controlling Zoonotic Diseases' (World Bank 2010) and the second entitled 'The Economics of One Health' (World Bank 2012). They provide advice not only for economists in determining funding arrangements and effectiveness evaluation but also for project planners and policy makers on understanding a rigorous economic framework for OH. These documents are a significant step forward in developing economic policy in considering the multidisciplinary nature and multiple outcomes of the OH approach. The importance of the application of the principles to least developed country funding cannot be under-estimated. Furthermore, in taking this forward, the World Bank is now supporting projects in individual countries to progress this approach at the nation level e.g. Mongolia ((South Asia Regional One Health Symposium, 2–6 December, 2013 Paro, Bhutan; personal communication, M Jeggo).

International regional instrumentalities have also supported OH approaches with respect to potential threats from zoonoses. In Europe, the European Union has been a key supporter of the IMCAPI and, since 2008, has promoted the OH approach and integrated it into certain EU strategy documents (European Union 2013). In Asia the Association of South East Asian Nations (ASEAN) and the Asia Pacific Economic Cooperation (APEC) have played active roles in capacity building and preparedness for influenza pandemic response, and developed and strengthened cross-sectoral networks and operations against the threat of emerging and zoonotic infectious diseases (ASEAN 2008; APEC 2011).

8.2.2 National, Multidisciplinary and Cross-Border Activities

Many regions and nations, particularly developing nations, have now taken up the principles of One Health and are applying them to the extent that their own capabilities and capacities permit. As examples, nations as diverse as the United Kingdom (UK), France, Sweden, Laos, Thailand, Mongolia, Canada and Tanzania, to name but a few, have set up national OH entities within or supported by government and varying in scope and resources. France has endorsed support for the Tripartite Collaboration and has committed to align its strategies and activities with the Tripartite Collaboration (French Government 2011; Hall and Coghlan 2011). In the UK, a cross-government approach chaired by Public Health England entitled 'Human Animal Infections and Risk Surveillance (HAIRS) group' provides a One Health approach to zoonoses and emerging infections across Government, and has been operating since 2004. It acts as a horizon-scanning group to identify emerging zoonoses which may pose a future threat, and brings together representatives from human and animal health from the Departments and Agencies across the UK (Public Health England 2014). Developing countries appear to be more successful in implementing OH programs than developed countries, due in large part to the fact that limited resources make it necessary for different sectors to work closely together, whereas more developed nations have older, established and more rigid public health structures with well-defined responsibilities met using a fairly narrow range of disciplines.

OH programs include elements of field epidemiology, the sharing of laboratory resources and health information, the use of multi-sectoral teams and the application of an evaluation framework. A good example of how this is working in practice can be seen in Mongolia. Coordination mechanisms established between veterinary and public health sectors in Mongolia are proving crucial in addressing issues of food safety, emergency disease management and the impact of climate change on disease emergence. Tuberculosis, anthrax, rabies and brucellosis are managed through the National Emergency Management Agency (NEMA) which adopts the OH approach involving both veterinarians and medical professionals. It is proposed to establish a "Centre" or "National Hub" for the further development of coordination mechanisms between relevant sectors, including animal and human sectors, food safety authorities, wildlife management authorities, border quarantine services and environmental health personnel, at all levels (sub-national, national, sub-regional, regional). This Centre will deal not only with the immediate threats but longer time risk reduction strategies. Increased and improved national surveillance, more comprehensive risk assessment, and the development of a broader based response capability will be further developed based on multilevel, multidisciplinary and multi-sectoral coordination through this national Center. This example in Mongolia provides a snapshot of what is being developed in countries in both Asia and Africa (Batsukh et al. 2013; Mazet et al. 2009; Hall and Coghlan 2011; Hueston et al. 2013).

Laos presents another example of a nationally operating One Health framework. Laos has an ideal platform for the spread of trans-boundary and emerging zoonotic pathogens through a complex web of wildlife, livestock and human interfaces, extending from remote protected areas to the center of the capital city, Vientiane. Wildlife is still widely hunted across the country, eaten by villagers for subsistence, traded locally in wet and medicinal markets, and transported widely for domestic and international trade. The extensive and often geographically rugged borders of land-locked Laos are extremely porous and movement of wildlife products, livestock, and other items are pervasive, both into and out of the country, representing a potentially significant conduit for the movement of novel pathogens. Bovine cysticercosis, tuberculosis, Q fever, brucellosis and leptospirosis are of particular concern (Vongxay et al. 2012). Laos has seen high levels of economic growth in the last decade and along with this growth have come new roads, many extractive and hydroelectric industries, and an increase in intensive farming resulting in large swathes of deforestation and dramatic land-use changes. All of these activities have the potential to alter disease ecology and in order to respond effectively to these challenges, Laos has reorganized to ensure effective collaboration between the animal and human health sectors. With support from a number of international agencies and NGOs, there is a significant shift towards a whole of Government and OH approach to tackling health issues, whether it be human, livestock or the environment.

The impact of these events, although often difficult to quantify, is at the community and environmental level. The European Union in the document '*Implementation of a One Health Approach in Asia and Europe*' (Hall and Coghlan 2011), noted over 90 OH programs and over 750 dedicated OH practitioners in Europe and Asia involved in emerging infectious diseases. In detailing ten diverse case studies ranging from the Swiss national antibiotic research program to controlling hydatid disease in Nepal, they noted firstly the need to take a systems approach when dealing with so many diverse players; secondly, that a One Health team must be trans-disciplinary; and thirdly, that there must be involvement of the community.

This trend is reflected in the changing face of health funding with large philanthropic donors such as the Bill and Melinda Gates Foundation leaning away from the more conservative vertical programs aimed at diseases which affect humans primarily, HIV, AIDS and malaria (Liden 2013) to ones involving a OH approach. The recent One Health call by the Gates Foundation through the Grand Challenge program (Gates Foundation 2013) espouses many One Health principles of multidiscipline cross-sectoral activity and in particular looks to fund innovations that benefit health but are developed in a non-health sector and vice versa.

Innovative OH partnerships have been set up in Africa in collaboration with various institutions in industrialized countries. These include the South African Centre for infectious Disease Surveillance (SACIDS) with the remit to increase capacity in zoonotic disease surveillance and to strengthen Africa's capacity to detect, identify and monitor infectious diseases of humans and animals (Rweyemamu et al. 2013a, b; SACID 2011); the One Health Central and East Africa (OHCEA 2013), which brings together 14 institutions of public health and veterinary medicine in Ethiopia, Democratic Republic of Congo, Kenya, Tanzania and Uganda; and East African Integrated Disease Surveillance Network (EAIDSNet) which brings together government departments, academic institutions, and health organizations in Kenya, Tanzania and Uganda (Ope et al. 2013); and Afrique One, the African Research Consortium on Ecosystem and Population Health, which brings together organizations in Côte D'Ivoire, Senegal, Chad, Ghana, Tanzania, and Uganda (Afrique One 2013). All of these organizations have demonstrated the ability to work across borders in a transparent and collaborative way.

Similarly, in southern Asia, the One Health Alliance of South Asia (OHASA) was launched initially by the Wildlife Trust to form a cohesive regional network of scientists and policy makers from Ministries of Health, Agriculture and Environment, as well as NGOs and universities in India, Bangladesh, Pakistan and Nepal to address zoonotic diseases (EcoHealth Alliance 2013).

There are a many other national and regional initiatives similar to those described above, some for specific diseases and some much broader to encompass zoonoses generally, as reflected in presentations at the First and Second International One Health Congresses. These range from dealing with the direct interface of animal and human disease such as the collaboration in regard to Hendra virus in Australia (Crawford et al. 2012) to looking comprehensively at food security through the development of Animal Health clubs in Sub Saharan Africa (One

Health Congress 2011; Prince Mahidol Conference 2013). The increased reporting of such initiatives through the One Health websites, the One Health Newsletter, and in journals dedicating issues to One Health such as Microbiology Australia (Volume 33, No 4; available on download from: http://www.theasm.org.au) and the EcoHealth Journal (Volume 7 Supp 1) reflects a gradual but increasing global diffusion of the One Health philosophy and approach,

8.2.3 Surveillance, Detection and Response

Surveillance is an essential component of the OH approach, especially for the early detection of novel threats arising at the human-animal interface. Key international surveillance networks are the Global Early Warning System for Major Animal Diseases (GLEWS), developed by FAO, OIE and WHO (GLEWS 2013), and also the Global Alert and Response Operations (GAR) of WHO. Coordinated rapid outbreak response is provided by WHO's Global Outbreak Alert and Response Network (GOARN), a network of technical institutions, research institutes, universities, international health organizations and technical networks willing to contribute and participate in international coordinated responses to infectious disease outbreaks (Mackenzie et al. 2014). FAO and OIE have also collaborated to develop a network of Expertise on Animal Influenza (OFFLU 2013). While there is no mechanism presently available to permit early and rapid detection of potential human pathogens at the human–animal interface, the PREDICT program of the USAID's Emerging Pandemic Threats has the potential to do this by detecting spillovers of pathogens from wildlife using their SMART (Strategic, Measurable, Adaptive, Responsive, and Targeted) surveillance method (Morse et al. 2012; PREDICT 2013). Animal disease surveillance or outbreak investigation is, in general, fragmentary, and tends to be highly focused on domestic livestock and supporting trade in livestock and livestock products. New platforms for pathogen discovery integrated with new surveillance procedures targeted at the animal–human interface are needed. In response to these needs, innovative molecular diagnostic platforms have been developed which have changed the face of pathogen discovery and detection with enormous sensitivity and specificity (Lipkin 2013), and novel procedures for targeted surveillance have been generated through the PREDICT program (Morse et al. 2012). For non OIE-listed emerging and reemerging diseases, the very recently launched OIE web application named WAHIS-wild interface will potentially assist in surveillance of wild animal diseases (OIE 2014).

The PREDICT project of the USAID's Emerging Pandemic Threats program is an exciting and far-reaching development with the aim of detecting the transmission of novel infectious agents from wildlife to humans. Coordinated by Dr. Jonna Mazet at the University of California at Davis and Dr. Stephen Morse at Columbia University, and with a number of partners including the EcoHealth Alliance and Metabiota, it supports and coordinates projects in countries in Asia, Africa and

South America. The SMART surveillance method developed by PREDICT is designed to detect novel diseases with pandemic potential early, giving health professionals the best opportunity to prevent emergence and spread. It also targets sentinel animal species at active human interfaces in specific hotspots to increase potential of success (PREDICT 2013). The Emerging Pandemic Threats program has additional projects linked to PREDICT, which are PREVENT, IDENTIFY, and RESPOND (USAID 2013). Importantly, the PREDICT project builds on a broad coalition of partners and thus provides a capacity building component in a collaborative framework. These surveillance systems are now established and improving and to date have received dedicated funding.

Regional surveillance networks, a key strategy of the Tripartite Collaboration, have been set up and progressively improved in Europe, Southern Africa, South and South East Asia, and in the Pacific Islands. A non-government platform for connecting and sharing information between regional disease surveillance networks was established in 2010 entitled ‘Connecting Organizations for Regional Disease Surveillance (CORDS: http://www.cordsnetwork.org). Surveillance networks included are the Mekong Basin Surveillance Network, Middle East Consortium on Infectious Disease Surveillance, South-Eastern European Health Network, Asian Partnership on Emerging Infectious Disease Research, Southern Africa Centre for Infectious Disease Surveillance, and East African Integrated Disease Surveillance Network (Gresham et al. 2011, 2013). The regional networks are generally staffed by national governments with assistance from WHO and philanthropic organizations such as the Rockefeller Institute. The networks are more than surveillance agencies, with the four key aims being: improving capacity, advancing One Health, promoting innovation and building sustainable networks. CORDS is currently undergoing an evaluation which includes assessing the value of community activity, knowledge capital, and the sustainability and value of change in practices.

In South–East Asia and Western Pacific, a plan known as the Asia-Pacific Strategy for Emerging Diseases (APSED) was developed to assist member states with achieving their core capacities for compliance with the new IHR, and to reduce the risk of emerging diseases by strengthening early detection and response to outbreaks, including zoonoses. This included a ‘Guide to Establishing Collaboration between Animal and Human Health Sectors at the Country Level’ as a joint collaboration between WHO, FAO and OIE covering surveillance and information sharing, response, and risk reduction (World Health Organization 2008b). The countries report back to a technical group on an annual basis.

8.2.4 Wildlife Interface

The wildlife-human interface is a crucial and pivotal step in the emergence of zoonotic diseases, and being able to detect a novel pathogen before or shortly after a spillover event and then to respond early would be a ‘game-changing’ development in the OH paradigm. As described previously, the emergence of new infectious diseases is increasing, as is the pathogenicity and distribution of re-emergent infectious diseases, with zoonotic infections predominating and primarily from wildlife. McFarlane et al. (2012) note the association of land conversion and global decline in diversity is associated with a rise of EIDs of wildlife origin. Underscoring this is that 'synanthropic' mammal species, those wildlife species which adapt well in human-modified environments, are 15 times more likely to be the source of EIDs. This would strongly argue that surveillance and control of diseases in wildlife should be at the same level as that for in farm animals as exchange of pathogens within and between the two populations are increasing.

General surveillance for diseases in wildlife that can impact on trade, human health and biodiversity occurs in most countries. However, the approach is often ad hoc, uncoordinated, or with lines of reporting that do not recognize the broad impact that diseases with wildlife as part of their ecology may have across different sectors. Different circumstances in different countries have led to a variety of different approaches, with some countries being more organized than others. Wild animals are of high social and economic importance to human societies. It is important to consider the value of wildlife in terms of ecology, species diversity, and, many would argue, the contribution to mankind's art and spirituality. Diseases of wildlife threatening whole species include: plague, chytrid fungus, avain cholera chronic wasting disease, mycoplasma spp., West Nile virus and avian malaria.

In many countries, however, wildlife surveillance is non-existent, and even where some ad hoc surveillance occurs, resources are extremely poor or unavailable. Systems often cited as being useful include those adopted by Australia (the Australian Wildlife Health Network; AWHN 2013), Canada (the Canadian Cooperative Wildlife Health Centre; CCWHC 2013), and France (The SAGIR Network; LaMarque and Artois 1997). In an effort to identify and respond to new zoonotic diseases before they spread to humans, partners in the PREDICT program, described above, locate their research in geographic "hotspots" and focus on wildlife that is most likely to carry zoonotic diseases—animals such as bats, rodents, and nonhuman primates. The USAID PREDICT program with its focus on human population densities and wildlife sources of pathogens may be a useful model for individual countries.

In the last 5 years, OIE, WHO and FAO have all increased their interest and activities associated with wildlife disease surveillance in recognition of the impact that these disease can have on global trade, human health and biodiversity. The OIE has an information network supported since 1994 by its international working group composed of high level scientists with an expertise in wildlife. The network relies on national Focal Points appointed by Member Countries Delegates for relaying information to the working group and to the OIE. In response to the need for improved knowledge of diseases in wildlife as well as in domestic animals, OIE is introducing wildlife species of significance in each of the disease specific chapters in the Terrestrial Code (OIE 2013a) and has developed a training manual on wildlife diseases and surveillance (OIE 2013b). The latter points out that the components of a National Wildlife Health Program should contain the same components as any population health program: prevention, including border management and environmental management to prevent emergence, early detection, timely decisions and responses, and effective pathogen management.

Despite increasing interest, wildlife programs remain poorly resourced even though the majority of emerging diseases arise from wildlife and spillovers from wildlife to domestic animals and humans continue to occur. The law of diminishing returns suggests that the greatest improvement in communicable disease control will occur from investment in important areas with the least resources. At the moment this is animal health, specifically wildlife health.

8.2.5 One Health Research

Research programs in OH are being developed and undertaken in various ways, however funding is precarious and if able to be summed up would easily be miniscule compared to agricultural and animal funding and non-existent to medical care funding. Most current funding organizations are targeted at specific disciplines or sciences and research projects on OH do not fit well into this model.

Recognition that a OH research methodology needs to be vastly different, including many more players and having different evaluation methods and indeed multiple hypotheses related not only to outcome but to processes and relationships.

Areas of excellence of OH research are nevertheless developing. The International Development Research Centre of Canada (IDRC) is one of the leaders in this field. The IDRC strives for a trans-disciplinary approach in all its research in developing countries and provides an excellent model not just for research but for intervention programs, education, and for strategy and policy development, arguably at all levels. The IDRC has methodology for measuring and evaluating relationships and the impact of relationship strength to project success (IDRC 2013). Much more work is needed in this field. A meeting held in 2012 to explore issues in the development of a research and education framework for OH identified two broad OH research issues: (1) the need to develop and evaluate interventions with the potential to provide economic benefits to human or animal health to demonstrate a return on investment; and (2) the need to engage behavioral science researchers and social marketers in conducting research to better understand consumer perspectives on OH issues. Other areas included the need to obtain longitudinal environmental data acquisition to support predictive modeling focused on the implications of changes at the human–animal-environment interfaces, and the development of rapid diagnostics for field and point of care use (Gargano et al. 2013).

8.3 Essential Issues and Challenges for Putting One Health into Practice

While there has been wide-ranging commitment to the One Health approach for addressing complex health problems by a large number of national and international organizations and professional bodies, its operationalization has so far proved to be challenging. Implementation is often a complex issue requiring collaboration between diverse and multi-disciplinary partnerships (Conrad et al. 2013). At a local or national level it often might be a matter of breaking down professional barriers through improved communication and incorporating information on OH and its benefits into professional training and university courses. At the international level it is usually much more difficult and can be hindered by dysfunctions which characterize current forms of global health governance (Lee and Brumme 2012).

8.3.1 Leadership

Leadership is a crucial issue in the development of One Health approaches. It is essential for building relationships and trust, both vertically within an organization and from community to international levels but most importantly horizontally between disciplines and within communities. In many countries, the sectors and disciplines needed to collaborate for a OH approach are based in different institutes, departments or ministries. Offering leadership in this complex environment is often seen as threatening or "empire building". On the contrary, the OH approach is alliance building, and leadership is crucial if the benefits of a One Health approach are to be realized.

Despite their divergent roles and resources the Tripartite collaboration between WHO, FAO and OIE is being looked at to provide international leadership. The history of international governance and leadership indicate that such collaboration needs dedicated resources and close attention to quality of key relationships (Fidler 2010), and this may be difficult in the absence of a crisis such as the influenza pandemic threat. There has been arguably a lull since the H1N1 pandemic with a false perception perhaps that sufficient preparation and infrastructure is in place for infectious disease crisis. The World Bank has continued to take a leadership role by stating that there are clear economic returns and benefits in driving a One Health Approach. There is also a swell of interest in OH from the professional and lay communities as evidenced by the burgeoning of national and international forums and collaborations. This is driven by the recognition that there are not yet enough mechanisms in place, nor are the wider drivers being given sufficient attention. There is also a demand for consideration of the wider scope of OH and a call for system thinking in all aspects of human, animal and environmental health. It may be that the global leadership required is in the form of a collaborating body such as

UNSIC formed for influenza. The demand for this is not clear, loud or certain enough.

Leadership within nations requires at least political agreement if not clear positioning such as public OH policies, strategies if not physical offices. The Canadian Food Protection Agency has delineated a key priority for the Office of the Chief Food Safety Officer to 'steward' OH nationally (Canadian Food Inspection Agency 2012). Thailand has emphasis on OH within its Public Health Ministry and has led the South East Asia region in promoting surveillance network and university OH networks.

8.3.2 Building Strong Relationships

Multi-disciplinary is defined as many disciplines being involved, *inter-disciplinary* as disciplines working very closely with one another and plans fully integrated, *trans-disciplinary* implies an exchange of knowledge and skills between disciplines in working together for a desired outcome. It is therefore important to recognize that OH requires a *trans-disciplinary* approach to be successful.

An examination of using a One Health approach to canine rabies control in Africa provides a good example of the need to build strong trans-disciplinary relationships. As considered by Landford and Nunn (2012), a broad and diverse number of entities were revealed to be important for success, involving in addition to Health and Agriculture Ministries, the Ministry of Justice (legal assistance, advice on laws and regulations) and non-governmental organizations (resources, programs), private sector (skills, capacity) and international organizations (guidelines, national, regional planning) to name a few.

The movement to true *trans-disciplinary* effort is seen mainly at a research level with this being one of the key drivers and outcomes of the research programs. Relationships have been developed at an international level through shared systems such as GOARN, GLEWS and OFFLU and through international law as with the IHR. Initiatives advanced through the Tripartite Concept rely on relationships, and to some extent good relationships are a by-product, but the strength of relationships needs to be recognized as an essential resource. Similarly while promising innovative partnerships have been set up, such as that between the southern African nations, London International Development Centre, Welcome Trust, the Rockefeller Foundation and the Google Foundation forming the South African Centre for infectious Disease Surveillance (SACIDS), care needs to be taken to nurture relationships and in particular to develop *trans-disciplinary* capacity and skills.

The One Health movement in the last decade has been driven primarily by veterinarians. HIV, tuberculosis and malaria are some of the infectious disease priorities for physicians, and as a profession, medical leaders in some countries appear unconvinced that emerging infectious diseases and zoonoses have a large impact on human health. Engaging the human health profession, and the public health sector in particular in developed countries is one of the challenges facing

One Health. Veterinarians have generally been much more willing to embrace the concept and value of OH than their human medical colleagues.

Community is an integral part of OH, being both the informer and the beneficiary. Collaboration in developing countries is in many ways easier than in developed countries because of clear need and fewer administrative barriers, however often a lack of skills and knowledge has also to be addressed. Techniques that maximize the use of community knowledge, provide additional knowledge, and impart skills, are needed. Participatory epidemiology using participatory rural appraisal techniques (PRA) is based on the understanding that the local population usually have of prevailing diseases and also conditions that might give rise to diseases, and is used in concert with other surveillance methodology. The use of PRA had contributed importantly to the successful eradication of rinderpest disease (Marcotty et al. 2013).

Success in OH also needs cultural and societal values to be an integral part of the process. Calvin Schwabe in his book Veterinary Medicine and Human Health (1984) stated that 'the critical needs of man include the combating of diseases, ensuring enough food, adequate environmental quality, and a society in which humane values prevail' (Schwabe 1984). This was supported by Dr Suwit Wibulpolprasert, senior advisor in disease control to the Ministry of Health, Thailand, who emphasized the need for social equity and justice in his keynote speech at the International Congress for One Health in Melbourne in 2011. He stated that a more sustainable world is a more just world (Wibulpolprasert 2011). Some of the ways in which One Health approaches can learn from many tenets of the paradigm of social-ecological systems was discussed by Ross (2013).

The effectiveness of relationships in OH endeavors is a key aspect but is not fully recognized as such. A methodology for assessing effectiveness of relationships in OH is sorely needed.

8.3.3 *Infrastructure*

Infrastructure for OH has developed pragmatically around surveillance and laboratory systems and networks of expertise, but infrastructure also implies dedicated services for development of skills and capacity, communication and information channels and organizational and policy frameworks to support OH. There is much to do in this area as we operationalize the concepts.

Most attention to date has been in the development of surveillance and laboratory systems, and those of expert networks. Monitoring of environmental health occurs separately to human and animal health activities, and there are very few linkages. This is the area that remains least well developed in terms of a OH approach. On the positive side, environmental perturbations are now being incorporated into risk predictions, as seen for example in temperature and humidity trends reflecting potential spread of mosquitos, and in floods indicating a possible increase in leptospirosis in infected areas. Complex predictive models are being developed for human health—for example Healthscape in King County, Washington, which takes into account land use, transport, air quality and potential climate change to predict impact on public health (King County HealthScape 2009).

8.3.4 Education and Training

It is essential to recognize that One Health requires thinking and acting across disciplines as a basic tenant. Ideally skills are 'trans-disciplinary' that is, not just to know how other disciplines work but sharing skills and agreed goals. One of the ways to demonstrate the necessary skills and benefits is to use examples and case studies of success stories, such as those described in the 'One Health for One World: a Compendium of Case Studies' assembled by Veterinarians Without Borders (Veterinarians Without Borders 2010). A number of other case studies have been reported, including those in Asia and Europe described by Hall and Coghlan (2011), Africa (Rweyemamu et al. 2013a), and various studies in One Health Talk (2013).

Practitioners working in resource poor countries see a clear benefit in trans-disciplinary training being introduced at an undergraduate level (Marcotty et al. 2013). As yet it seems that 'western society' veterinary and medical faculties have not recognized the benefit of pre-graduate training—but this is changing. Global health is now reported to be included in 40 % of US and Canadian medical schools either as mandatory or elective programs. Many of the Masters and graduate training courses, and other professional training modules available in support of OH are listed on One Health Global Network website (see below).

There are currently no standardized guidelines for competencies or curricula for OH, or accreditation across professions.

8.3.5 Communication and Technology

Communication channels and networks are numerous, with many websites and informal and formal reporting tools concerned with OH and other arenas relevant to the animal–human-ecosystems interfaces such as disease emergence. The two dedicated OH websites, the One Health Global Network (www.onehealthglobal.net) and the One Health Initiative (www.onehealthinitiative.com/index.php) have different emphases although they overlap in a number of areas. The objectives of the former are on developing collaborative international programs and projects, on the promotion of OH education, and on ensuring a coherence of messages regarding OH strategies, communication and advocacy, whereas the latter has a broader remit as a means of communication, news and advocacy for all and any OH activities, but with a special interest in educational efforts across the OH spectrum and cross-species surveillance, transmission, prevention. A shared language is developing informally as is consensus definitions. One Health Sweden (formerly Infection Ecology and Epidemiology) has a focus on OH research but also works to provide open and transparent access to research articles, new methodologies and technological advances (One Health Sweden 2013). The EcoHealth Alliance, with an emphasis on holistic environmental concerns and ecosystem stability is a strong supporter of One Health and collaborates on language and definition (Zinsstag 2012).

The rapid growth and sophistication of new technologies, particularly information technology, has greatly assisted many areas underlying the One Health approach, including communication, data management, pathogen detection, risk analysis, and modelling. One such new technology has been the development of Geographic Information Systems (GIS) is an element in numerous projects enabling activities such as herd mobilization mapping, and research related to mixing of wildlife and livestock. On the ground easy to use Personal Digital Assistants (PDAs) and spreading availability of Wi-Fi are making collection and transfer of information easier and faster. In the world of molecular sciences, the availability of complex sequence data on a single pathogen is leading to the widespread use of molecular epidemiology to map and understand more closely the spread of viruses both within and between epidemics. All this science can be equally applied to human or animal health and the skills and knowledge to utilize these technologies have a clear One Health dimension.

8.4 The Way Forward

It is 10 years since the Manhattan Principles were promulgated and international concerns began to be raised about the risk posed by HPAI H5N1, and while the OH approach has been greeted widely with enthusiasm, it can also be argued that it has now reached a cross-road. To go forward, it is essential that professional silos are broken down to present a common, cross-disciplinary approach to health issues at the animal–human-ecosystems interfaces, and that the three major international organizations together with other international partners must begin to provide the global leadership which is presently lacking. This might best be done by a process similar to that taken by the United Nations Secretary-General in forming the UNSIC—that is by high-level international United Nations System coordination.

It is also essential that resources be made available to support research in the OH arena, and particularly in developing a better understanding of the human–animal-ecosystems interfaces including wildlife disease surveillance. The majority of emerging diseases arise from wildlife but the vast majority of funds are spent on understanding and controlling them in humans. There is an immediate need to invest in the frameworks, policies and processes required to better identify, articulate and manage risks posted to trade, human health and biodiversity by diseases with wildlife as part of their ecology.

Globally accepted frameworks and standards for research, education and accepted core competencies are required along with the need for an identified career path. However, there is little interest in any form of 'global governance': OH is a concept or approach not an association or society. Identifying a body that can lead relationship development between major disciplines and foster a true transdisciplinary approach, develop global guidelines and strategies, and ensure sustainable funding is needed, probably in an advisory capacity. The recent establishments of groups doing this at the regional level in developing countries (e.g. SACID) do seem to be achieving success in operationalizing OH at the national level and a similar approach could be emulated at the international level. In its most recent guise, OH is still just an approach for detecting and managing global health at the animal–human-environment interfaces, albeit relatively recent, and requires nurturing and further development to realize its full potential. It has been suggested that there should be a One Health Global Guidance Group as one of the outcomes of Stone Mountain, but this has still to be widely accepted. To some, this should be part of the leadership devolved to FAO, OIE and WHO, with an invitation list of international experts added to form an 'oversight' group. The importance of food safety and security is reason enough to support their leadership. However, there is no doubt that there is increasing global support for the usefulness and value of the OH approach for the detection and response to diseases at the animal–human-ecosystems interfaces in a transparent, collaborative, and multi-disciplinary way, and that this approach will become an essential component to future global health, and to the provision of safe food and water for an increasingly hungry world.

References

Afrique One (2013) African research consortium on ecosystem and population health: expanding frontiers in health. http://www.afriqueone.net. Accessed 12 June 2013

Alagaili AN, Briese T, Mishra N, Kapoor V, Sameroff SC, de Wit E, Munster VJ, Hensley LE, Zalmout IS, Kapoor A, Epstein JH, Karesh WB, Daszak P, Mohammed OB, Lipkin WI (2014) Middle East respiratory syndrome coronavirus infection in dromedary camels in Saudi Arabia. mBio 5(2):e00884-14. doi:10.1128/mBio.00884-14

APEC (2011) APEC one health action plan. http://did.go.th/dcontrol/th/images/stories/OneHealth_2011/doc-onehealth/apec_one_health_action_plan__september_2011.pdf. Accessed 6 June 2013

ASEAN (2008) ASEAN secretariat established working group for one health. http://www.aseanplus3-eid.info/newsread.php?nid=125&node=3&gid=3. Accessed 10 Jan 2014

AWHN (2013) The Australian wildlife health network. http://www.wildlifehealth.org.au/. Accessed 18 Dec 2013

Batsukh Z, Tsolmon B, Otgonbaatar D, Undraa B et al (2013) One health in Mongolia. Curr Top Microbiol Immunol 366:123–137

Canadian Food Inspection Agency (2012) Chief food safety office. www.inspection.gc.ca/abouth-the-cfia/organizational-information/president/chief-food-safety-officer/eng/1319478474855/1319478664830

CCWHC (2013) Canadian cooperative wildlife health network. http://www.ccwhc.ca/. Accessed 18 Dec 2013

CDC (2010) Operationalizing "one health": a policy perspective - taking stock and shaping an implementation roadmap. http://www.cdc.gov/onehealth/pdf/atlanta/meeting-overview.pdf. Accessed 6 June 2013
Centre for Health Protection (2014) Avian influenza report. Hong Kong Centre for Health Protection 10(17):3. http://www.chp.gov.hk/files/pdf/2014_avian_influenza_report_vol10_wk17.pdf. Accessed 1 May 2014
Chen Y, Liang W, Yang S et al (2013) Human infections with the emerging avian influenza A H7N9 virus from wet market poultry: clinical analysis and characterisation of viral genome. Lancet 381:1916–1925
Conrad PA, Meek LA, Dumit J (2013) Operationalizing a one health approach to global health challenges. Comp Immunol Microbiol Infect Dis 36:211–216
Crawford B, Roth I, Grillo T (2012) One health and Hendra virus: a collaborative approach in action. NSW Public Health Bull 23:160
EcoHealth Alliance (2013) One health alliance of South Asia. www.ecohealthalliance.org/programs/24-one_health_alliance_of_south_asia_ohasa. Accessed 12 June 2013
European Union (2013) One health: addressing health risks at the interface between animals, humans and their environments. http://eeas.europa.eu/health/pandemic_readiness/index_en.htm
FAO (2011) One health: food and agriculture organization of the United Nations strategic action plan. http://www.fao.org/docrep/014/al868e/al868e00.pdf. Accessed 11 Jan 2014
FAO (2014) ECTAD: FAO's response to animal disease emergencies. http://www.fao.org/ag/againfo/programmes/en/empres/ah1n1/Ectad.html. Accessed 10 Jan 2014
Fidler DF (2010) The challenges of global health governance working paper. International institutions and global governance program. Council on Foreign Relations, New York
French Government (2011) French position on the one health concept: for an integrated approach to health in view of the globilization of the health risks. http://www.diplomatie.gouv.fr/fr/IMG/pdf/Rapport_One_Health.pdf. Accessed 7 June 2013
Gargano LM, Gallagher PF, Barrett M et al (2013) Issues in the development of a research and education framework for one health [Conference summary]. Emerg Infect Dis. doi:10.3201/eid1903.121103
Gates Foundation (2013) Grand challenges in global health. The "one health" concept: bringing together human and animal health for new solutions. http://www.grandchallenges.org/Explorations/Topics/Pages/OneHealthRound11.aspx and http://www.grandchallenges.org/Explorations/Topics/Pages/OneHealthRound12.aspx
Ge XY, Li JL, Yang XL et al (2013) Isolation and characterization of a bat SARS-like coronavirus that uses the ACE2 receptor. Nature 503:535–538. doi:10.1038/nature12711
GLEWS (2013) Global early warning system for major animal diseases including zoonoses. www.glews.net. Accessed 12 June 2013
Gresham L, Wibulpolprasert S, Rweyemamu M, Leventhal A (2011) Connecting health organizationsfor regional disease surveillance (CHORDS): transforming international dialogue to counter biological threats. Ecohealth 7(Suppl 1):S53–S54
Gresham LS, Smolinski MS, Suphanchaimat R et al (2013) Creating a global dialogue on infectious disease surveillance: connecting organizations for regional disease surveillance. Emerg Health Threats J. doi:10.3402/ehtj.v6i0.19912
Guan Y, Zheng BJ, He YQ et al (2003) Isolation and characterization of viruses related to the SARS coronavirus from animals in southern China. Science 302:276–278
Hall D, Coghlan B (2011) Implementation of the one health approach in Asia & Europe: How to set-up a common basis for action and exchange of experience. European Union, Asia-Europe Meeting. http://eeas.europa.eu/health/docs/2011_09_eeas_operationalizing_one_health_report_en.pdf. Accessed 7 June 2013
Hueston W, Appert J, Denny T et al (2013) Assessing global adoption of one health approaches. Ecohealth 10:228–233

IDRC (2013) Evaluation. http://www.idrc.ca/EN/Programs/Evaluation/pages/default/aspx. Accessed 06 June 2013

INFOSAN (2013) The international food safety authorities network. http://www.who.int/foodsafety/fs_management/INFOSANOverview2013.pdf. Accessed 18 Dec 2013

IOM (1992) Emerging infections. Microbial threats to health in the United States. The National Academy Press, Washington DC

IOM (2003) Microbial threats to health. Emergence, detection, and response. The National Academy Press, Washington DC

IOM (2012) Improving food safety through a one health approach. The National Academies Press, Washington DC

Kelly AM, Ferguson JD, Galligan DT et al (2013) One health, food security, and veterinary medicine. J Am Vet Assoc 242:739–743

King L, Anderson L, Blackmore C (2008) Executive summary of the AVMA one health initiative task force. J Am Vet Assoc 233:259–260

King County HealthScape (2009) IPLACES health and climate enhancements and their application in king county. http://your.kingcounty.gov/kcdot/planning/ortp/HealthScape/I-PLACE3S-FINALREPORT%2006-01-09.pdf. Accessed 14 Aug 2014

Lamarque F, Artois M (1997) Surveillance of the wildlife diseases in France: the SAGIR network. Epidémiol Santé Anim 31–32:07 B31

Landford J, Nunn M (2012) Good governance in 'one health' approaches. Rev Sci Tech Off Int Epiz 31:561–575

Lau SK, Woo PC, Li KS et al (2005) Severe acute respiratory syndrome coronavirus-like virus in Chinese horseshoe bats. Proc Natl Acad Sci U S A 102:14040–14045

Lee K, Brumme Z (2012) Operationalizing the one health approach: the global governance challenges. Health Policy Plan 28:778–785

Li W, Shi Z, Yu M et al (2005) Bats are natural reservoirs of SARS-like coronaviruses. Science 310:676–679

Liden J (2013) The grand decades for global health, 1998–2008. Working group paper. Chatham House, London

Lipkin WI (2013) The changing face of pathogen discovery and surveillance. Nat Rev Microbiol 11:133–141

Lubroth J (2013) FAO and the one health approach. Curr Topics Microbiol Immunol 366:65–72

Mackenzie JS, Jeggo M (2013) Reservoirs and vectors of emerging diseases. Curr Opin Virol 3:170–179

Mackenzie JS, Drury P, Arthur RR et al The global outbreak alert and response network. Global Publ Health. 2014. doi:10.1080/17441692.2014.951870

Marcotty T, Thys E, Conrad P et al (2013) Intersectoral collaboration between the medical and veterinary professions in low-resource societies: the role of research and training institutions. Comp Immunol Microbiol Infect Dis 36:233–239

Mazet JA, Vlifford DL, Coppolillo PB et al (2009) "One health" approach to address emerging zoonoses: the HALI project in Tanzania. PLoS Med 6(12):e1000190. doi:10.1371/journal.pmed.1000190

McFarlane R, Sleigh A, McMichael T (2012) Synanthropy of wild mammals as a determinant of emerging infectious diseases in the Asian-Australasian region. Ecohealth 9:24–35

Memish ZA, Mishra N, Olival KJ et al (2013) Middle East respiratory syndrome coronavirus in bats, Saudi Arabia. Emerg Infect Dis 19:1819–1823

Meyer B, Mueller MA, Corman VM et al (2014) Antibodies against MERS coronavirus in dromedary camels, United Arab Emirates, 2003 and 2013. Emerg Infect Dis 20(4):552–559. doi:10.3201/eid2004.131746. Accessed 6 Jan 2014

Morse SS, Rosenberg BH, Woodall J (1996) ProMED global monitoring of emerging diseases: design for a demonstration program. Health Policy 38:135–153

Morse SS, Mazet JAK, Woolhouse M et al (2012) Prediction and prevention of the next pandemic zoonosis. Lancet 380:1956–1965

Nabarro D (2012) One health: towards safeguarding the health, food security and economic welfare of communities. Onderstepoort J Vet Res 79(2):450. doi:10.4102/ojvr.v79i2.450. Accessed 12 Jan 2014

OFFLU (2013) OIE/FAO Network of expertise on animal influenza. www.offlu.net. Accessed 12 June 2013

OHCEA (2013) One health Central and Eastern Africa, a regional network of pubic health and veterinary institutions. http://www.ohcea.org. Accessed 10 June 2013

OIE (2013a) Terrestrial animal code (2013). http://www.oie.int/international-standard-setting/terrestrial-code/access-online/. Accessed 6 Jan 2014

OIE (2013b) Training manual on wildlife diseases and surveillance. http://www.oie.int/fileadmin/Home/eng/Internationa_Standard_Setting/docs/pdf/WGWildlife/A_Training_Manual_Wild life.pdf. Accessed 6 Jan 2014

OIE (2014) The OIE worldwide monitoring system for wild animal diseases. http://oie.int/wahis_2/public/wahidwild.php. Accessed 11 Jan 2014

One Health Congress (2011) First international one health congress abstracts, Feb 14–16, 2011, Victoria, Australia. EcoHealth 7(Suppl 1)

One Health Sweden (2013) http://www.infee.se/infeecommunity/. Accessed 12 June 2013

One Health Talk (2013) Past discussions and case studies. www.onehealthtalk.org/index.cfm/pastdiscussions/one-health-capacity-building-case-studies/. Accessed 12 June 2013

Ope M, Sonoiya S, Kariuki J et al (2013) Regional initiatives in support of surveillance in East Africa: the East Africa integrated disease surveillance network (EAIDSNet) experience. Emerg Health Threats J. doi:10.3402/ehtj.v6i0.19948

PREDICT (2013) PREDICT: building global surveillance to detect and prevent spillover of pathogens of pandemic potential that can move between wildlife and people. www.vetmed.ucdavis.edu/ohi/predict/. Accessed 14 June 2013

Prince Mahidol Award Conference (2013) Report on the 2013 conference on a World United against infectious disease: cross-sectoral solutions. www.pmaconference.mahidol.ac.th. Accessed 10 June 2013

ProMED (2014) Avian influenza, Human – China (95): H7N9, WHO. ProMED Post 28 April 2014

Public Health England (2014) Human animal infections and risk surveillance group. http://www.hpa.org.uk/Topics/InfectiousDiseases/InfectionsAZ/EmergingInfections/HAIRS/. Accessed 10 Jan 2014

Public Health Agency of Canada (2009) One World One Health: from ideas to action. Report of the expert consultation, Winnipeg (Canada), March 16–19, 2009. http://aitoolkit.org/site/DefaultSite/filesystem/documents/OWOH%20Winnipeg%20July%202009%20version.pdf. Accessed 6 June 2013

Rabinowitz PM, Kock R, Kachani M et al (2013) Toward proof of concept of a one health approach to disease prediction and control. Emerg Infect Dis 19(12). doi:10.3201/eid1912.130265. Accessed 6 Jan 2014

Ross H (2013) One health from a social-ecological systems perspective: enriching social and cultural dimensions. Curr Top Microbiol Immunol 366:217–229

Rubin CS (2013) Operationalizing one health: stone mountain and beyond. Curr Top Microbiol Immunol 366:173–183

Rweyemamu M, Kambarage D, Karimuribo E, Karimuribo E, Wambura P, Matee M, Kayembe JM, Mweene A, Neves L, Masumu J, Kasanga C, Hang'ombe B, Kayunze K, Misinzo G, Simuunza M, Paweska JT (2013a) Development of a one health national capacity in Africa: the Southern African centre for infectious disease surveillance (SACIDS) one health virtual centre model. Curr Top Microbiol Immunol 366:73–91

Rweyemamu MM, Mmbuji P, Karimuribo E, Paweska J, Kambarage D, Neves L, Kayembe JM, Mweene A, Matee M (2013b) The Southern African centre for infectious disease surveillance: a one health consortium. Emerg Health Threats J. doi:10.3402/ehtj.v6i0.19958

SACID (2011) Southern African centre for infectious disease surveillance. www.sacids.org/kms/frontend/. Accessed 12 June 2013

Schwabe C (1984) Veterinary medicine and human health. Williams and Wilkins, Baltimore
Shomaker TS, Green EM, Yandow SM (2013) One health: a compelling convergence. Acad Med 88:49–55
Stone Mountain Workgroups (2011) Operationalizing one health. Stone Mountain meeting workgroups updates. Centers for Disease Control and Prevention, Atlanta. http://www.cdc.gov/onehealth/pdf/workgroups/newsletter-june-2011.pdf. Accessed 6 June 2013
Stone Mountain Workgroups (2012) Operationalizing one health. Stone mountain meeting workgroups updates. Centers for Disease Control and Prevention, Atlanta. http://www.cdc.gov/onehealth/pdf/workgroups/newsletter-april-2012.pdf. Accessed 6 June 2013
UNSIC (2008) Contributing to One World, One Health. A strategic framework for reducing risks of infectious diseases at the human-animal-ecosystems interface. www.un-influenza.org/files/OWOH_14Oct08.pdf. Accessed 6 June 2013
UNSIC (2010) Hanoi declaration, IMCAPI Hanoi, Vietnam. www.un-influenza.org/node/4040. Accessed 6 June 2013
USAID (2013) Emerging pandemics threat program. www.usaid.gov/news-information/fact-sheets/emerging-pandemic-threats-program. Accessed 16 June 2013
Veterinarians Without Borders (2010) One health for One World: a compendium of case studies. https://www.vetswithoutborders.ca/get-involved/resources/73-one-health-case-studies. Accessed 6 June 2013
Vongxay K, Conlan JV, Khounsy S et al (2012) Seroprevalence of major bovine-associated zoonotic infectious diseases in the Lao People's Democratic Republic. Vector Borne Zoonotic Dis 12:861–866
Wibulpolprasert S (2011) http://www.slideshare.net/melinhle/one-health-1st-congress-report. Accessed 12 June 2013
Wielinga PR, Schlundt J (2013) Food safety: at the center of a one health approach for combating zoonoses. Curr Topics Microbiol Immunol 366:3–17
Wildlife Conservation Society (2004) One World-One Health. www.wcs.org/conservation-challenges/wildlife-health/wildlife-humans-and-livestock/one-world-one-health.aspx. Accessed 5 June 2013
World Bank (2010) People, pathogens and our planet, vol 1. Toward a one health approach for controlling zoonotic diseases, Washington DC http://siteresources.worldbank.org/INTARD/Resources/PPP_Web.pdf. Accessed 10 June 2013
World Bank (2012) People, pathogens and our planet, vol 2. The economics of one health. World Bank, Washington DC. https://openknowledge.worldbank.org/handle/10986/11892. Accessed 10 June 2013
World Health Organization (2008a) International health regulations, 2nd edn. World Health Organization, Geneva
World Health Organization (2008b) Zoonotic diseases: a guide to establishing collaboration between animal and human health sectors at the coutry level. World Health Organization South-East Asia and Western Pacific Regional Offices, Manila, pp 1–19
World Health Organization (2010) The FAO-OIE-WHO collaboration. Sharing responsibilities and coordinating global activities to address health risks at the animal-human-ecosystems interfaces. A tripartite concept note. http://www.who.int/influenza/resources/documents/tripartite_concept_note_hanoi_042011_en.pdf. Accessed 6 June 2013
World Health Organization (2014a) Influenza at the human-animal interface. Summary and assessment as of 24th March 2014. http://www.who.int/influenza/human_animal_interface/Influenza_Summary_IRA_HA_interface_24March14.pdf?ua=1. Accessed 1 May 2014
World Health Organization (2014b) WHO risk assessment. Human infections wth avian influenza A (H7N9) virus. http://www.who.int/influenza/human_animal_interface/influenza_h7n9/140225_H7N9RA_for_web_20140306FM.pdf?ua=1. Accessed 01 May 2014
World Health Organization (2014c) WHO vigilant on new Middle East respiratory syndrome developments. http://www.emro.who.int/media/news/mers-developments.html. Accessed 1 May 2014
Zinsstag J (2012) Convergence of EcoHealth and one health. Ecohealth 9:371–373

Chapter 9
Field Epidemiology and One Health: Thailand's Experience

Sopon Iamsirithaworn, Karoon Chanachai, and David Castellan

Abstract Field epidemiology has become a major principle in public health and animal health service. Field epidemiologists are the primary group of professionals responding to outbreaks and other health emergencies who provide evidence-based recommendations for decision makers. In outbreak investigations, their duties are to identify the disease etiology, risk factors or source of an outbreak, and to contain the spread of the disease. Field epidemiology training programs (FETP) recruit professionals with training in medical, veterinary medical and other related health sciences to deal with real-life outbreaks and health problems. Joint training in surveillance and outbreak investigation has led to improved surveillance and control of zoonotic diseases by young professionals from human health and animal health sectors. Field epidemiology training programs for veterinarians (FETPV) is a living branch of a mature FETP which could support the development of wildlife and ecosystem needs under a broad One Health canopy. FETP can be a practical means of actualizing the One Health approach based upon shared needs and mutual benefit. One such model for sustainable, joint capacity development in field epidemiology under a One Health approach has been initiated in Thailand and is being adopted in other Asian countries. The Field Epidemiology Training Network of FETP in Southeast Asia is a useful platform for further strengthening regional disease surveillance and improving response to both public and animal health problems of international concern.

Keywords Epidemiologist • Field epidemiology • Outbreak • Training

---

S. Iamsirithaworn (✉)
Department of Disease Control, Ministry of Public Health, Bangkok, Thailand
e-mail: sopon@ddc.mail.go.th

K. Chanachai
Department of Livestock Development, Ministry of Agriculture and Cooperatives, Bangkok, Thailand

D. Castellan
Food and Agriculture Organization, Emergency Center for Transboundary Animal Diseases, Bangkok, Thailand

A. Yamada et al. (eds.), *Confronting Emerging Zoonoses*,
DOI 10.1007/978-4-431-55120-1_9

9.1 Needs for Field Epidemiologists and Field Epidemiology Training

9.1.1 Function and Role of the Field Epidemiologist

In many public health emergencies, effective prevention and control of health threats requires appropriate response by health professionals who have been trained in field epidemiology. Prompt investigation and proper response to major disease outbreaks by a team of field epidemiologists can minimize negative impacts on the health and economies of affected populations. Outbreak investigation is commonly recognized as an important function of field epidemiologists who work in a variety of field settings in different countries. Outbreaks of communicable and non-communicable diseases can occur in various settings ranging from a big city or rural town to a sophisticated healthcare facility (Ungchusak and Iamsirithaworn 2009). During outbreak events, field epidemiologists work closely with health professionals from different disciplines to gather relevant information from patients, family members, close contacts and residents of the affected area. Common challenges to field epidemiologists include working with data sources that are incomplete and using investigation protocols that are not well planned (Goodman and Buehler 2008). The nature of the work of field epidemiologists demands a timely response and travel to the field site which can sometimes require difficult access to remote areas, in order to solve the problem. The extent of the investigation is often limited because of the need for timely intervention as well as situational constraints on study methods.

During public health crises (e.g., major outbreaks or events of illness with unexpected deaths), field epidemiologists are usually the first group of professionals deployed to investigate the outbreaks and initiate control measures to protect at-risk populations. The primary function of field epidemiologists is to use findings obtained from an investigation to define the disease etiology, to identify the possible source(s) of an outbreak, and risk factors in order to contain its spread, and to implement prevention and control measures. Key objectives of the investigation generally include creating a case definition, implementing methods for case-finding, and establishing the occurrence of cases by time, place, and attributes of the affected population including, gender and age. The data generated are used to formulate hypotheses regarding etiology, mode of transmission, and other key facets of the disease. An important aspect of the investigation is the collection of samples from patients, the environment, and other sources for diagnostic tests and other laboratory investigations. An investigation of a foodborne *Streptococcal suis* outbreak attributable to consumption of fresh pig blood is an example of field epidemiologists identifying an implicated food item in this zoonotic disease outbreak (Khadthasrima et al 2008). The associated epidemic curve shows a pattern of common point source outbreak following a funeral banquet in Thailand in 2007 (Fig. 9.1). Roughly 75 % of recently emerging diseases and 60 % of human pathogens are zoonoses, outbreak response required coordinated and field

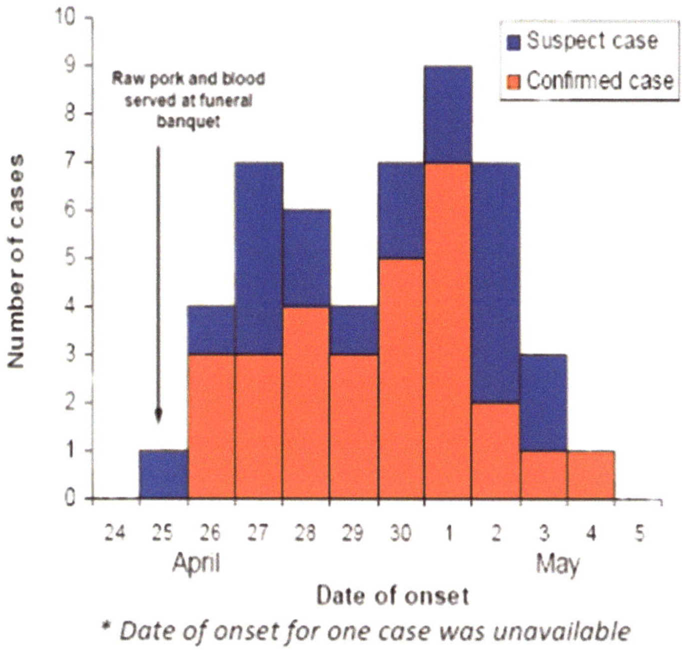

Fig. 9.1 Epidemic curve of of *S. suis* infections (N = 49*) in Phu Sang District, Phayao Province, Thailand, April to May 2007

investigations (U.S. CDC 2013). Therefore harmonized, yet adapted approaches for field investigation are essential components of a holistic approach to understand, prevent and control zoonoses in a comprehensive manner.

Surveillance is a major epidemiological intelligence tool and serves as the brain of health authorities and workers. Effective surveillance systems provide timely information to guide appropriate interventions for solving health problems. Field epidemiologists are often the key persons in establishing and maintaining surveillance systems since they design the system as well as collecting, analyzing, interpreting and sharing surveillance information with stakeholders for necessary action. Regular evaluations of surveillance systems by epidemiologists are necessary for quality improvement and ensuring utilization of the surveillance information.

In recent decades, the principle of field epidemiology has become increasingly applied to studying and solving complex chronic health threats, e.g., cancer in humans as well as both animal production (health) and disease. Field epidemiologists play a role in conducting field studies to identify causes and risk factors for the diseases, then select, design and implement appropriate intervention(s) to tackle the problems. Furthermore, field epidemiologists get involved with many interventional studies that are often carried out in the field setting to test the effectiveness of health interventions.

9.1.2 Concept of Field Epidemiology Training

In 1951, a pioneer training program for field epidemiology was first established in the USA (Gregg 2008). The U.S. Centers for Disease Control and Prevention (CDC) program, Epidemic Intelligence Service or EIS, is the first model that has been progressively adopted by many countries around the world. The EIS program aims to produce competent epidemiologists to combat infectious diseases and to mitigate the non-communicable disease burden in communities. The program has trained nearly 3,000 professionals in this discipline and assisted other nations to establish a number of field-based training programs. Since its onset, the US CDC's EIS program included primarily physicians but also recruited a variety of other health professionals such as statisticians, sanitary engineers, microbiologists, and veterinarians.

The term "Field Epidemiology Training Program: FETP" was first used by the Thai FETP at the time of its establishment in 1980. In the first two decades of the program, only young medical doctors (physicians) who were dedicated and enthusiastic to pursue their career in public health and epidemiology service were enrolled into the field-based training program. Beginning in 2005, veterinarians were regularly recruited through close collaboration between human health and animal health sectors including a formal memorandum of understanding signed by both the Ministries of Public Health and of Agriculture and Cooperatives in 2008.

In most countries, field epidemiology training courses are specifically designed for health professionals as applied post-graduation education. Graduates with a background in medicine, veterinary medicine or other health sciences are recruited and enrolled into a 2-year, non-degree program. Some programs closely collaborate with universities or academic institutes as a part of a joint diploma-degree program. In Thailand, the curriculum of FETP is incorporated as a part of the 3-year residency training for human medical doctors who aim to build their expertise in preventive medicine and epidemiology. As a second component of the residency training, physicians pursue a Master's of Public Health (MPH) degree in universities of their choice. This 1-year graduate degree provides broad knowledge for working in public health. On-the-job, in-service training in applied epidemiology provides trainees opportunities to learn from actual disease outbreaks and direct experience in responding to important health problems. Close supervision by mentors and learning through providing service are essential features and important challenges of delivering field epidemiology training.

FETP has a unique curriculum that intends to provide opportunities for trainees to learn epidemiology while they provide public health services. Transferable skills and competencies are built which improve the level of government services required to meet the challenges of emerging infectious diseases (EID) (TEPHINET 2013). Trainees learn how to conduct an outbreak investigation starting from verifying its existence, then organizing and traveling to the field to collect, manage and analyze data and interpret the findings in order to provide relevant recommendations. Most FETP curricula are designed for 2-year training though some are shorter (6 months to 1 year) and are referred to as field epidemiology training (FET). Trainees learn basic concepts of epidemiology and biostatistics and how to conduct field surveys and surveillance. In addition they also develop basic skills related to laboratory diagnosis, communication and outbreak management. Major requirements for graduation in most FETPs include an evaluation of a surveillance system, conducting outbreak investigations and epidemiologic field research.

In general, FETPs are based in ministries of health where trainees can legally access surveillance data and participate in outbreak response with no delay and with full legal authority. It should be noted that formal education in epidemiology has existed in a variety of Master's degree programs for decades but field-based practice involving real-life outbreak investigation is less emphasized in the universities due to time constraints, difficult field management and a limited mandate to deliver public health service. In contrast, the on-the-job epidemiology training provides trainees opportunities to learn from field practice. During the 2-year, in-service training in field epidemiology, the trainees' time spent conducting field work activities is placed at the highest priority and accounts for 70 % or more of their time. The remaining time allocation is for non-field activities, such as didactic teaching and workshops. This training strategy has proven to be successful in producing competent field epidemiologists for effective response to disease outbreaks and other public health challenges.

As of June 2013, 55 training programs in field epidemiology have been developed around the world (TEPHINET 2013) and the FETPs can address the needs for capable epidemiologists in health service and to meet a requirement of the International Health Regulations 2005 by building surveillance and response capacity in the country.

9.1.3 Development of Field Epidemiology Training in Thailand and Southeast Asia

The Field Epidemiology Training Program (FETP) in Thailand was initiated following a demand by health authorities for competent field epidemiologists who can effectively respond to acute health problems in a timely manner. The Thai program is the first field-based epidemiology training outside the American continent that was modelled after the EIS in the U.S. (Malison et al 1989). The curriculum of the Thai FETP was jointly developed by a collaborative effort of the Thai Ministry of Public Health (MoPH), the World Health Organization (WHO) and the U.S. CDC. The first cohort of trainees consisted of five medical doctors who began on-the-job training in June 1980. The label "FETP" has been recognized as applied epidemiology training that is directly linked to providing public health services including the surveillance and outbreak investigation mandates of the MoPH. An international training course was established in 1998 under FETP-Thailand with the goal of addressing needs for capable field epidemiologists in controlling public health problems in Asian countries that include Bhutan, Cambodia, China, Lao PDR, Malaysia, Myanmar and Vietnam.

The philosophy of Thai FETP is "Learning through Providing Services". The Thai FETP recruits young medical doctors with high interest in public health into the 2-year in-service training at the Bureau of Epidemiology in the MoPH. Trainees develop competence in operating surveillance systems by analyzing and interpreting the data and disseminating key information to support policy decisions. They also have opportunities to conduct a surveillance system evaluation or become involved with establishment of a new surveillance system. Each trainee investigates at least three outbreaks as a principal investigator and reports major findings and recommendations to local authorities and relevant policy makers. One of the requirements, an epidemiologic study, is carefully carried out to understand priority health issues and disease epidemiology, and to solve those health problems. In situations of health emergencies, FETP trainees are the front line investigators in the field. Close supervision by well-trained mentors, the majority of whom are FETP graduates, is a vital component of the successful field training.

As of June 2013, the program has graduated over 200 field epidemiologists to serve in various public health programs. Approximately 70 % of graduates have served in the MoPH by providing epidemiological services at national and local levels. Others graduates work in universities, military service and international organizations including WHO, US CDC and United Nations Populations Fund (UNFPA).

During the first decade of FETP-Thailand in 1980s, infectious diseases were major health challenges. Trainees and mentors worked with responsible agencies to control the public health problems through outbreak investigation and providing recommendations for improving vaccination strategies in the Expanded Program on Immunization. In the early 1990s, the Human immunodeficiency virus infection/acquired immunodeficiency syndrome (HIV/AIDS) epidemic provided another opportunity for the Thai FETP to demonstrate its usefulness. Trainees and graduates from FETP assisted in investigating the first 100 AIDS cases in the country and revealed epidemics among drug users and sex workers. The FETP graduates played a key role in establishing a sentinel surveillance system to monitor the HIV prevalence and risk behaviours among target populations (Weniger et al 1991). A famous intervention called "100 % Condom Use Program" was initiated and advocated and implemented by a program graduate. This program resulted in slowing down the HIV epidemic in Thailand (Rojanapithayakorn and Hanenberg 1996).

With a demonstration of the prominent 2-year field epidemiology training, a policy to establish and train 1,030 "Surveillance and Rapid Response Teams", so called SRRTs to cover all districts in Thailand was initiated following the emergence of the global severe acute respiratory syndrome (SARS) epidemic and outbreaks of Highly Pathogenic Avian Influenza (HPAI) H5N1 in Southeast Asia

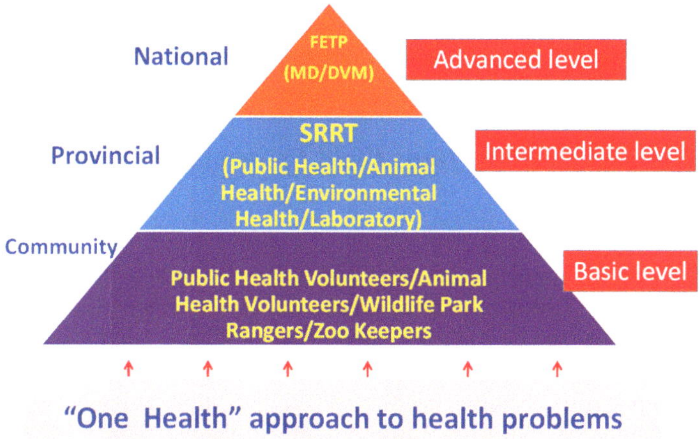

Fig. 9.2 Target of Field Epidemiology Training in Thailand

including Thailand in the early 2000s. SRRTs provide timely surveillance and are an effective response mechanism for containment of the epidemics. Thai FETP was asked to design a curriculum and training materials for medical doctors and other health professionals working at the district level. Each SRRT has approximately five members each of whom was trained for 6 months in surveillance and outbreak investigation, under supervision of more experienced epidemiology officers at the provincial level (Fig. 9.2). In recent years, a short course for basic epidemiology and event-based surveillance was developed for training personnel in communities including public health and animal health volunteers under the One Health approach.

In 1998, a new track for training of international participants was established under FETP-Thailand. On average, two to three trainees from Southeast Asian nations are enrolled into the 2-year program. After an 8-week introductory training module of basic epidemiology and biostatistics in classroom and field studies, the trainees perform actual surveillance studies and outbreak investigations in their home countries under the supervision of a local advisor. Every 6-months, they return to the training center in Thailand to work with their program advisor to share their work and experiences with colleagues from other countries. By 2013, the cumulative number of international graduates was 39. In late 2011, a network of field epidemiology training programs, "ASEAN+3 Field Epidemiology Training Network: ASEAN+3 FETN" was formed to strengthen field epidemiology training capacity in ten Asian countries plus China, Japan and Republic of Korea.

9.2 Establishment of FETPV in Thailand: A Collaboration Between Human and Animal Health Sections Through an Applied Field-Based Training Model

9.2.1 *The Demand for Joint Field Epidemiology Capacity in Thailand*

In Thailand, animal zoonoses control and prevention is mainly the responsibility of the Department of Livestock Development (DLD), Ministry of Agriculture and Cooperatives while the human zoonoses control and prevention task is under the authority of the Department of Disease Control (DDC), Ministry of Public Health. Effective response to zoonoses needs close collaboration, at least, between these two organizations. Many zoonotic diseases are notifiable diseases required for the surveillance system under the responsibility of DDC, i.e., rabies, leptospirosis, anthrax, influenza, trichinosis, Japanese encephalitis and streptococcosis caused by *Streptococcus suis*. However, the most effective approach to fight against these zoonoses is timely control at their origin. Thus, DLD and DDC have worked collaboratively to tackle those important zoonoses for decades.

Human resource development, especially capacity building for field epidemiologists, is an essential component of preparedness and response for emerging, re-emerging and endemic infectious zoonotic diseases under the concept of "One Health". In Thailand, the evidence shows that animal health and human health sectors working closely together can effectively control many important zoonotic diseases. For example, during the avian Influenza crisis in Thailand between 2004 and 2006, staff from DLD and DDC joined hands to fight against this emerging disease and successfully controlled its spread in the country.

Considering differences in human resources, infrastructure, expertise and mandate, collaboration in the area of joint human capacity development has begun in 2005. The Ministry of Public Health FETP recruited two veterinarians from DLD into the 2-year training program at the Bureau of Epidemiology, DDC, Ministry of Public Health. Since then, FETP has recruited one to two veterinarians annually to participate in the FETP. It has been demonstrated that the collaboration results in improved response to zoonoses as well as creating a working network between public health and animal health sectors. Thus, DLD and DDC agreed to strengthen the collaboration by the signing of a Memorandum of Understanding in January 2008 with the goal to produce 30 veterinary epidemiologists for DLD within the next 10 years. Graduate veterinary epidemiologists are now core persons responsible for human capacity development in DLD and strengthening networks from central to local levels.

9.2.2 Development of FETPV

The Field Epidemiology Training Program for Veterinarians (FETPV) curriculum was developed under close consultation with FETP faculty and veterinary experts from DLD, universities and FAO for the most effective application of the FETP concept, "Learning through Providing Services". Thai veterinarians and physicians are trained and work together to respond to any possible zoonotic disease outbreaks, thereby contributing to the public health sector mandate. The program aims not only to strengthen capacity of animal health personnel but also to promote self-sufficiency and multi-disciplinary response capacity for animal diseases that improves technical services for DLD.

In the first year of the training program, FETPV trainees are assigned for health situation analysis and outbreak investigation and they are focused equally on exclusive animal diseases and zoonotic diseases. This approach ensures that the animal health situation will be analyzed and critiqued for a better understanding it and improvement of the surveillance system. Trainees are expected to generate hypotheses and research questions to identify gaps in the response system through health situation analysis and outbreak investigation assignments in the first year of the program. A field epidemiological study is developed during the second year of the training program to further assess and provide recommendations to fill gaps identified previously.

The program shares financial and technical resources among partners, particularly DLD and DDC in order to train veterinarians. Trainees spend 30 % of the time in formal classroom training and 70 % of the time conducting field work to fulfil the requirements for graduation. During the 2-year training program, veterinarians participate in various short training courses and workshops such as basic epidemiology and biostatistics, field surveillance and outbreak demonstration and basic computer software and laboratory workshops to ensure that trainees are competent to carry out field epidemiological activities. Some of these short training courses and workshops are organized in conjunction with the regional aspect of FETPV and the animal health professional development programs of DLD. A summary of the FETPV curriculum is shown in Table 9.1.

During the training course, veterinarians and physicians have a chance to share their experiences and expertise and work collaboratively to solve animal and human health problems. Veterinarians and physicians in the program are trained to understand both animal and public health service systems. Since the beginning of FETPV, a number of collaborative health situation analyses, outbreak responses and field epidemiological studies of both human and animal health problems have been conducted. These include human leptospirosis, human streptococcosis caused

Table 9.1 Summary of the 2-year FETPV curriculum

Time[a]	Duration	Classroom training	Description	In-service assignments
Year 1				
June	1 month	Introductory course in field epidemiology and biostatistics	Formal starting of two-year training program. Participants: Veterinarians and physicians Objectives: Provide an integrated training under a One Health approach	1. Health situation analysis (either animal or human) 2. Minimum three outbreak investigations as principal investigator (zoonoses and non-zoonotic diseases) • One animal disease outbreak investigation • Two zoonotic disease outbreak investigations 3. Receive and verify outbreak report 4. Participate in weekly meeting 5. Lead scientific seminar 6. Journal club
July	1 week	Computer software	Participants: Veterinarians and physicians Objectives: Trainees shall be able to apply computer software as a tool for their investigation. Software: Excel, EpiInfo, Stata, etc.	
July	1 week	Surveillance evaluation demonstration workshop	Participants: Veterinarians and physicians Objective: Understand surveillance system and demonstrate how to conduct surveillance evaluation, i.e. avian influenza and Japanese encephalitis.	
July	1 week	Animal disease outbreak investigation demonstration workshop	Participants: Veterinarians Objective: Conduct a systematic animal disease outbreak investigation during a real disease event.	
January	1 week	Advance data analysis workshop	Participants: Veterinarians and physicians Objective: Apply advanced data analysis methods, i.e. multivariate logistic regression.	
January	3 days	Computer software for animal disease surveillance	Participants: Veterinarians Objective: Gain proficiency with computer	

(continued)

Table 9.1 (continued)

Time[a]	Duration	Classroom training	Description	In-service assignments
			software for animal disease surveillance.	
			Software: Win Episcope, Survey toolbox, C survey, etc.	
February	3 days	Laboratory surveillance workshop	Participants: Veterinarians Objective: Understand and apply sampling procedures, and interpretation of laboratory results.	
Year 2				
August	1 week	Scientific writing and presentation skills workshop	Participants: Veterinarians and physicians Objective: Improve communication skills with scientific information.	1. Conducting field epidemiological study
				2. Participate in outbreak investigation as co-investigator
				3. Participate in weekly meeting
				4. Give presentation in international conference
				5. Teaching assistant for animal and public health

[a]Time may be adjusted year by year to fit with overall program

by *S. suis type 2* from pigs, human and animal brucellosis caused by *Brucella melitensis* and *B. suis*, human and animal rabies, Porcine Reproductive and Respiratory Syndrome (PRRS) in pigs, zoonotic Influenza A in humans and animals, and Methicillin-resistant *Staphylococcus aureus* (MRSA) in pigs and farm workers.

It has been 8 years since veterinarians joined the FETP—tangible proof of the program sustainability and commitment from both DLD and DDC. As of June 2013, a total of 13 Thai veterinarians graduated from the program. Further, as trainees have graduated and returned to their own organizations, several animal and public health collaboration activities have been initiated by an alumni network at the local and central level. These include joint field investigations, and joint surveillance database programs for avian influenza and rabies. Our experience with FETPV development reinforces the concept that effective response to zoonoses and emerging animal-health related problems requires a broad range of competencies from various disciplines.

9.3 Collaboration Between Human and Animal Health Sectors at Regional Level

9.3.1 The Challenge

The emergence and spread of Nipah virus, SARS and highly pathogenic avian influenza (HPAI) H5N1 subtype in Asia since 2004 provided a stress test of both country and regional capacity to deal with emerging infectious diseases (EIDs) and transboundary animal diseases (TADs). Common constraints were identified in addressing an emerging zoonosis including the lack of resources and the difficulties in establishing a seamless interaction among human health, animal health and environmental health agencies (Schelling et al. 2005). In order to fully support a One Health approach to zoonoses, it was evident that the veterinary health infrastructure needed to be strengthened in order to deal with EIDs which will continue to pose a significant threat to human, animal and environmental health, food security, livelihoods, sustainability of food production systems and the economies of developing countries (FAO 2011).

9.3.2 The Response

A regional needs assessment was conducted in nine countries from Southeast Asia between May and August 2008 to establish a baseline measurement of veterinary epidemiology capacity. It included a semi-structured interview and questionnaire and a gap analysis of 32 skills related to field epidemiology for veterinarians (FAO 2008b). Stakeholders from animal health and public health sectors at national, regional and international agencies in each country were interviewed over several days to assess capacity development needs for veterinarians working to fulfil the mandates to address TADs and zoonotic diseases in the region.

The needs assessment revealed that the level of cooperation between animal and human health sectors among countries in the region varied greatly and was generally limited. In 2008, eight of nine countries surveyed had established cooperative relationships in dealing with zoonotic diseases, however in 2013 all countries have established cooperative relationships related to zoonoses. In general, the lead agency for H5N1 was designated according to the species first affected, related to whether the index species was human or animal. The needs assessment in 2008 also revealed that joint surveillance rapid response teams (SRRTs) existed only in Thailand. Similar efforts to build SRRTs in China and Bangladesh as well as several other countries in the region are now underway.

Most notably, epidemiologists from both human and animal health agencies favored joint training of human medical and veterinary medical staff in applied field epidemiology training programs such as FETP and FETPV. Joint training was considered as a practical way to establish trust and working relationships among local stakeholders for zoonotic diseases. However the needs assessment concluded that institutional barriers prevent greater linkage among human and animal health sectors. Removal of institutional barriers at middle and upper management levels of human and animal health agencies remains an essential target to address in order to promote integration of joint training and implementation of functional SRRT in the region for zoonotic diseases.

9.4 Regional Application of the Field Epidemiology Training Program for Veterinarians (FETPV)

9.4.1 The Need for Veterinary Field Epidemiologists

Veterinary infrastructure will require improvement to support a successful One Health approach. The 79th General Session of the World Assembly of the World Organization for Animal Health (OIE) cited that more than half the countries of the world have fewer than 35 public sector veterinarians per million inhabitants and with a median number of 5.4 veterinarians per million with an animal health mandate (including zoonotic and animal specific diseases (Bonnet et al. 2011)). In addition, over half the countries of the world invest less than US$ 2 per capita per year of public investment on veterinary services. The functional state of animal health surveillance systems is limited since even though 86 % of countries reported having the capacity for early detection by the animal health sector, 30 % of them had not reported suspected disease events during the previous 5 years (Bonnet et al. 2011). However diagnostic capacity has greatly improved since the advent of avian influenza H5N1 subtype through the creation of an active regional laboratory network which is increasingly capable of detecting a wider array of zoonotic diseases, in partnership with public health laboratories and enabled through donor support. "Two-way linking" of field and laboratory services within the animal health sector is being strengthened with the aim of enabling the eventual "four-way linking" between laboratory and field services of animal health and public health services.

Veterinarians have been engaged in public health training and service in the U. S. through the epidemiology intelligence service (EIS) since 1951 (Pappaioanou et al. 2003). Engagement of veterinarians in field epidemiology training and service has evolved more recently in Asia. Its development is also coincident with wider recognition of the general need for field training in veterinary epidemiology (Salman 2009). In 2008, three strategic areas were identified for the development of FETPV, which include establishing partnerships, designing a programmatic framework and developing a relevant curriculum to meet country needs within the region.

Table 9.2 Epidemiology related training parameters and findings, January 2006 to June 2008 (FAO 2008a)

Parameter	Findings
Training Topic	59 % related to vaccinator training
	12 % related to epidemiology
	29 % other areas including disease response
Didactic Versus Applied	66 % didactic
	25 % combined didactic and applied
Duration of Training	59 % less than 1 week
	29 % 1–4 weeks
	6 % 1–24 months
Target Group	18 % national level
	13 % subnational level
	28 % local
	24 % combination
Training Emphasis	26 % focused specifically on surveillance and outbreak investigation

9.4.2 *Assessing Veterinary Epidemiology Capacity Development in Asia*

Prior to the development of FETPV, FAO conducted an inventory of training related to epidemiology for veterinarians in the Region of Asia and the Pacific in 2008 for training conducted in the region that was largely supply-driven. The training included an 30-month period between January 2006 and June 2008 associated with efforts to prevent and control the panzootic of HPAI H5N1 in Asia. The review included 356 training courses in 10 countries, which assessed the quantity and quality of efforts at country level (FAO 2008a). A summary of these parameters and findings is presented in Table 9.2.

It was necessary to adjust capacity development to provide demand-driven training and to address country and regional ownership and stakeholder needs. From its inception, FETPV was envisioned as a model that addresses both public health and animal health sector needs for both zoonotic and animal-specific diseases in order to be relevant and sustainable.

FETPV is a living branch of a mature FETP which has been developed in many countries in the region. This joint collaboration can also support the development of wildlife and ecosystem needs under a broad One Health canopy.

9.4.3 Regional Needs Assessment for Veterinary Field Epidemiology Training

Vision and mission statements, six core competencies and 32 skills were defined through a consultative workshop facilitated by FAO that included technical inputs from Thailand MoPH, Thailand DLD and university representatives (FAO 2008c). A list of skills and competencies for FETPV is presented in Table 9.3.

The stated objective of FETPV was "training through providing services" as with FETP. The required skills, competencies and concepts from the Training Programs in Epidemiology and Public Health Interventions Network (TEPHINET) were adapted for veterinarians based on the public health needs as well as the nature of the animal health work and services provided (Traicoff et al. 2008) (U.S. CDC 2006).

The output from this consultative workshop provided the basis for a needs assessment tool including a gap analysis in nine member countries of the Region of Asia and the Pacific including Cambodia, China, Indonesia, Lao PDR, Malaysia, Mongolia, Myanmar, Thailand and Viet Nam (FAO 2008b). The needs assessment included a semi-structured questionnaire using a Delphi approach which created a profile of country level animal health services and their relationship with public health services and needs. An analysis of skills identified for veterinary field epidemiologists was produced for zoonotic and animal specific disease priorities at the country level. The gap analysis and country profiles served as the basis for the development for the FETPV pre-requisite short course and for the 2-year modular curriculum. The process of conducting a needs assessment promoted trust and ownership for countries that continue to support FETPV. Key outcomes of the 2008 regional needs assessment are listed in Table 9.4.

The assessments uncovered the need to develop short, medium and long-term approaches for capacity development at local, subnational and national levels that include vocational, in-service and academic training models.

In 2010, seeing the potential use of field epidemiology to solve animal disease problems in their country the China Ministry of Agriculture (MoA) requested FAO to assist in developing a roadmap to develop a 2 year national FETPV that would complement the China FETP that which was established in 2003. After 2 years, China MoA now provides the majority of funding to support China FETPV (Fusheng 2012, personal communication). Further, coordination between the Thailand based FETPV and China FETPV includes sharing lessons learned, resource persons and training materials.

9.4.4 Application of FETPV in the Region

Demand for veterinary field epidemiology training at the country and sub-regional level has grown since 2010 through application of the pre-requisite short course

Table 9.3 FETPV core competencies and skills

Core competencies
1. Conduct Outbreak Investigations
2. Develop Surveillance Systems
3. Effective Communication
4. Conduct Field Research
5. Produce Scientific Publications
6. Conduct Surveys
Skills
1. Conducting Outbreak Investigation Steps
2. Teamwork
3. Leadership
4. Computer Software Applications
5. Project Management and Logistics
6. Laboratory Knowledge
7. Oral and Written Communication (including English)
8. Basic Epidemiology Principles
9. Human Relations
10. Biosafety and Biosecurity
11. Animal Health Management
12. Animal Diseases and Control Methods
13. Laws and Regulations
14. Questionnaire Design
15. Data Analysis
16. Surveillance Evaluation
17. Data Management
18. Data Quality
19. Data Analysis
20. Principles of Market (Value) Chains
21. Public Relations Presentation and Public Speaking
22. Effective Interviewing, Negotiating
23. Study Proposals and Design
24. Identifying Research Questions (generating hypothesis)
25. Critical Literature Review
26. Ethical issues
27. Financial Management
28. Manuscript Preparation & Submission
29. Perseverance and Life-Long Learning
30. Sampling Design
31. Sample Size Calculations
32. Diagnostic Test Evaluation

Table 9.4 Findings of the 2008 regional needs assessment for the development of FETPV in the Region of Asia and the Pacific (FAO 2008b)

Area	Finding
Gap Analysis	• Overall, 44 % of the 32 epidemiology skills had significant gaps (score 2 out of a maximum of 3) between current capacity and needs
	• Significant gaps between current capacity and needs for skills ranged from 16 to 72 % among countries
Priority areas identified by countries	• Application of epidemiological concepts
	• Data management
	• Outbreak investigation and surveillance
	• Field epidemiology research and report writing
Institutional challenges	• Lack of functional epidemiology units
	• Cross-utilization of epidemiology staff
	• Variable and limited interaction with public health agencies
Primary target group in-country	• Subnational group identified as highest priority by all countries
Selection criteria proposed by countries	• Merit based and age restricted
	• Gender equality
	• English language competency

including 1 week of field activities based on country needs assessment profiles. The FETPV pre-requisite course has been adapted for country level and regional contexts including China, Lao PDR, Cambodia, India and the eight member countries of the South Asian Association for Regional Cooperation (SAARC). The Food Agricultural Organization (FAO) has developed working partnerships with national, international and other public health agencies to provide joint training under a One Health umbrella through memoranda of understanding that include country FETP, the World Health Organization and the U.S. CDC.

To further create an enabling environment for field epidemiology training in the region, strategic planning and regional institutionalization are required to sustain FETPV. FAO has conducted advocacy with country Chief Veterinary Officers (CVO) to raise awareness of the need for epidemiology and related institutional changes required to enable and support field epidemiologists following their return to the workplace (FAO 2010). In 2011 and 2012, member country representatives from the Association of South East Asian Nations (ASEAN) developed a strategic framework and plan for epidemiology capacity development (FAO 2012a). As of May 2013 both FETPV and the strategic plan are now formally recognized by ASEAN as regional platforms. There is similar potential to support the needs of the SAARC member countries.

Since 2008, FETPV training has expanded to include both ASEAN and SAARC member countries with 2-year programs established in Thailand and China. In addition, a total of 12 FETPV pre-requisite short courses have been conducted in six countries since January 2009 in Thailand (4), China (2), Lao PDR (2), Cambodia (1), India (2) and Nepal (1). In consultation with country decision makers, trainees attend the FETPV short course in order to prepare in-country field studies upon their return from the training. Fifty field studies have been conducted for seven of the 12 short courses conducted since 2009 despite limitations in funding and mentoring. Constraints related to travel, laboratory testing and mentoring limit the number of eligible field surveys that can be implemented by FETPV short course participants. Follow up workshops to report results are held in Thailand and are currently under development in Cambodia and Nepal.

Veterinary participants in the FETPV short course demonstrate a strong interest in conducting surveys for zoonotic diseases in their countries. Fifty-four percent (27/50) of all field surveys conducted by FETPV short course participants involve zoonotic diseases including brucellosis, rabies, anthrax, leptospirosis, and Japanese encephalitis.

Institutional barriers such as lack of regional enabling mechanisms, lack of funding and the limited availability of resource persons restrict input from national and international public health experts for veterinary field epidemiology training programs. However, well-established FETPs have provided significant inputs in three of the six training centers. Training contributions from national health agencies occurred at three of five training centers noted above and one of five training centers received training inputs from international health agencies for FETPV short course training across the region. Resource constraints limit input from some public health agencies in addition to existing institutional barriers in some countries or sub-regions.

9.4.5 Visible Implementation of One Health

FETP and FETPV are practical means of actualizing the One Health approach based upon shared needs and mutual benefit. Integration of One Health through joint training is progressing, however institutional barriers and financial constraints must be overcome to increase the benefits of training investments in some countries. Mechanisms are needed to transfer One Health knowledge and approaches beyond the conceptual stage within the existing institutional frameworks of each country in order to conduct field studies that are trans-sectoral. The development of joint surveillance and rapid response teams is limited and incipient in most countries in the region with some exceptions including Thailand, Bangladesh and Cambodia.

Inter-sectoral as well as inter-disciplinary efforts are essential components of FETPV that include training modules linking these various aspects. Inter-sectoral commitment includes joint classroom and field training by instructors and mentors to demonstrate disease relationships at the human-animal-environment interface. Similarly, FETPV trainees consider disease events at the social-economic-communications interface to broaden the scope and understanding of the role and interrelationships of field epidemiology with other disciplines.

9.5 Opportunities for Collaboration and Future Challenges

Historically, the need for collaboration under a One Health approach has been forced upon the global community by biological and political imperatives (Zinsstag and Weiss 2001). The succession of large-scale disease events illustrates the need to develop common platforms such as field epidemiology capacity in order to respond to emerging zoonotic diseases and livestock diseases comprehensively. The collaboration between FETP and FETPV represents a practical vehicle for One Health in action through training, working and sustainably building together at individual, institutional and country levels.

9.5.1 Joint Training

Joint training and fieldwork raises awareness of the benefits of working together to overcome differing institutional cultures and mandates. Joint training creates a common language, culture and experience that can lead to operationalized responses to pathogens and disease agents that are not constrained within institutional walls.

Joint training through FETP and FETPV in Thailand fosters a culture of mutual respect and appreciation among countries, sectors and disciplines. Individuals in applied epidemiology training are exposed to a performance (student-centered) model as opposed to an acquisition (teacher-focused) model (Taylor 2009). This learning model builds critical thinking skills that are transferrable and can adjust to the demands of dealing with any emergent disease agent that encompasses multiple sectors, disciplines and geo-political boundaries. Joint training from a broad range of instructors further expands understanding about how to implement surveillance or outbreak investigation jointly where the disease agent interacts at the interface of humans, livestock, wildlife and the environment. Quality assurance, particularly in an operational sense, is essential for continual improvement and adaptation of training to meet evolving challenges (White et al. 2001).

The FETPV is complementary to other training models needed for training paraveterinarians as well as undergraduate and post-graduate academic training for veterinarians. Veterinary Public Health needs must also be addressed including meat inspection, food safety and zoonotic disease control in live animals (Zamri-Saad et al 2009). Joint training of participants from many countries raises awareness of transboundary and regional perspectives that affect prevention and control of diseases at the country level (White et al. 2001).

9.5.2 Working Together

Service delivery models for joint surveillance and investigation of zoonoses when present will be specific for each disease event and within each country (Schelling et al. 2005). The relative scarcity of human resources to conduct field assessment and response in most countries requires a clear strategic approach to optimize these scarce resources. The SRRTs model in Thailand reflects a technically sound and financially practical option for disease response and surveillance from local communities to the national level.

Field epidemiology training programs focus on producing high quality graduates through an in-service training model at similar cost to a conventional academic training program. Balancing this cost, it has been demonstrated that governments also directly benefit from the "training through services model" (U.S. CDC 2006). Funding and cost sharing for training and field investigation and research greatly leverages scarce funds available, particularly for animal health agencies. Cost sharing by the Thai national government and international development partners has been used to support field activities and international fellowships including sponsoring attendance at international conferences to present findings of field research.

The increasing demand for food will require the use of systems-based approaches in order to mitigate risks to human health (King 2008). The occurrence of H7N9 in China in 2013 illustrated the opportunity to combine value chain analysis with epidemiological information in order to characterize and mitigate risks within large systems related to human exposure. These integrated systems-based approaches are greatly needed as food systems become larger and more complex. There are opportunities to extend the FETPV to building linkages with animal industry that have been promoted in Indian veterinary curricula, (Rahman 2004).

9.5.3 Building Sustainably Together

At the country level, applied training through FETP has contributed greatly to designing and implementing health programs since their initiation (White et al. 2001) (Yin et al 2006) (Petersen et al. 2000). This need for applied field training is now being extended to support the inclusion of both veterinary and non-veterinary wildlife professionals through joint FETP and FETPV training such as the wildlife training module (WILD) (FAO 2012b). TEPHINET has included the participation of FETPV at regional and global meetings to discuss curriculum development, program assessment and to share findings of field studies at conferences.

Building functional regional networks of alumni and mentors is will ensure the sustainability of FETP, FETPV and Field Epidemiology Training Network (FETN), which is a regional platform, recognized by 10 ASEAN nations plus China, Japan and the Republic of Korea. These integrated networks will support effective front line capacity to meeting the ongoing challenge of controlling emerging infectious diseases.

References

Bonnet P, Lancelot R, Seegers H, Martinez D (2011) Contribution of veterinary activities to global food security for food derived from terrestrial and aquatic animals. In: 79th General Assembly of the World Organization for Animal Health, Paris, 22–27 May 2011

FAO (2008a) Assessment of epidemiology training in the region of Asia and the Pacific, January 2006 to June 2008, D Castellan internal FAO report

FAO (2008b) Needs assessment for the development of a field epidemiology training program for veterinarians (FETPV) in the region of Asia and the Pacific, D Castellan internal FAO report, August 2008

FAO (2008c) FETPV workshop to develop a vision, mission competencies and skills for the FETPV program and curriculum, April 2008, Bangkok

FAO (2010) Report of the regional advocacy meeting of chief veterinary officers, August 2010, Phuket

FAO (2011) Approaches to controlling, preventing and eliminating H5N1 Highly pathogenic avian influenza in endemic countries. Animal Production and Health Paper, No. 171. Rome, 2011

FAO (2012a) Report of the regional meeting to develop a strategic plan and framework for the development of epidemiology in Asia, August 2012, Chiang Mai

FAO (2012b) WILD training press release. http://www.fao.org/ag/againfo/programmes/en/empres/news_241012.html

Gregg M (ed) (2008) Field epidemiology, 3rd edn. Oxford University Press, New York, 572 pp

Goodman RA, Buehler JW (2008) In: Gregg MB (ed) Field epidemiology. Oxford University Press, New York, pp 1–5

Khadthasrima N, Hannwong T, Thammawitjaya P, Pingsusean D, Akkanij B, Jaikhar A et al (2008) Human streptococcus suis outbreak in Phayao province, Thailand. OSIR 1 (1):4–7. http://www.osirjournal.net/issue.php?id=3

King L (2008) Collaboration in public health: a new global imperative. J Vet Med Educ 35 (2):264–265

Malison MD, Dayrit MM, Limpakarnjanarat K (1989) The field epidemiology training programmes. Int J Epidemiol 18(4):995–996

Pappaioanou M, Garbe PL, Glynn MK, Thacker SB (2003) Veterinarians and public health: the epidemic intelligence service of the centers for disease control, 1951–2002. J Vet Med Educ 30 (4):383–391

Petersen LR, Ammon A, Hamouda O, Breuer T, Keibling S, Bellach B, Nimer U, Bindert FJ, Ostroff S, Kurth R (2000) Developing national epidemiologic capacity to meet the challenges of emerging infections in Germany. Emerg Infect Dis 6(6):576–584

Rahman SA (2004) The history of veterinary education in India. J Vet Med Educ 33(2)

Rojanapithayakorn W, Hanenberg R (1996) The 100 % condom program in Thailand. AIDS 10 (1):1–7

Salman MD (2009) The role of veterinary epidemiology in combating infectious animal diseases on a global scale: the impact of training and outreach programs. Prev Vet Med 92 (2009):284–287

Schelling E, Wyss K, Bechir M, Doumagoum D, Zinsstag J (2005) Synergy between public health and veterinary services to deliver human and animal health interventions in rural low income settings. BMJ 331:1264–1267
Taylor RM (2009) Defining, constructing and assessing learning outcomes. Rev Sci Tech 28 (2):779–788
TEPHINET (2013) Atlanta. http://www.tephinet.org/
Traicoff DA, Walke HT, Jones DS, Gogstead EK, Imtiaz R, White ME (2008) Replicating success: developing a standard FETP curriculum. Public Health Rep 123(Suppl 1):28–34
Ungchusak K, Iamsirithaworn S (2009) Principles of outbreak investigation. In: Detels R (ed) Oxford textbook of public health, 5th edn. Oxford University Press, New York, pp 484–497
U.S. CDC (2006) Field epidemiology training program development handbook. U.S. CDC, Atlanta
U.S. CDC (2013) http://www.cdc.gov/ncezid/
Weniger BG et al (1991) The epidemiology of HIV infection and AIDS in Thailand. AIDS 5(Suppl 2):S71–S85
White ME, McDonnell SM, Werker DH, Cardenas VM, Thacker SB (2001) Partnerships in international applied training and service, 1975–2001. Am J Epidemiol 154(11):993–999
Yin J, Li G, Ren X (2006) An overview of veterinary medical education in China: current status, deficiencies and strategy for improvement. J Vet Med Educ 33(2):238–243
Zamri-Saad M, Romziah S, Kunavongkrit A, Valdez CA (2009) Integrating the world of animal health and world public health into the veterinary curriculum: a southeast Asian perspective. Rev Sci Tech 28(2):719–725
Zinsstag J, Weiss WG (2001) Livestock diseases and human health. Science 294:477

Chapter 10
One Health and Food Safety

Peter R. Wielinga and Jørgen Schlundt

Abstract Many, if not most, of all important zoonoses relate in some way to animals in the food production chain. Therefore food becomes an important vehicle for many zoonotic pathogens. One of the major issues in food safety over the latest decades has been the lack of cross-sectoral collaboration across the food production chain. Major food safety events have been significantly affected by the lack of collaboration between the animal health, the food control, and the human health sector.

One Health formulates clearly both the need for, and the benefit of cross-sectoral collaboration. Here we will focus on the human health risk related to zoonotic microorganisms present both in food animals and food derived from these animals, and typically transmitted to humans through food. Some diseases have global epidemic—or pandemic—potential, resulting in dramatic action from international organizations and national agricultural- and health authorities in most countries, for instance as was the case with avian influenza. Other diseases relate to the industrialized food production chain and have been—in some settings—dealt with efficiently through farm-to-fork preventive action in the animal sector, e.g. *Salmonella*.

Finally, an important group of zoonotic diseases are 'neglected diseases' in poor settings, while they have been basically eradicated in affluent economies through vaccination and culling policies in the animal sector, e.g. *Brucella*. Here we will discuss these three different foodborne disease categories, paying extra attention to the important problem of antimicrobial resistance (AMR). In addition, we present some of the One Health inspired solutions that may help reduce the threat of several of the foodborne diseases discussed.

Keywords Antimicrobial resistance • Farm-to-fork • Food safety • Foodborne zoonoses • One health

---

P.R. Wielinga • J. Schlundt (✉)
National Food Institute, Technical University of Denmark (DTU), Copenhagen, Denmark
e-mail: peter.wielinga@gmail.com; jors@food.dtu.dk

A. Yamada et al. (eds.), *Confronting Emerging Zoonoses*,
DOI 10.1007/978-4-431-55120-1_10

10.1 Introduction

People in ancient times already understood they could get sick from consumption of infected meat. By keeping their animals healthy, and by using dedicated methods of food preparation and conservation, ancient farmers learned how to improve health and prevent disease. Probably the oldest written document about it is '*On Airs, Waters, and Places*', written by Hippocrates,[1] which describes how human health is influenced by its interaction with the environment. Since then, our health situation has improved by applying these simple rules of thumb, and even more through improved technologies such as good animal management, hygiene and biosecurity, vaccination programs and prudent animal drug treatment. Nowadays, some of the most feared zoonotic diseases such as anthrax or brucellosis are absent in many countries. However, there are still many important diseases that threaten human health and which have animals as their reservoir. These animal reservoirs range from wildlife to domestic animals, both in companionship and agricultural settings. By the obvious close contact and the sheer number of animals needed for consumption, the animals produced for food form the largest reservoir and production grounds for emerging zoonotic pathogens.

Actions of authorities to protect society from zoonotic diseases differ significantly according to socio-economic status and the zoonotic pathogen in question. Basically, zoonotic diseases related to food animals can be separated into three groups. In the first group are diseases with a potential for global spread and with a dramatic public relations potential, often these diseases have a significant human reservoir showing human-human transmission, e.g. SARS, avian influenza and certain types of AMR bacteria. The second group is constituted by persistent zoonotic diseases related to the industrialized food production chain, such as *Salmonella* and *Campylobacter,* which are broadly distributed in the farm-to-fork chain. These human pathogens are often non-pathogenic in animals and seem to be distributed in all countries, both rich and poor. In the third group are the 'neglected zoonotic diseases'. They are zoonotic diseases which have been eradicated (or drastically reduced) in affluent economies through vaccination and culling policies, and through introduction of hygienic and animal biosafety management practices. However, in many poor settings these diseases are 'neglected diseases' and receive very little attention from national authorities or even international organizations. This group includes *Brucella*, bovine TB (tuberculosis), i.e. Mycobacterium bovis, and many parasitic diseases, e.g. leishmaniasis and cysticercosis. In addition to these traditional infectious diseases there is a new threat of antimicrobial resistant (AMR) bacteria. Caused by the use of antimicrobials both in human and veterinary medicine this problem has emerged and is now to be recognized as one of the most important threats to human health.

Although in the detail the control of these groups of diseases differ, they are all most efficiently prevented by a One Health approach which considers the full

[1] http://classics.mit.edu/Hippocrates/airwatpl.mb.txt.

farm-to-fork chain. Such preventive and holistic approaches may reduce both the disease burden to human health and the economic burden to developing economies, and therefore represent a significant *potential for improvement* related to food safety as seen in a One Health perspective.

10.2 Transmission Routes

Through food and feed, direct contact, and via the environment, the human- and the animal microbial flora are in contact with each other. Figure 10.1 outlines the most important routes of transmission for infectious diseases between humans and animals. Via these routes infectious diseases from (food-)animals may enter the human reservoir and vice versa. The foodborne transmission route is probably the most important gateway for this contact, and the vast majority of human infections with enteric zoonotic bacterial pathogens, such as *Salmonella enterica*, *Campylobacter coli/jejuni*, and *Yersinia enterocolitica*, occur through this route. For other diseases there is evidence that transmission also occurs via direct contact between (food) animal and humans, e.g. live-stock associated methicillin resistant *Staphlylococcus aureus* (MRSA) (Graveland et al. 2011). Next, there is transmission via the environment (e.g. surface water or water used to irrigate plants) mainly as a result of spreading of manure into the environment (Spencer and Guan 2004; Hutchison et al. 2005). And though much less frequently reported, there may be transmission of pathogens from humans to animals, which most probably was the

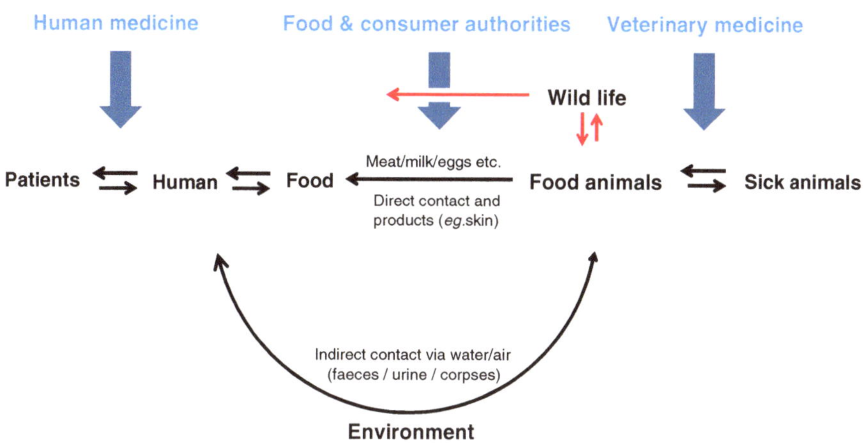

Fig. 10.1 Schematic presentation of important microbial transmission routes via which the human and (food) animals are in contact with each other. In *blue* control mechanisms are shown, and in *red* some of the transmission routes that are more difficult to control. Via the environment transmission may take place of microorganisms present in excretion products, and in diseased animals and people. In addition, wildlife constitutes a risk, as it holds a broad spectrum of diseases, including many highly pathogenic diseases

case for the *Staphylococcus aureus* CC398 (Price et al. 2012). In many developing countries, wildlife forms an additional important reservoir for foodborne pathogens, not only through consumption of wildlife. Because of often lower bio-safety levels in these countries, direct contact between humans and food animals is generally more frequent, and diseases from the wildlife community may cross over more easily to domestic animals. For instance, the general understanding now is that the SARS epidemic in 2003 originated in direct human contact with and/or consumption of wildlife, or indirectly through contact between wildlife and domestic animals (Guan et al. 2003; Shi and Hu 2008).

Wildlife holds a broad spectrum of diseases including some of the most feared, such as Ebola, rabies and anthrax and, and in contrast to other food sources, much of the consumption of wildlife goes undetected by food controlling agencies. For these reasons, and because of the global trade in wildlife derived food and other items (Pavlin et al. 2009), consumption of wildlife animals, and the spillover of infectious diseases from wildlife to food/production animals, should not be overlooked.

10.3 Food Animal Zoonoses in General

The spread of foodborne zoonoses through the food production chain is often referred to as the 'farm-to-fork' (or 'farm-to-table' or 'boat-to-throat') chain. It should be noted that risk mitigation solutions under this framework typically have focused on a consideration of the full food production continuum, involving all relevant stakeholders, i.e. a typical One Health framework invented before the One Health paradigm was defined. Figure 10.2 tries to capture a generalized picture of a farm-to-fork chain, starting with animal feed and ending in human consumption of animal food products.

Although a number of very important zoonoses are related to wildlife—and in some cases directly transmitted from wildlife animals—the vast majority of zoonotic disease cases in the world relate to animals that are bred for food purposes. Such zoonotic pathogens include bacteria such as *Brucella*, *Salmonella*, *Campylobacter*, verotoxigenic *Escherichia coli* and *Leptospira*; parasites, such as *Taenia*, *Echinococcus* and *Trichinella*; and viruses, such as Influenza A H5N1 (Avian influenza) and Rift Valley Fever virus. Next to these infectious diseases, derived agents such as (microbial) toxins and prions (Prusiner 1997) form another important zoonotic subgroup.

Diseases originating on the farm can in many cases most efficiently be dealt with on the farm itself, thereby eliminating more complex measures or cross-contamination down the farm-to-fork chain. For example, brucellosis in animals (mainly cattle, sheep and goats) has been eliminated in many countries, thereby virtually eliminating the human disease burden (Godfroid and Käsbohrer 2002). Also, some of the main parasites can be effectively controlled at the farm level, and this could work for both *Taenia solium* in pigs (defined by WHO/FAO/OIE as a 'potentially eradicable parasite'), as well as, *Trichinella spiralis* which is found in

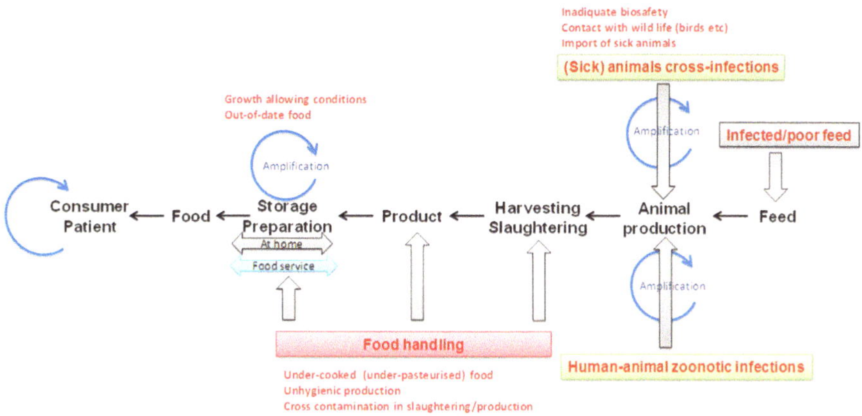

Fig. 10.2 Farm-to-fork scheme showing how infectious diseases may travel through the food chain. Along the chain we have indicated some of the most prevalent causes of infectious disease introduction

many wild animals and importantly in pigs for human consumption. Both parasites have essentially been eliminated from farmed pigs in most northern European countries (WHO/FAO/OIE 2005; Gottstein et al. 2009). However, both diseases still form a serious 'neglected diseases' threat in settings where there is potential for contact between wild and domestic animals.

10.3.1 Zoonoses with a Dramatic Public Relations Impact and Potential for Global Spread

It was primarily the outbreaks of SARS and zoonotic influenza, AMR (dealt with separately) and BSE (Bovine Spongiform Encephalopathy) which alerted the world to the need for a One Health approach. Outbreaks of viral diseases in humans, originating in or spreading through farm animals (avian flu—H5N1 and swine flu—H1N1) have caused major global alerts in the last decade. These influenza outbreaks spread very quickly, either in the animal population (H5N1) or directly in the human population (H1N1), and formed a global threat for human health. H1N1 was therefore characterized by the World Health Organization (WHO) as a pandemic. Although in total the human disease burden related to the endemic bacterial zoonoses is probably many fold higher than these influenza outbreaks, it is basically these relatively few but fast spreading outbreaks that have put One Health on the global agenda. In addition, the failure to predict, monitor and control the spread of these diseases in animals presented regulators and politicians with a wake-up call, and made them demand (better) cross-sectoral collaboration between the animal and human health sectors.

Prions, non-living infectious agents, have been a significant burden of disease in animal and man. The most well-known zoonotic prion disease is probably the one causing Bovine Spongiform Encephalopathy (BSE) in cows, and new variant Creutzfeldt-Jakob disease in humans, as represented by the massive outbreak of 'Mad Cow Disease' in the UK in the 1980s and 1990s. This agent, a mutant protein, which mainly sits in the brain, got into the (beef) food chain by the feeding of ruminant derived meat and bone meal to ruminants. Prions (Prusiner 1997), Scrapie (the disease in goats), spongiform encephalopathy of Rocky Mountain elk, transmissible mink encephalopathy, kuru and Creutzfeldt–Jakob disease were known before the large outbreak of bovine spongiform encephalopathy in the UK. It, however, took some time and great efforts, and an early One Health approach, to establish the links between the different animal diseases and the human disease (Hill et al. 1997; Prusiner 1998; Ghani et al. 1998). This insight created a background to efficiently stop the spread of this prion disease, by banning the use of animals in animal feed, and seeing a subsequent decrease of the disease in humans (Hoinville 1994).

In the Western world prion diseases have attracted much attention, and their control has resulted in a large economic burden to society. In developing countries, often with less strict rules about the re-use of dead animals, and more direct contact with wildlife, prion diseases may still be endemic though unrecognized.

10.3.2 Endemic Zoonoses Related to Industrialized Food Production

In contrast to the dramatic outbreaks discussed above, many food-related zoonoses are endemic in farm animals and some of the most important of these do actually cause disease in the animals. It should be realized that most countries—including most developing countries—produce large amounts of food animals, and most of the production takes place in some sort of industrialized setting. Such settings are invariably linked to a number of important zoonotic pathogens. Table 10.1 shows three lists of the most important food pathogens, as reported in studies published by the CDC in the USA and by RIVM (Havelaar et al. 2012) in the Netherlands, as well as a list of pathogens recognized by ECDC as focus organisms for the EU.

Although widespread, these pathogens are often not recognized as important human pathogens because of their often mild disease syndromes in healthy persons (e.g. limited to diarrhea and vomiting) and because of the complexity of source attribution. However, they do form a serious threat to the vulnerable segments of our societies (i.e. the young, the elderly, the immune-compromised and recovering patients), and some patients may develop long-lasting chronic disorders (e.g. arthritis and neurological disorders) (McKenna 2012). These facts, together with the sheer number of infections they cause, results in a substantial total burden of disease for these pathogens as expressed in Disability-Adjusted Life Years

Table 10.1 The most important foodborne infections as experienced in the US, EU and the Netherlands

Ranking of the top five foodborne diseases in the USA, ranked according to incidence (regular), hospitalizations (*italic*) deaths (bold) (CDC 2011)	Ranking of top 14 foodborne pathogens and toxins in the Netherlands, ranked according to their burden of disease in DALYs[a] (Havelaar et al. 2012)	Focus panel of food- and waterborne diseases in the EU, without ranking (ECDC 2010)
Norovirus 1, *2*, ***4***	*Norovirus 4*	
Salmonella (nontyphoidal) 2, *1*, **1**	*Salmonella* spp. *5*	Salmonellosis
Clostridium perfringens 3,..,..	*perfringens* toxin *7*	
Campylobacter spp. 4, *3*, **5**	*Campylobacter* spp. *2*	Campylobacteriosis
Staphylococcus aureus 5, ..,..	*Staphylococcus aureus* toxin *6*	
Toxoplasma gondii .., *4*, **2**	*Toxoplasma gondii 1*	Toxoplasmosis
E. coli (STEC) O157 .., *5*, ..	STEC O157 *10*	Infection with VTEC/STEC
Listeria monocytogenes ..,.., **3**	*Listeria monocytogenes 11*	Listeriosis
	Cryptosporidium spp. *13*	Cryptosporidiosis
	Giardia spp. *8*	Giardiasis
	HepA *9*	*Hepatitis* A
	HepE *12*	
	Rotavirus 3	
	Bacillus cereus toxin *12*	
		Shigellosis
		Variant Creutzfeldt–Jakob disease

[a]Disability-adjusted life years (DALYs), are a combined estimate of the burden of disease due to both death and morbidity. One DALY can be thought of as 1 year of healthy life lost and is often expressed in years of life lost on the population level, and can be thought of as a measure of the gap between current health status and an ideal situation where each individual in the population lives to old age, free from disease and disability (Murray 1994)
For the list of the CDC and the Netherlands the ranking in terms of incidence, hospitalizations, deaths or DALY is given. The list from ECDC was generated by expert consultation and for use as an EU focus list in future disease burden studies

(DALYs) (see Murray 1994). For instance, the study of Havelaar et al. (2012) showed that the total burden of the 14 diseases he studied was 13,500 DALYs, for a total of 1.8 million cases and 233 deaths caused by these diseases (in 16 million people). Source attribution estimates showed that one-third could be attributed to foodborne transmission. Similarly large numbers were reported for the USA (CDC 2011), where surveillance studies of 31 known pathogens gave an estimated total of 9.4 million illnesses, 55,961 hospitalization and 1,351 deaths attributed to foodborne diseases (in 315 million people). Importantly, the latter report also showed that the 31 pathogens studied make up only 44 % of the foodborne diseases, and the majority of 56 % is caused by unknown agents. This situation is probably not unique for the USA, and thus indicates that there is still much health to be gained from improved food safety. Table 10.1 shows that an almost identical list of important foodborne pathogens are found in Europe, and that *Toxoplasma gondii*, *Listeria monocytogenes*, *Campylobacter*, rotaviruses, noroviruses and *Salmonella*, should probably form the key targets for interventions. Except for norovirus, these pathogens have been shown to be zoonotic, and find their way to humans via food and the environment (Fig. 10.1). A One Health approach ensuring efficient cross-sectoral collaboration and data-sharing, could lay the foundation for a realistic description of the situation, and could help implement sensible cross-sector solutions.

Building on the idea of One Health to control these diseases, there are several countries (especially in northern Europe and North America) that have instituted cross-sectoral data collection and collaboration. This is typically done through the construction of zoonosis centers or their equivalents. These centers aim to stimulate and facilitate the collaboration between human, veterinary and food institutes. Some examples of such specialized centers are the US National Center for Emerging and Zoonotic Infectious Diseases (http://www.cdc.gov/ncezid), the British National Centre for Zoonosis Research (http://www.zoonosis.ac.uk/zoonosis) and Danish Zoonosis Centre which is part of the Danish National Food Institute (http://www.food.dtu.dk/English/Research/Research_Groups/Zoonosis_Centre.aspx). Two clear examples of what such centers can accomplish are: the reduction of *Salmonella* in food animals in Denmark, and the Danish integrated approach to combat AMR (described below in a separate section). In the Danish *Salmonella* reduction program, data sharing across animal, food and human health sectors has enabled science-based solutions, and has most noticeably resulted in significant reductions in human salmonellosis through lowering *Salmonella* prevalence in animals (Wegener et al. 2003). In relation to laying hens the program started with a simple and inexpensive serological surveillance of egg producers. Flocks found positive were either culled and repopulated, or used to produce heat-processed eggs, Danish eggs are now considered free of *Salmonella*. Next to this arm, a program of surveillance and eradication of infected broiler flocks was setup. The effect of the whole program was measured in term of cases of human salmonellosis, which were found to be significantly reduced as the project progressed in time.

The construction and solutions of this program clearly followed One Health principles. Food, veterinary and human health sciences worked together, using similar detection and (geno)typing techniques, which enabled comparison and sharing of data. This top-down selection of *Salmonella*-free poultry could work in other countries with industrialized food animal production as well. In other countries—including most likely most developing countries, *Salmonella*-positive animals have been imported from big producers in industrialized countries.

One such documented example was the import into Zimbabwe of *Salmonella enteritidis* via live animals. *Salmonella* entered the country through import of infected poultry in the commercial national production system around 1993, and thereafter spread quickly to the communal sector (small-scale farming), as well as to the human population (Matope et al. 1998). The most likely reason for the spread within Zimbabwe was that old animals from the commercial sector were sold to small-scale communal production systems. As the trade of live animals is done on a global level and does not take into account whether the traded animals carry any of the diseases from Table 10.1, reducing the prevalence in the commercial sector in producing countries, may also lower the global spread and human disease burden in the rest of the world.

10.3.3 Neglected Zoonoses Related to Poverty

The spectrum of neglected diseases is broad and includes diseases caused by bacteria, viruses and parasites. Many are found world-wide but their prevalence in the human and animal populations varies according to the local agricultural, demographic and geographic conditions and tradition. For many of the neglected diseases solutions to dramatically decrease the disease burden are well-known, but action is lagging, this is the case for example for many of the parasitic zoonoses. This is the reason why the WHO refers to these diseases as 'Neglected Diseases' (WHO 2006; Molyneux et al. 2011). Neglected diseases may be categorized into two (strongly overlapping) categories. In the first category are the neglected tropical diseases which include Chagas disease, trypanosomiasis, leprosy, rabies, schistosomiasis and others, many of which are zoonotic and parasitic diseases.[2] The second category are the neglected zoonotic diseases, covering many of the diseases above and also some bacterial diseases such as anthrax, bovine tuberculosis (TB), brucellosis, and also cysticercosis and echinococcosis.[3]

Many of the neglected diseases are carried by wildlife and in poor and rural settings by livestock (e.g. brucellosis, anthrax, leptospirosis, Q-fever and bovine TB). In addition, many are food- and waterborne (e.g. brucellosis, cysticercosis/taeniasis and echinococcosis). In particular, the prevalence of bovine TB appears to be increasing in many poor settings and has been linked to HIV infections as an important factor for progression of a TB infection to an active TB disease (LoBue et al. 2010). Brucellosis and bovine TB in cattle cause lowered productivity in the animal population, but seldom death, and both have largely been eradicated from the bovine population in the developed world by test-and-slaughter programs, which in effect has eliminated the human health problem (Godfroid and Käsbohrer 2002).

Some of the parasitic diseases (e.g. schistosomiasis, cysticercosis, trematodiasis and echinococcosis) have high mortality rates and long-term sequelae including cancer and neurological disorders. Cysticercosis is emerging as a serious public health and agricultural problem in poor settings (García et al. 2003). Humans acquire *Taenia solium* tapeworms when eating raw or undercooked pork contaminated with cysticerci. The route of transmission is, pigs infected through *Taenia*

---

[2] http://www.who.int/neglected_diseases/diseases/en/.

[3] http://www.who.int/neglected_diseases/zoonoses/en/.

eggs shed in human faeces, and the disease is thus strongly associated with pigs raised under poor hygienic conditions. This means that the cycle of infection can be relatively easily broken by introducing efficient animal management, as has been done in most developed countries.

Given that 70 % of the rural population in poor countries is dependent on livestock as working animals to survive (FAO 2002), the effect of these animals carrying a zoonotic disease can be dramatic, both relative to human health directly, but also as it affects the potential to earn an income. This also affects the potential mitigation action; for instance the large-scale culling of animals, which can be a viable solution in rich countries, might be problematic in the poorest countries. Such solutions would not only mean loss of food, but also a serious socio-economic disruption, in some cases leading to national instability.

10.4 AMR in Food Animals

Some of the recent serious outbreaks of antibiotic resistant (AMR) foodborne disease, such as EHEC in Germany (Mellmann et al. 2011), have shown us a new problem. There seems to be a global trend with the prevalence of AMR rising (WHO 2001; DANMAP 2010; ECDC 2010; Aarestrup 2012). Especially dangerous is the emergence of resistance against antimicrobials that are considered critically important in human medicine, and in multidrug resistant (MDR) infections (Potron et al. 2011; Kumarasamy et al. 2010).

In the early 1940s antibiotics were first introduced to control bacterial infections in humans. The success in humans led to their introduction in veterinary medicine in the 1950s, where they were used in both production and companion animals. Nowadays, antibiotics are also used with intensive fish farming and to control some infectious diseases in plants. Their use is thus widespread. Antibiotics in animals are mainly used in three ways: (1) for therapy of individual cases, (2) for disease prevention (prophylaxis) treating groups of animals, and (3) as antibiotic growth promoters (AGP) treating groups of healthy animals with sub-therapeutic concentrations to promote animal growth. When first introduced, the use of antibiotics led to improved animal health, and subsequently higher levels of both food safety and food security. All use, but in particular the use as AGP, resulted in a dramatic rise in the use of antibiotics, and for instance, between 1951 and 1978 the use in the United States alone went from 110 to 5,580 t (WHO 2011).

However, the use of antibiotics in animals has over the years also resulted in a selective pressure for AMR microorganisms, contributing significantly to the human health problem of AMR bacteria. Notably a number of bacterial strains that were previously susceptible to antibiotics are now in very high frequencies becoming resistant to various antibiotics, some of which are very important as last resort treatment potential for humans (Bonten et al. 2001). In particular the use as AGP is questionable, as the concentrations used are sub-therapeutic which result in the selection for resistance but do not efficiently kill microorganisms. Nowadays there are serious efforts by national authorities and some international organizations to reduce the antibiotic overuse in animals (Food Agricultural Organization of the United Nations (FAO)/World Organization for Animal Health (OIE)/WHO 2003; WHO 2011; U.S. Food and Drug Administration (FDA) (2012)), especially—but not only—through abolishing their use as AGP. However, there seem to be major problems in ensuring cross-sectoral understanding, since the veterinary and medical professions are still in debate about how the AMR problem has emerged (Phillips et al. 2004; Karp and Engberg 2004; Smith et al. 2005; Price et al. 2012). To achieve a science-based understanding of the problem, data on AMR from both the animal and the human side should be compared, and both risk assessments and source attributions performed in an integrated way. In other words, a One Health approach in which human and animal health sectors, including food and environmental sectors, work together, may help to deliver answers needed and suggest ways to reduce problems in both human and animal reservoirs (Figs. 10.1 and 10.2).

10.5 Global One Health Efforts

In addition to the factors described above, food production and food trade are now more and more global, and thus some of the food related problems are also global food problems. On the positive side, globalization has helped with some of the important global food issues: it raised food security, made our food more varied and tastier, and even including transport costs in the equation, still has global financial advantages. However, together with the food also the foodborne diseases now travel the globe. And if we do not stay on top of the problem, disease outbreaks might affect large parts of the global food sector negatively, in the end leading to negative health—but will also have financial and socio-economical consequences. A more holistic and pro-active approach to food safety may help prevent future food disasters and build healthy economies.

10.5.1 Global Initiatives to Contain Foodborne Zoonoses

One Health approaches to combat zoonotic foodborne diseases need to consider at least three levels, the international level, national level and the farm level where the actual production takes place. To facilitate the work at all these levels, many countries have established specialized zoonosis centers. These centers focus their work on zoonotic diseases and promote collaboration between different sectors, and between different countries. They examine the prevalence of zoonotic diseases in humans and (food-)animals, their routes of transmission, the risk associated with their presence in our food chain, and the relation between human disease and zoonotic transmission. In addition, as our food production system has become increasingly dependent on global trade, the approaches taken by these zoonoses centers (should) also include a global angle. National zoonosis centers may also help tackle the global problems associated with zoonotic diseases.

However, at the moment most of this work is done by international and global organization, such as the WHO, OIE, and FAO. These three international organizations have recognized that combating zoonoses is best achieved via a One Health approach, as stated in their seminal paper 'A Tripartite Concept Note' (FAO/OIE/WHO 2011), in which they express the need to collaborate for a common vision. Given the impact zoonotic diseases have in socio-economical terms and on the vulnerable sectors in our societies, a One Health vision is also endorsed by the World Bank and the United Nations Children's Fund (UNICEF) (World Bank/WHO/UNICEF/OIE/FAO/UNSIC 2008). In their common vision, they say that a One Health approach may lead to novel and improved solutions, including solutions that have not been considered before because of the high costs involved.

For instance, while in some cases vaccination is the ultimate tool to prevent disease, it is not always considered because the costs of mass vaccination are higher than the public health benefit savings, or because of global trade regulations. Under a One Health approach sharing of costs, as well as other mitigation strategies could likely enable novel ways of reaching sensible solutions (Narrod et al. 2012).

For global infectious disease safety national authorities report to WHO important outbreaks of human disease which have the potential of cross-border spread, under the auspices of the International Health Regulations (IHR) (WHO 2005). These regulations also cover foodborne diseases associated with globally traded food. However, given the major impact that such announcements may have on global food trade, such as was the case with BSE in the UK or the more recent trade barriers put up after the EHEC outbreak in Germany, national authorities may have become more careful and restricted in what they report.

A global One Health approach which both considers human health aspects and socio-economic consequences would therefore be a welcome improvement to the IHR of 2005. Next to WHO, other organizations are active in reporting global infectious disease outbreaks, most notably ProMED-mail (http://www.promedmail.org), which is an internet based Program for Monitoring Emerging Diseases worldwide, set up by the International Society for Infectious Diseases. The program is dedicated to rapid global dissemination of information on outbreaks of infectious diseases and acute exposures to toxins that affect human health, including those in animals and in plants grown for food or animal feed, and thereby supports the One Health principles.

Many of the (international) organizations and governing bodies named above have generated guidelines to control – and disseminate information about—food related zoonoses, such as for instance WHO's Global Foodborne Infections Network (GFN) (www.who.int/gfn), the European Food Safety Authority (EFSA) (www.efsa.europa.eu/en/topics/topic/zoonoticdiseases), Foodnet from the US Centers for Disease Control and Prevention (www.cdc.gov/foodnet) and others.

The goal of these networks is essentially the same: *To help capacity-building and promote integrated, laboratory based surveillance and intersectional collaboration among human health, veterinary and food-related disciplines to reduce the risk of foodborne infections.*

10.5.2 Efforts to Contain AMR Zoonoses

The emergence of AMR in food animals is a serious threat for modern human medicine. The risks exist that both (i) the overuse by mass prophylaxis and AGP in animals, and (ii) the misuse of human critically important antibiotics in animals, will lead to the emergence of new AMR organisms which may spread to the human reservoir, and via global food trade spread around the world. In the most critical scenario this will make our arsenal of antibiotics unfit to treat previously treatable infectious disease, and it might take us back to a situation as before World War II, when antibiotics were not yet used in human medicine.

One Health principles may help mitigate this risk and deal with the AMR problem in an efficient way. Collaboration between the FAO/WHO Codex Alimentarius Commission and the OIE have generated important guidance on how an integrated approach and the prudent use of antimicrobials may reduce the emergence of AMR in (food-)animals and subsequently in humans.[4] Previous to this, in 2000 the WHO published the 'Global Principles for the Containment of Antimicrobial Resistance in Animals Intended for Food' (WHO 2000) which all countries should follow to reduce the risk of AMR. The three major principles are:

- Use of antimicrobials for prevention of disease can only be justified where it can be shown that a particular disease is present on the premises or is likely to occur. The routine prophylactic use of antimicrobials should never be a substitute for good animal health management.
- Prophylactic use of antimicrobials in control programs should be regularly assessed for effectiveness and whether use can be reduced or stopped. Efforts to prevent disease should continuously be in place aiming at reducing the need for the prophylactic use of antimicrobials.
- Use of antimicrobial growth promoters that belong to classes of antimicrobial agents used (or submitted for approval) in humans and animals should be terminated or rapidly phased-out in the absence of risk-based evaluations.

These Global Principles have been supplemented with, (1) guidance on the prudent use of antibiotics from the Codex Alimentarius Commission together with OIE, and (2) six priority recommendations from WHO to reduce the overuse of antibiotics in food animals for the protection of human health (WHO 2001), being:

---

[4] See: www.codexalimentarius.org; www.who.int/foodborne_disease/resistance; www.oie.int/our-scientific-expertise/veterinary-products/antimicrobials.

(1) Require obligatory prescriptions for all antibiotics used for disease control in (food) animals;
(2) In the absence of a public health safety evaluation, terminate or rapidly phase out the use of antibiotics for growth promotion if they are also used for treatment of humans;
(3) Create national systems to monitor antibiotic use in food-animals;
(4) Introduce pre-licensing safety evaluation of antibiotics [intended for use in food animals] with consideration of potential resistance to human drugs;
(5) Monitor resistance to identify emerging health problems and take timely corrective actions to protect human health;
(6) Develop guidelines for veterinarians to reduce overuse and misuse of antibiotics in food animals.

A recent publication (WHO 2011) covers the broader scope of AMR in relation to both animals and humans. Thus, a 'One Health' approach has explicitly been proposed by these international organizations to mitigate the risk of AMR.

Since the occurrence of AMR in the food production sector, different programs to contain zoonoses and AMR zoonoses have been developed following these Principles and Guidelines. The Danish program to contain AMR zoonoses, DANMAP, has in particular gained international attention and has been analyzed in different publications (WHO 2003; Hammerum et al. 2007; Aarestrup et al. 2010). The reason for this was the early One Health approach that the Danish government and stakeholders proposed to combat AMR. In 1995, after publication of the finding that 80 % of *Enterococci* in all industrial produced chickens in Denmark were highly resistant to vancomycin (a last resort drug for human therapy) the government decided that actions had to be taken (Wegener et al. 2003) and set up the Danish Integrated Antimicrobial Resistance Monitoring and Research Program (DANMAP). Figure 10.3 shows the organization of DANMAP and how the animal health, food safety and public health sectors work together.

The objectives of DANMAP are: (1) to quantitatively monitor the consumption of antimicrobials used in (food) animals and humans, (2) to quantitatively monitor the occurrence of AMR in (zoonotic) bacteria in animals, food and humans, (3) to study and describe the associations between antimicrobial consumption and antimicrobial resistance, and (4) to identify routes of transmission and areas for further research. Next to this an automated/ICT program, called Vetstat, was introduced to collect quantitative data on all prescribed medicine for animals from veterinarians, pharmacies and feed mills (Stege et al. 2003).

Vetstat data on drug usage proved important for understanding the different aspects of the antibiotic usage problem, and to provide tools to control the use. For instance, with the information from Vetstat it has been possible for the Danish Veterinary and Food Authority (DVFA) to introduce "The Yellow Card Initiative" (DVFA 2012). This initiative works similarly to that in football, and farmers and veterinarians get a yellow card when their antimicrobial use is excessive as compared to similar farms. Only by reducing the antibiotic use, which may be done for instance by adopting management practices from low users, the card can be

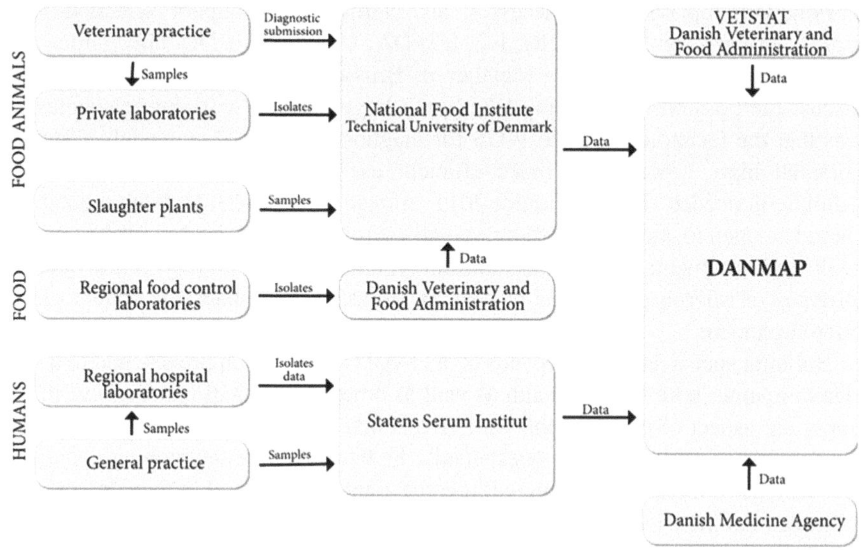

Fig. 10.3 Organization of DANMAP showing the different collaborating institutes and agencies, and how human, animal and food information is brought together (taken from: www.danmap.org)

retracted. This has not only worked as a stick, it also gives the farmers a sense of how they are doing compared to their colleagues. In the European Union several countries have now also started to collect antibiotic usage data and to compare antibiotic use at country level (EMA 2011).

10.5.3 *Global Microbial Identifier*

Surveillance of foodborne infections, and infectious diseases in general, is important to understand the transmission of infectious diseases and identify risks. To do this efficiently data collection should be done in a harmonized way, so data can be compared and integrated. Until now, this has been difficult because different human medical, veterinary medical, food and environmental laboratories have been using different techniques for surveillance, making it often almost impossible to compare data.

With the introduction of whole genome sequencing (WGS) this problem may be solved. Its unbiased way of detecting DNA and its single platform (the DNA code) for comparing genomic information gives WGS the potential to take disease diagnostics to a new level. Some early uses showing the value of WGS for diagnostic and epidemiological purposes were the tracking of the massive cholera outbreak in Haiti in 2011 (Hendriksen et al. 2011) and by the EHEC outbreak that was first detected in Germany and later also found in other countries and which could be traced back to Egyptian imported fenugreek seed using WGS (Mellman et al. 2011).

Following up on these successes, an international group of scientists with representatives from OIE, WHO, EC, USFDA, US CDC, ECDC, universities and public health institutes, came together in Brussels, September 2011, to further discuss the possibilities of using WGS on a larger scale. Their simple conclusion was that the technology to use WGS for diagnostic purposes is available, and its potential high, however, to make efficient use of the data, a global genomic database is needed (Kupferschmidt 2011; Aarestrup et al. 2012). Such a database should be open to, and supported by, scientists from all fields: human health, animal health, environmental health and food safety, and should include genomic data for all types of microorganisms as well as meta-data to trace back the source of the microorganism.

Building such a database depends on a global One Health approach, and in a One Health manner both human health as well as other sectors will benefit from it. An important aspect of pursuing this initiative is that it will not only be beneficial for the developed world, but it may especially be beneficial for developing countries. For them genomic identification will mean a giant leap forward as they do not need to implement the wide variety of specialized methods that are nowadays used in the developed world. If set up in a sensible, inclusive, open-source framework WGS analysis will provide the world with a strong weapon in the fight to combat infectious diseases in all sectors.

10.6 Discussion

One health approaches may be synergistic in controlling foodborne zoonotic diseases to support both sufficient food safety, and sustainable food security. Clearly, because of the unique situation of transmissibility between humans and animals, zoonoses control relies on the control of the microorganisms in (1) animals, (2) the food chain and in (3) humans. In addition, as zoonoses originate in animals before being transmitted to humans, the most effective interventions may be achieved at the source, i.e. at the farm. To be most effective, approaches to reduce the risk of foodborne zoonoses should include all stakeholders from the human as well as the animal health side. At the transmission level, it will be of major importance to involve food and consumers authorities and related stakeholders (e.g. environmental specialist), to make sure the spillover from the animal reservoir is kept as low as possible.

The exact solution will differ per country and type of disease (e.g. in many developing countries neglected diseases may still be of importance). Given that 70 % of the rural population in poor countries is still dependent on livestock as working animals to survive, the effect of these animals carrying a zoonosis will work out differently than in the industrialized settings. A number of the most important zoonoses relate directly to food production systems in poor settings which could be reduced dramatically through well-known interventions, such as has been the case for *Brucella*, bovine TB and cysticercosis.

Furthermore, it is important to realize that much of the One Health efforts until now have focused on zoonotic pathogens with a potential for dramatic global spread (such as avian influenza and BSE). However, major health gains can be obtained with the endemic zoonotic pathogens. For instance *Salmonella* causes a dramatic global disease burden because of the sheer number of cases and the global spread via food and live animal trade. For *Salmonella* there are efficient methods to reduce the prevalence in food animals.

One Health approaches in the food sector are complex and involve both public and commercial stakeholders, which may put limitations on what can be done. On the one hand food should be nutritious and adding to one's health. On the other hand most food is commercially produced and traded. As food is a commercial product, one of the ways to make food producers (the supply-side) produce more healthy food, is if the public demands this (the supply–demand balance). Therefore, educating the public to buy healthier food may be a way to make food manufacturers produce healthier food. Next, there are other stakeholders and fields of science (e.g. industrial sciences or logistics) and policy (e.g. economics) that contribute to the food chain, but which may focus on other aspects than healthy food alone, and their conclusions may conflict with the food safety principles (e.g. the use of AGP to improve animal growth).

Clearly food safety is a complex issue, and integration of all its problems and data is difficult and should best be limited to its essential components. For this reason, countries should learn from experiences abroad that have documented success. There are many examples of One Health approaches that have helped lower the risk of zoonotic foodborne disease. Key to all approaches has been surveillance of the farm-to-fork chain.

Surveillance should be done at relevant levels of the chain, at the farm level by the veterinary system, and at the food production stage by food-scientist. Findings should be shared and compared with the findings in human medicine, to be able to make decisions about potential risks for human health. Thus the animal, food and human sectors need working together, to collect and to share data in such a way that they may be compared. As there are still many different techniques used in all three fields, it is still difficult to compare data. An important development in infectious disease diagnostics will be the introduction of WGS techniques, and the construction of a global, open-access genomic database for microorganisms. In a One Health manner, the latter would take diagnostics to a new level, and will greatly improve human, animal and food safety.

References

Aarestrup F (2012) Sustainable farming: get pigs off antibiotics. Nature 486(7404):465–466

Aarestrup FM, Jensen VF, Emborg HD, Jacobsen E, Wegener HC (2010) Changes in the use of antimicrobials and the effects on productivity of swine farms in Denmark. Am J Vet Res 71 (7):726–733

Aarestrup FM, Brown EW, Detter C, Gerner-Smidt P, Gilmour MW, Harmsen D, Hendriksen RS, Hewson R, Heymann DL, Johansson K, Ijaz K, Keim PS, Koopmans M, Kroneman A, Lo Fo

Wong D, Lund O, Palm D, Sawanpanyalert P, Sobel J, Schlundt J (2012) Integrating genome-based informatics to modernize global disease monitoring, information sharing, and response. Emerg Infect Dis. doi:10.3201/eid/1811.120453
Bonten MJ, Willems R, Weinstein RA (2001) Vancomycin-resistant *enterococci*: why are they here, and where do they come from? Lancet Infect Dis 1:314–325
CDC (2011) CDC estimates of foodborne illness in the US, CDC 2011 estimates. http://www.cdc.gov/foodborneburden/2011-foodborne-estimates.html. Accessed 2 April 2013
DANMAP (2010) The Danish integrated antimicrobial resistance monitoring and research programme 2010. http://danmap.org/Downloads/Reports.aspx. Accessed 2 April 2013
DVFA (2012) The Yellow Card initiative. http://www.foedevarestyrelsen.dk/english/Animal/AnimalHealth/Pages/The-Yellow-Card-Initiative-on-Antibiotics.aspx. Accessed 2 April 2013
ECDC (2009) Annual report of the European Antimicrobial Resistance Surveillance Network (EARS-Net). Antimicrobial resistance surveillance in Europe 2009. http://ecdc.europa.eu/en/publications/Publications/1011_SUR_annual_EARS_Net_2009.pdf. Accessed 2 April 2013
ECDC (2010) Methodology protocol for estimating burden of communicable diseases. http://www.ecdc.europa.eu/en/publications/Publications/1106_TER_Burden_of_disease.pdf. Accessed 2 April 2013
EMA (2011) Trends in the sales of veterinary antimicrobial agents in nine European countries (2005–2009) http://www.ema.europa.eu/docs/en_GB/document_library/Report/2011/09/WC500112309.pdf. Accessed 2 April 2013
FAO (2002) Improved animal health for poverty reduction and sustainable livelihoods. FAO animal production and health paper 153. http://www.fao.org/DOCREP/005/Y3542E/Y3542E00.HTM#Contents. Accessed 2 April 2013
FAO/OIE/WHO (2011) The FAO-OIE-WHO collaboration – a Tripartite concept note. http://www.fao.org/docrep/012/ak736e/ak736e00.pdf. Accessed 2 April 2013
Food Agricultural Organization of the United Nations (FAO)/World Organization for Animal Health (OIE)/WHO (2003) Joint FAO/OIE/WHO expert workshop on non-human antimicrobial usage and antimicrobial resistance: scientific assessment, Geneva. December 1–5 2003. http://www.who.int/foodsafety/publications/micro/en/amr.pdf. Accessed 2 April 2013
García HH, Gonzalez AE, Evans CA, Gilman RH (2003) Cysticercosis working group in Peru. Taenia solium cysticercosis. Lancet 362(9383):547–556
Ghani AC, Ferguson NM, Donnelly CA, Hagenaars TJ, Anderson RM (1998) Epidemiological determinants of the pattern and magnitude of the vCJD epidemic in Great Britain. Proc Biol Sci 265(1413):2443–2452
Godfroid J, Käsbohrer A (2002) Brucellosis in the European Union and Norway at the turn of the twenty-first century. Vet Microbiol 90(1–4):135–145
Gottstein B, Pozio E, Nöckler K (2009) Epidemiology, diagnosis, treatment, and control of trichinellosis. Clin Microbiol Rev 22(1):127–145
Graveland H, Waagenar JA, Bergs K (2011) Persistence of livestock associated MRSA CC398 in humans is dependent on intensity of animal contact. PLoS One 6(2):1
Guan Y, Zheng BJ, He YQ, Liu XL, Zhuang ZX, Cheung CL, Luo SW, Li PH, Zhang LJ, Guan YJ, Butt KM, Ong KL, Chan KW, Lim W, Shortridge KF, Yuen KY, Peiris JS, Poon LL (2003) Isolation and characterization of viruses related to the SARS coronavirus from animals in southern China. Science 302(5643):276–278
Hammerum AM, Heuer OE, Emborg HD, Bagger-Skjøt L, Jensen VF, Rogues AM, Skov RL, Agersø Y, Brandt CT, Seyfarth AM, Muller A, Hovgaard K, Ajufo J, Bager F, Aarestrup FM, Frimodt-Møller N, Wegener HC, Monnet DL (2007) Danish integrated antimicrobial resistance monitoring and research program. Emerg Infect Dis 13(11):1632–1639
Havelaar AH, Haagsma JA, Mangen MJ, Kemmeren JM, Verhoef LP, Vijgen SM, Wilson M, Friesema IH, Kortbeek LM, van Duynhoven YT, van Pelt W (2012) Disease burden of foodborne pathogens in the Netherlands, 2009. Int J Food Microbiol 156(3):231–238
Hendriksen RS, Price LB, Schupp JM, Gillece JD, Kaas RS, Engelthaler DM, Bortolaia V, Pearson T, Waters AE, Upadhyay BP, Shrestha SD, Adhikari S, Shakya G, Keim PS, Aarestrup

FM (2011) Population genetics of *Vibrio cholerae* from Nepal in 2010: evidence on the origin of the Haitian outbreak. MBio 2(4):e00157-11. doi:10.1128/mBio.00157-11
Hill AF, Desbruslais M, Joiner S, Sidle KC, Gowland I, Collinge J, Doey LJ, Lantos P (1997) The same prion strain causes vCJD and BSE. Nature 389(6650):448–450
Hoinville LJ (1994) Decline in the incidence of BSE in cattle born after the introduction of the 'feed ban'. Vet Rec 134(11):274–275
Hutchison ML, Walters LD, Avery SM, Munro F, Moore A (2005) Analyses of livestock production, waste storage, and pathogen levels and prevalences in farm manures. Appl Environ Microbiol 71(3):1231–1236
Karp BE, Engberg J (2004) Comment on: does the use of antibiotics in food animals pose a risk to human health? A critical review of published data. J Antimicrob Chemother 54(1):273–274
Kumarasamy KK, Toleman MA, Walsh TR, Bagaria J, Butt F, Balakrishnan R, Chaudhary U, Doumith M, Giske CG, Irfan S, Krishnan P, Kumar AV, Maharjan S, Mushtaq S, Noorie T, Paterson DL, Pearson A, Perry C, Pike R, Rao B, Ray U, Sarma JB, Sharma M, Sheridan E, Thirunarayan MA, Turton J, Upadhyay S, Warner M, Welfare W, Livermore DM, Woodford N (2010) Emergence of a new antibiotic resistance mechanism in India, Pakistan, and the UK: a molecular, biological, and epidemiological study. Lancet Infect Dis 10(9):597–602
Kupferschmidt K (2011) Outbreak detectives embrace the genome era. Science 333(6051):1818–1819
LoBue PA, Enarson DA, Thoen CO (2010) Tuberculosis in humans and animals: an overview. Int J Tuberc Lung Dis 14(9):1075–1078
Matope G, Schlundt J, Makaya PV, Aabo S, Baggesen DL (1998) *Salmonella enteritidis* in poultry: an emerging zoonosis in Zimbabwe. Zimbabwe Vet J 29:132–138
McKenna M (2012) Food poisoning's hidden legacy. Sci Am 306(4):26–27
Mellmann A, Harmsen D, Cummings CA, Zentz EB, Leopold SR, Rico A, Prior K, Szczepanowski R, Ji Y, Zhang W, McLaughlin SF, Henkhaus JK, Leopold B, Bielaszewska M, Prager R, Brzoska PM, Moore RL, Guenther S, Rothberg JM, Karch H (2011) Prospective genomic characterization of the German enterohemorrhagic *Escherichia coli* O104:H4 outbreak by rapid next generation sequencing technology. PLoS One 6(7): e22751. doi:10.1371/journal.pone.0022751
Molyneux D, Hallaj Z, Keusch GT, McManus DP, Ngowi H, Cleaveland S, Ramos-Jimenez P, Gotuzzo E, Kar K, Sanchez A, Garba A, Carabin H, Bassili A, Chaignat CL, Meslin FX, Abushama HM, Willingham AL, Kioy D (2011) Zoonoses and marginalised infectious diseases of poverty: where do we stand? Parasit Vectors 4:106. doi:10.1186/1756-3305-4-106
Murray CJ (1994) Quantifying the burden of disease: the technical basis for disability adjusted life years. Bull World Health Organ 72:429–445
Narrod C, Zinsstag J, Tiongco M (2012) A one health framework for estimating the economic costs of zoonotic diseases on society. Ecohealth 9(2):150–162
Pavlin BI, Schloegel LM, Daszak P (2009) Risk of importing zoonotic diseases through wildlife trade, United States. Emerg Infect Dis 15:1721, Available from http://wwwnc.cdc.gov/eid/article/15/11/09-0467.htm. Accessed 2 April 2013
Phillips I, Casewell M, Cox T, De Groot B, Friis C, Jones R, Nightingale C, Preston R, Waddell J (2004) Does the use of antibiotics in food animals pose a risk to human health? A critical review of published data. J Antimicrob Chemother 53(1):28–52
Potron A, Kalpoe J, Poirel L, Nordmann P (2011) European dissemination of a single OXA-48-producing *Klebsiella pneumoniae* clone. Clin Microbiol Infect 17(12):E24–E26. doi:10.1111/j.1469-0691.2011.03669.x
Price LB, Stegger M, Hasman H, Aziz M, Larsen J, Andersen PS, Pearson T, Waters AE, Foster JT, Schupp J, Gillece J, Driebe E, Liu CM, Springer B, Zdovc I, Battisti A, Franco A, Zmudzki J, Schwarz S, Butaye P, Jouy E, Pomba C, Porrero MC, Ruimy R, Smith TC, Robinson DA, Weese JS, Arriola CS, Yu F, Laurent F, Keim P, Skov R, Aarestrup FM (2012) *Staphylococcus aureus* CC398: host adaptation and emergence of methicillin resistance in livestock. MBio 3(1):e00305–e00311. doi:10.1128/mBio.00305-11

Prusiner SB (1997) Prion diseases and the BSE crisis. Science 278(5336):245–251
Prusiner SB (1998) Prions. Proc Natl Acad Sci U S A 95(23):13363–13383
Shi Z, Hu Z (2008) A review of studies on animal reservoirs of the SARS coronavirus. Virus Res 133(1):74–87
Smith DL, Dushoff J, Morris JG (2005) Agricultural antibiotics and human health. PLoS Med 2(8): e232
Spencer JL, Guan J (2004) Public health implications related to spread of pathogens in manure from livestock and poultry operations. Methods Mol Biol 268:503–515
Stege H, Bager F, Jacobsen E, Thougaard A (2003) VETSTAT-the Danish system for surveillance of the veterinary use of drugs for production animals. Prev Vet Med 57:105–115
U.S. Food and Drug Administration (FDA) (2012) 35 Years of resistance. Nat Rev Microbiol 10 (6):373
Wegener HC, Hald T, Lo Fo Wong D, Madsen M, Korsgaard H, Bager F, Gerner-Smidt P, Mølbak K (2003) *Salmonella* control programs in Denmark. Emerg Infect Dis 9(7):774–780
WHO (2000) WHO global principles for the containment of antimicrobial resistance in animals intended for food: report of a WHO consultation with the participation of the Food and Agriculture Organization of the United Nations and the Office International des Epizooties, Geneva. 5–9 June 2000. http://whqlibdoc.who.int/hq/2000/WHO_CDS_CSR_APH_2000.4.pdf. Accessed 2 April 2013
WHO (2001) WHO global strategy for containment of antimicrobial resistance. http://www.who.int/drugresistance/WHO Global Strategy English.pdf. Accessed 2 April 2013
WHO (2003) Impacts of antimicrobial growth promoter termination in Denmark. The WHO international review panel's evaluation of the termination of the use of antimicrobial growth promoters in Denmark. http://www.who.int/gfn/en/Expertsreportgrowthpromoterdenmark.pdf. Accessed 2 April 2013
WHO (2005) International health regulations 2005, 2nd edn. http://www.who.int/ihr/9789241596664/en/index.html. Accessed 2 April 2013
WHO (2006) Neglected tropical diseases: hidden successes, emerging opportunities. http://whqlibdoc.who.int/hq/2006/WHO_CDS_NTD_2006.2_eng.pdf. Accessed 2 April 2013
WHO (2011) Tackling antibiotic resistance from a food safety perspective in Europe. http://www.euro.who.int/__data/assets/pdf_file/0005/136454/e94889.pdf. Accessed 2 April 2013
WHO/FAO/OIE (2005) Guidelines for the surveillance, prevention and control of taeniosis/cysticercosis. http://www.oie.int/doc/ged/D11245.PDF. Accessed 2 April 2013
World Bank/WHO/UNICEF/OIE/FAO/UNSIC (2008) Contributing to one world, one health: a strategic framework for reducing risks of infectious diseases at the animal–human–ecosystems interface. http://www.fao.org/docrep/011/aj137e/aj137e00.htm. Accessed 2 April 2013

Chapter 11
The Clinical Biomedical Research Advances Achievable Utilizing One Health Principles

James L. Cook and B. Sonny Bal

Abstract Translational biomedical research is an excellent example of the evolving paradigm of the "one health" concept. The multidisciplinary collaborations between physicians, veterinarians, physical therapists, engineers, molecular biologists and a host of other scientific disciplines have yielded, and will continue to yield, significant advancements in the quality of healthcare in both animals and man.

Keywords Animal model • Biomedical research • FDA • Orthopaedic

The concept of One Health is almost as old as medicine itself. The ancient Greeks understood that information about the process of life could be gained by dissecting and studying animals (Olsson 1969). From the comparative anatomical and physiological studies of Galen and William Harvey to the discovery of insulin by Frederick Banting and Charles Best, the careful and detailed studies performed on animals were often responsible for significant advances in human medicine (Lyons and Petrucelli 1978). Today, there is an entire industry dedicated to biomedical research that is intended to effectively translate data from animal models and veterinary research to clinical use in humans. Companies in every area of the healthcare industry understand the necessity for ethical animal-based data for determining safety and efficacy of their diagnostics, pharmaceuticals, implants, and techniques in order to satisfy the FDA, investors, stock-holders, physicians and the public. This necessity provides the impetus and funding for quality translational biomedical research utilizing One Health principles.

For biomedical research to become clinically applicable, there are a number of facets pertaining to the "end product" that should be considered prior to initiation of the work. First, the end product must address a true need in clinical medicine. It is not effective to "make a hammer and then figure out what nail you can hit with it." Numerous innovative and efficacious technologies have failed to see clinical success because of this problem. A very sensitive and specific diagnostic test, a

J.L. Cook, DVM., PhD (✉) • B.S. Bal, MD., JD., MBA
Comparative Orthopaedic Laboratory and Department of Orthopaedic Surgery, University of Missouri, Columbia, MO, USA
e-mail: cookjl@health.missouri.edu

A. Yamada et al. (eds.), *Confronting Emerging Zoonoses*,
DOI 10.1007/978-4-431-55120-1_11

pharmaceutical with a well-defined mechanism of action and few side effects, or a biomaterial that perfectly mimics function of the native tissue may not attain clinical utility because it addresses only a niche diagnosis, benefits a small number of patients, or is useful for only a narrow indication. While it is certainly admirable and important to try to address all clinical problems through quality biomedical research, an already-difficult pathway to clinical application becomes nearly impossible when the need it addresses is not both widespread and currently unmet.

Importantly, the diagnostic test, therapy, or technique must be proven safe (http://www.fda.gov/). Demonstrating safety is the primary pre-requisite for FDA approval for clinical use of any new product or technique. To gain FDA approval, demonstration of safety typically requires use in a relevant animal model or corollary of disease. As such, the One Health concept can be a valuable tool in this process. In many areas of medicine, dogs, cats, or horses suffer from orthopaedic, metabolic, oncologic, infectious, and neurologic disorders that share similar signs, symptoms, and clinical courses seen in their human counterparts. Once initial safety has been demonstrated in vitro, in rodents, and/or in large animal models of disease, application to veterinary clinical patients can be an effective way to validate safety while also gaining data regarding efficacy. This pathway to US Food and Drug Administration (FDA) approval and clinical application then becomes a "two-way bridge" in that the safety and efficacy data can be applied to veterinary patients as well, benefiting both fields (O'Connell et al. 2012; Stoker et al. 2012; Garner et al. 2011; Kuroki et al. 2010; Arnoczky et al. 2010; Bal et al. 2010; Janicek et al. 2010; Luther et al. 2009; Anz et al. 2009; Cook et al. 2008a, b; Mahn et al. 2005).

Another often overlooked but critically important facet in translational research that should be considered from the onset is the financial aspect. For biomedical research to reach widespread clinical use, the benefits associated with use of product must have a positive benefit to the treated individuals and/or to society. On the cost side, the components include research and development (R&D) costs, the indication(s) for the technology and the achievable market share for the indication(s), all associated costs to the healthcare provider, all associated costs to the patient, and to what degree associated costs will be reimbursed. While the R&D costs are often addressed before initiation of the research, the latter components are often not, especially in the context of small business where the assumption may be that a strategic partner will assume downstream development responsibilities. Like it or not, the financial part of the equation is a major driving force governing clinical applicability and should be taken into account during the initial planning stages of research.

The factors discussed above contribute to the acceptability of the technology, which is the final determinant for successful clinical application. The diagnostic test, new medication, new implant, or new surgical procedure must be "accepted" by the profession, patients, industry, regulators, and policy-setting organizations in order to be used clinically. In addition to the factors discussed above, acceptability is also determined by the ability of the "average" practitioner to effectively apply the technology in their practice, ease of access and use, and patient perception.

Even a safe, effective and highly applicable technology can fail if it does not gain acceptance by all interested parties. The cost: benefit equation (pharmacoeconomics) is especially critical to policy setters and third-party payers, and may determine recommendations for use and government-sponsored programs.

So, how can all of this be accomplished? The keys to successful translational biomedical research include establishing a direct clinical connection for the research team, involving "end users" in the R&D process, completing a pre-IDE or Pre-IND meeting with the FDA, and putting great science behind the technology being developed. The clinical connection can come in many forms and mechanisms, but should be set up to provide the much-needed "real life" perspective for the research. The clinical connection may be as simple as a single nurse who is on the research team or be as complex as an international advisory board of physicians, veterinarians, nurses, and salespeople. The key is that the clinical connection provides the perspective necessary to foster acceptability when the research is completed. To do so, key questions need to be asked and answered, such as: *Which patients will this technology truly benefit? Can the technology be incorporated into a busy clinical setting? What associated costs do we need to consider? Will it be accessible? What data will be convincing to clinicians? Patients? Regulatory bodies? Industry?*

While clinicians are one of the major end users for biomedical technologies, other interested parties should be incorporated into the process as well. Three parties that play key roles in successful clinical application of biomedical technologies are patients, industry, and regulatory bodies. These parties can be included in the initial stages of the work as well by engaging patient advisory groups, industry R&D representatives, and the appropriate regulatory body. Of course, regulatory body approval will be a requisite step prior to clinical use of any technology in humans. In the US, the FDA provides the mechanism for a pre-IDE and Pre-IND meeting that is intended to provide guidance to researchers proposing to study new technologies or new uses for existing technologies prior to the submission of an IDE or IND application (http://www.fda.gov/). This communication with the FDA at the start of the process can help in understanding FDA requirements, regulations, and guidance documents, and allows FDA personnel to familiarize themselves with the new technology. It can also help speed the regulatory process and minimize delays in the development of useful devices intended for human use.

The communication with FDA can be in the form of Informal or Formal Guidance Meetings. Informal Guidance Meetings are telephone conference calls, video conferences, or face-to-face discussions with the reviewing division so that they can provide advice and guidance for the development of supporting pre-clinical data or the investigational plan for incorporation into the IDE application. Formal Guidance Meetings are based on a written request to discuss the type of valid scientific evidence that will be necessary to demonstrate that the technology is safe and effective for its intended use. These meetings are designed to develop a broad outline of a clinical trial design. Applicants may submit a written request for a meeting to reach an agreement regarding FDA's review of an investigational plan. The written request should include a detailed description of the technology, a detailed description of the proposed conditions of use of the technology, a proposed plan (including a clinical protocol and a toxicology protocol) for determining whether there is a reasonable assurance of effectiveness, and information regarding the expected performance of the technology. If an agreement is reached between FDA and the applicant regarding the parameters of an investigational plan, the terms of the agreement are put in writing and made part of the administrative record by FDA.

In addition to telephone contacts and informal or formal meetings, applicants may submit preliminary information as a "pre-IDE" or Pre-IND submission. This is encouraged when preparing a formal IDE or IND submission that requires FDA guidance on troublesome parts of the application, e.g., clinical protocol design, pre-clinical testing proposal, pre-clinical test results, protocols for foreign studies when the studies will be used to support future marketing applications to be submitted to FDA. Upon completion of the review of the pre-IDE or pre-IND submission, the reviewing division will issue a response to the applicant to help guide the research accordingly. A similar procedure is followed in Europe, where "Scientific Guidance" may be obtained in advance of clinical testing.

Today, "one health" laboratories are located throughout the world and typically employ both a comparative and translational research approach in an effort to improve diagnostics, develop preventative and therapeutic strategies, and advance our understanding of disease mechanisms. While comparative research focuses on common pathological and regenerative pathways across species, translational research refers to studies that attempt to extend basic science discoveries into practical clinical applications. An example of a translational research study would be to determine if a drug's ability to increase cartilage cell metabolism could be translated into a clinical treatment of arthritis. In these laboratories, translational studies are usually carried out in the animal species deemed most appropriate by comparative research investigations and are the "pre-clinical" foundations on which clinical applications in humans are based.

Importantly, the comparative and translational avenues of "one-health" research must always be considered as *"two-way bridges"*. While the knowledge gained, and the technologies developed, through comparative research are often initially focused on human applications, many of these advancements can be (and have been) brought back to the animal species in which they were originally studied in for veterinary clinical application. Because many of the comparative biomedical research laboratories are directed by veterinarians who have had extensive clinical, as well as research, training, such clinical applications are becoming more common. In addition, several of the major pharmaceutical, orthopaedic, and supplement companies have established animal health divisions dedicated to development and use of technologies in clinical veterinary medicine. These factors provide the framework for the two-way bridges to be built, leveraging funding and resources in order to benefit veterinary patients as biomedical research advances healthcare for humans around the world.

While the regulatory pathway for animal health technologies is not as difficult as it is for human health technologies, the FDA's Center for Veterinary Medicine

(CVM) regulates the manufacture and distribution of drugs and food additives given to animals (http://www.fda.gov/AnimalVeterinary/default.htm). These include animals from which human foods are derived, as well as food additives and drugs for companion animals. CVM is responsible for regulating drugs, devices, and food additives given to, or used on, over one hundred million companion animals, plus millions of poultry, cattle, swine, and horses.

A new animal drug is any drug intended for use in animals other than man, including any drug intended for use in animal feed, the composition of which is such that the drug is not already recognized as safe and effective for use under the conditions prescribed, recommended, or suggested in the labeling of the drug. A new animal drug may not be sold into interstate commerce unless it is the subject of an approved new animal drug application (NADA), abbreviated NADA (ANADA), or there is a conditional approval (CNADA).

If requirements for exemption are met, unapproved investigational new animal drugs may be exempt from approval requirements, and may be shipped in interstate commerce for use by experts, qualified by scientific training and experience, to investigate their safety and effectiveness. FDA also has regulatory oversight over veterinary devices and can take appropriate regulatory action if a veterinary device is misbranded, mislabeled or adulterated. It is the responsibility of the manufacturer and/or distributor of these devices to assure that they are safe, effective, and properly labeled. FDA recommends that devices should meet or be equivalent to the performance standards for use in humans. This is especially important for devices that can be used both in humans and animals, such as examination gloves, sterile catheters, infusion pumps, orthopaedic implants, and other similar technologies. FDA recommends that manufacturers and/or distributors of veterinary medical devices request a review of their product labeling and promotional literature to ensure that it complies with regulations. This includes devices manufactured in another country and offered for importation into the U.S.

Numerous examples of clinical biomedical research advances that have already been achieved through the One Health concept exist in the areas of infectious disease, oncology, ophthalmology, cardiovascular disease, neurologic disorders, and many more (Koff et al. 2013; Patz and Hahn 2012; Mbugi et al. 2012; Henry and Bryan 2013; Essman et al. 2003; Ranieri et al. 2013; Nishida 2010; Delaney et al. 2011; Arendash 2012; Uccelli and Mancardi 2010). From our area of expertise, orthopaedics, an excellent example of this process is the work by Arnoczky and colleagues in the area of meniscal pathology, repair and replacement (Arnoczky et al. 2010; Arnoczky and Warren 1982; Arnoczky and Warren 1983; Crawford et al. 2004). From their seminal work describing the vasculature of the canine and human menisci (Arnoczky and Warren 1982; Arnoczky and Warren 1983), including their close similarity and the critical role of blood supply in treatment and healing, to their award-winning research addressing safety of allograft tissue preservation using a feline model (Crawford et al. 2004), these investigators have provided the data needed to validate diagnostic and therapeutic approaches for surgeons, radiologists, sports team physicians, trainers, and therapists, as well as a broad spectrum of the orthopaedic industry and non-profit tissue banks. Importantly, this work has also directly benefited the veterinary clinical world in the same way. Today, dogs, horses, and people receive function-saving treatments based on this One Health-One Medicine biomedical research.

Hopefully, this chapter provides both insight and inspiration into the mechanisms by which a multi-disciplinary, multi-species approach to solving clinical problems in all areas of medicine is accomplished. As purveyors and beneficiaries of One Health principles, the authors suggest that clinical advances are not only possible by utilizing this approach, but can be made more efficient, safe, and effective as a result.

References

Anz AG, Smith MJ, Stoker AM, Linville C, Markway H, Branson K, Cook JL (2009) The effect of bupivacaine and morphine in a co-culture model of diarthrodial joints. Arthroscopy 25:225–231

Arendash GW (2012) Transcranial electromagnetic treatment against Alzheimer's disease: why it has the potential to trump Alzheimer's disease drug development. J Alzheimers Dis 32:243–266

Arnoczky SP, Warren RF (1982) Microvasculature of the human meniscus. Am J Sports Med 10:90–95

Arnoczky SP, Warren RF (1983) The microvasculature of the meniscus and its response to injury. An experimental study in the dog. Am J Sports Med 11:131–141

Arnoczky SP, Cook JL, Carter T, Turner AS (2010) Translational models for studying meniscal repair and replacement. What they can and cannot tell us. Tissue Eng 16:31–39

Bal BS, Rahaman MN, Jayabalan P, Kuroki K, Cockrell MK, Yao JQ, Cook JL (2010) In vivo outcomes of tissue engineered osteochondral grafts. J Biomed Mater Res B Appl Biomater 93:164–174

Cook JL, Hudson CT, Kuroki K (2008a) Autogenous osteochondral grafting for treatment of stifle osteochondrosis in dogs. Vet Surg 37:311–321

Cook JL, Fox DB, Kuroki K, Jayo M, DeDeyne PG (2008b) *In vitro* and *in vivo* comparison of five biomaterials used for orthopedic soft tissue augmentation. Am J Vet Res 69:148–156

Crawford MJ, Swenson CL, Arnoczky SP, O'Shea J, Ross H (2004) Lyophilization does not inactivate infectious retrovirus in systemically infected bone and tendon allografts. Am J Sports Med 32:580–586

Delaney LE, Arco-Esquivel AA, Kuroki K, Laughlin MH (2011) Exercise training improves vasoreactivity in the knee artery. Int J Sports Med 33:114–122

Essman SC, Lattimer JL, Cook JL, Turnquist S, Kuroki K (2003) Effects of ^{153}Sm-ethylenediaminetetramethylene phosphonate on physeal and articular cartilage in juvenile rabbits. J Nucl Med 44:1510–1515

Garner BC, Stoker AM, Kuroki K, Evans R, Cook CR, Cook JL (2011) Using animal models in osteoarthritis biomarker research. J Knee Surg 24:251–264

Henry CJ, Bryan JN (2013) Not lost in translation: how study of diseases in our pets can benefit them and us. J Missouri State Med Assoc 110:144–147

Janicek JC, Cook JL, Wilson DA, Ketzner KM (2010) Multiple osteochondral autografts for treatment of a medial trochlear ridge subchondral cystic lesion in the equine tarsus. Vet Surg 39:95–100

Koff WC, Russell ND, Walport M, Feinberg MB, Shiver JW, Karim SA, Walker BD, McGlynn MG, Nweneka CV, Nabel GJ (2013) Accelerating the development of a safe and effective HIV vaccine: HIV vaccine case study for the Decade of Vaccines. Vaccine 31(Suppl 2):B204–B208

Kuroki K, Stoker AM, Sims HJ, Cook JL (2010) Expression of Toll-like receptors 2 and 4 in stifle joint synovial tissues of dogs with or without osteoarthritis. Am J Vet Res 71:750–754
Luther JK, Cook CR, Cook JL (2009) Meniscal release in cruciate ligament intact stifles causes lameness and medial compartment cartilage pathology in dogs 12 weeks post operatively. Vet Surg 38:520–529
Lyons AS, Petrucelli RJ (1978) Medicine: an illustrated history. Abradale, New York
Mahn MM, Cook JL, Cook CR, Balke MT (2005) Arthroscopic verification of ultrasonographic diagnosis of meniscal pathology in dogs. Vet Surg 34:318–323
Mbugi EV, Kayunze KA, Katale BZ, Kendall S, Good L, Kibik GS, Keyyu JD, Godfrey-Faussett P, Van Helden P, Matee MI (2012) 'One Health' infectious disease surveillance in Tanzania: are we all on board the same flight? Onderstepoort J Vet Res 79:E1–E8
Nishida T (2010) Translational research in corneal epithelial wound healing. Eye Contact Lens 36:300–304
O'Connell GD, Lima EG, Liming B, Chahine NO, Albro MB, Cook JL, Ateshian GA, Hung CT (2012) Toward engineering a biological joint replacement. J Knee Surg 25:187–196
Olsson S-E (1969) Comparative orthopaedics. Clin Orthop Relat Res 62:3–5
Patz JA, Hahn MB (2013) Climate change and human health: a one health approach. Curr Top Microbiol Immunol 366:141–171
Ranieri G, Gadaleta CD, Patruno R, Zizzo N, Daidone MG, Hansson MG, Paradiso A, Ribatti D (2013) A model of study for human cancer: Spontaneous occurring tumors in dogs. Biological features and translation for new anticancer therapies. Crit Rev Oncol Hematol 88:187
Stoker AM, Garrity JT, Hung CT, Stannard JP, Cook JL (2012) Improved preservation of fresh osteochondral allografts for clinical use. J Knee Surg 25:117–126
Uccelli A, Mancardi G (2010) Stem cell transplantation in multiple sclerosis. Curr Opin Neurol 23:218–225

Chapter 12
One Health Successes and Challenges

Peter M. Rabinowitz and Lisa A. Conti

Abstract The concept of One Health encompasses a wide range of issues and is subject to a variety of interpretations as a term. Central to the concept, however, is the interconnectedness of human, animal, and environmental health. This chapter reviews the experience of a number of disease detection and control efforts that have relevance to the One Health concept. These examples include rabies control, lead poisoning surveillance, animals as sentinels of environmental health hazards, parasitic disease control, and comparative clinical medicine research. The successes and shortcomings of these efforts illustrate both the potential of the One Health concept as well as the inherent challenges in its implementation. The chapter describes how integrated risk assessment and integrated disease control and prevention strategies involving human, animal, and environmental health components are central to the success of the One Health approach. It also outlines ways to overcome existing barriers to implementing such integrated strategies.

Keywords Animal sentinel • Comparative clinical medicine • Environmental intervention • Integrated assessment

As the different chapters in this book illustrate, the concept of One Health encompasses a wide range of issues and is subject to a variety of interpretations as a term. Central to the concept, however, is the interconnectedness of human, animal, and environmental health. This chapter reviews the experience of a number of projects, including rabies control, lead poisoning surveillance, parasitic disease control, and comparative medicine research, illustrating both the potential of the One Health concept as well as the inherent challenges in its implementation.

In judging the success or failure of "One Health" efforts, it is useful to have a working definition of what constitutes a successful One Health approach to a particular problem. For the purposes of this discussion, we have focused on particular disease prediction or control efforts that simultaneously considered the

A working definition of One Health: a transdisciplinary approach involving human, animal, and environmental health aspects.

P.M. Rabinowitz (✉) • L.A. Conti
e-mail: peterr7@uw.edu

A. Yamada et al. (eds.), *Confronting Emerging Zoonoses*,
DOI 10.1007/978-4-431-55120-1_12

spheres of human, animal, and environmental health. In particular, we propose that One Health involves *integrated assessment* of human, animal, and environmental health aspects of a disease problem, as well as *integrated interventions* across these spheres to prevent or manage such diseases.

12.1 Rabies Prediction and Control: A One Health Success and Challenge

Rabies is a classic zoonosis transmitted by the bites of domestic and wild animals and remains almost uniformly fatal to humans. While the disease has been a focus of veterinary and human public health efforts for centuries, it continues to cause thousands of human deaths yearly, mostly in developing countries, is resurgent in a number of regions, and would therefore seem to be a prime candidate for the attention of a One Health approach. Modern management techniques include effective vaccines for humans and animals, and in many countries there is ongoing and effective communication between animal health and human health professionals regarding rabies detection and prevention. Yet attempts to control the spread of rabies epizootics in wildlife by incorporating an environmental or ecosystem aspect such as ecological manipulation have largely been unsuccessful. Rabies control therefore illustrates some of the challenges to implementing One Health concepts.

The bullet-shaped, single stranded RNA rabies virus is spread typically via the bite of an infected animal to its next host. Non-bite transmission mechanisms such as inhalation of aerosols or organ and corneal transplantation are rarely reported (Constantine 1962; Winkler et al. 1973; Wohlsein et al. 2011). The neurotropic virus replicates in its new host at the entry point and quickly works centripetally through the nervous system. Once established in the central nervous system (CNS), the resulting encephalitis is responsible for the characteristic furor and rage seen among "mad" animals. This occurs concomitantly with further replication and viral shedding in the host's saliva, thus promoting viral transmission into subsequent hosts. However, the virus itself is fragile and short-lived outside a host. Indeed, washing a bite wound thoroughly to flush out the virus may aid in preventing infection.

The incubation period from infection to clinical signs varies, depending on how close the site of infection is to the CNS. From days to months following infection, the resulting disease is progressive from malaise, fever, headache, agitation, paralysis, and finally coma and death. People rarely survive rabies once they exhibit clinical signs; there are rare documented/reported exceptions including an unvaccinated Wisconsin girl who underwent extreme medical measures to survive (Porras et al. 1976; Willoughby et al. 2005; Health News 2011).

Rabies virus is found on all continents with the exception of Antarctica. While the virus can infect all mammals, different strains have been identified by monoclonal antibody analysis among certain (typically mesocarnivore and bat) species with "spillover" occurring when these reservoir hosts infect other mammals, including humans. Principal reservoir hosts vary worldwide from dogs in Latin America, Africa and Asia to wild mesocarnivores (especially raccoons, foxes, coyotes and skunks) and bats in the Americas and Europe (insectivorous species in North America and Europe and hematophagous species in South America). The Brazilian marmoset is the sole primate identified with a unique viral variant associated with human rabies cases (Favoretto et al. 2013).

12.1.1 History of Rabies Control

Louis Pasteur began the field of immunology by working on two zoonotic diseases: fowl cholera and bovine anthrax. In 1885, he developed the first vaccine against rabies, basing his work on that of Dr. Pierre-Victor Galtier, a veterinarian and professor at the Veterinary School of Lyon, France (Jackson 2013). Two years later, he established the interdisciplinary Pasteur Institute for the purpose of furthering translational research. Today, prophylactic vaccination among high-risk individuals, including animals, provides protection. In addition, post exposure prophylaxis (PEP), involving vaccine and rabies immune globulin (RIG) are critical to preventing rabies (http://www.cdc.gov/rabies/resources/acip_recommendations.html. Accessed July 2013; http://www.who.int/rabies/human/postexp/en/. Accessed July 2013). In the United States, public health veterinarians frequently provide consultation to human health clinicians regarding rabies management.

In North America and Europe, the public health approach to controlling rabies includes vaccinating dogs and controlling strays. The resulting reduction of the dog strain of rabies combined with PEP has led to human rabies becoming relatively rare in developed countries. In the USA, dog strains of rabies have been declared eradicated (Velasco-Villa et al. 2008).

Distributed into the environment, oral rabies vaccine (ORV) in baits targeted toward foxes, coyotes and raccoons has been a rabies control strategy used with some success in Europe, Texas and eastern North America, respectively (http://www.aphis.usda.gov/wildlife_damage/oral_rabies/. Accessed July 2013). Indeed, ORV development and distribution has been a fruitful partnership between wildlife biologists and public health professionals. ORV has also been used as a strategy to immunize dogs in developing countries.

NOTE: As another example of interdisciplinary cooperation between human, animal, and environmental health professionals, state wildlife health officials in Kentucky were recently facing budget limitations in their effort to track and control white nose syndrome, an emerging fungal disease of bats. Surveillance for white nose syndrome is labor intensive and expensive. The

(continued)

(continued)

National Park Service proposed a plan to make use of ongoing surveillance for rabies in bats to test the animals simultaneously for white nose syndrome. This successful pooling of resources provided cost-effective benefits for human and domestic animal health as well as wildlife disease control (Griggs et al. 2012).

12.1.2 One Health Challenges for Rabies Control

Over 50,000 human deaths, particularly in children bitten by rabid dogs, are estimated to occur in Asia and Africa, each year, demonstrating that, global rabies prevention and control remains an urgent challenge (http://www.who.int/mediacentre/factsheets/fs099/en/index.html. Accessed July 2013). Further, rabies spillover from domestic dogs to wildlife populations results in additional stresses to threatened and endangered species such as the Ethiopian wolf and the African wild dog (Randall et al. 2004). In some developing countries, dog control efforts are often underfunded or met with resistance (http://www.who.int/vaccine_research/diseases/zoonotic/en/index4.html. Accessed July 2013).

In North America and Europe, the shift from rabies in domestic to wild animals presents new challenges for disease control. Recent human cases in the United States are attributed most frequently to bat strain rabies with occult exposure (http://www.cdc.gov/rabies/bats/education/. Accessed March 2014). There are currently no oral rabies vaccines available for bats, and as with other reservoir species, widespread depopulation of bats is impractical, unethical and contraindicated ecologically due to their dual role in reducing noxious insect populations and pollinating fruit trees (Rupprecht et al. 1995). Rabies ecology is also influenced by human development encroaching into wildlife habitat or creation of "wildlife islands" which can concentrate susceptible animals.

12.1.3 Going Forward

International efforts to improve rabies control in developing countries have advocated for a One Health approach involving multiple sectors. Presently, these efforts are focusing on the domestic dog reservoir because that is the most direct threat to human health, and current canine vaccination levels are low in many countries (Lembo and Partners for Rabies Prevention 2012). While disease modeling suggests that focusing on domestic dogs could be effective for control at country level, even in countries where wildlife canines are abundant, there is the possibility that wildlife rabies reservoirs will become more important as the domestic dog reservoir is controlled, as has happened in developed countries (Fitzpatrick et al. 2012). Because rabies virus persists in diverse ecologic communities, the One Health approach must be enhanced in the future to pursue further understanding of enzootic maintenance mechanisms, in essence incorporating more consideration of environmental health and ecological systems into an integrated One Health assessment framework. This necessitates the partnership of multiple disciplines including wildlife biologists to provide rabies control programs that are efficacious, cost effective, publicly supported, and have negligible negative impacts on humans, animals, and the environment.

12.2 Animal Sentinels for Lead and Other Hazards

A promising application of One Health principles is the prediction of infectious and non-infectious disease events and risk in the environment. These prediction efforts are based on the concept that one species may serve as a "sentinel" for another species, just as the "canary in the coal mine" was used to warn miners of the presence of dangerous gases (Rabinowitz 2009). While non-human animals can serve as "sentinels" for human health threats in the environment, in some circumstances humans themselves may be the sentinels for such shared environmental health risks.

The usefulness of the sentinel concept is exemplified by case reports of dogs found to have lead toxicity, leading to the discovery that children sharing the same household were found to have lead poisoning as well (Dowsett and Shannon 1994). Similarly, cats and even cows have also been found to serve as sentinels for environmental lead poisoning risk in humans (Bischoff et al. 2010).

Sentinel disease events can be incorporated into formal prediction systems in order to provide more effective early warning. A study in California found that the most effective combination of warning signals for West Nile virus infection risk to humans involved environmental (climate) data as well as data on mosquitoes and animal cases of disease (Kwan et al. 2012). Other examples of sentinel events for human disease include:

- abundant wild piñon nut years in the Four Corners region of the USA predicting deermice population explosions associated with hantavirus outbreaks in humans (Mills et al. 1999),
- increased rainfall in East Africa predicting Rift Valley fever epizootics in livestock and epidemics in humans (Linthicum et al. 1999),
- Monkey deaths in India predicting Kyasanur Forest disease (Pattnaik 2006), and
- Rat deaths (rat-fall) in India predicting plague (Dennis and Staples 2009).

12.3 Schistosomiasis Prevention: Integrating human, animal and environmental health

To date, a limited number of disease intervention and control efforts simultaneously incorporated and documented the human, animal, and environmental health aspects of the intervention. One project in China that took such approach was an effort to prevent new cases of *Schistosoma japonicum*, a zoonotic form of Schistosomiasis with a bovine reservoir. The traditional approach to controlling the disease in humans was to treat human cases with anti-parasitic medication. Investigators found, in a controlled intervention study, that when they simultaneously treated humans and bovines (an animal reservoir for the disease), infection rates in both humans and the environment (the rate of infection in freshwater snails) was much lower than when only humans were treated (Gray et al. 2009). In a separate study, they found even better results when they carried out environmental sanitation measures including restricting cows from access to water bodies and installing latrines on boats to prevent human fecal contamination of the same bodies of water (Wang et al. 2009). While the authors of the study did not specifically label their approach as "One Health", it met the criteria described above for an integrated One Health intervention.

One Health approaches in agriculture have also shown promise. One study found that when humans and horses developed respiratory conditions related to the presence of high levels of organic dusts in a barn, better ventilation of the barns improved the air quality as well as the health of both species (Wålinder et al. 2011).

12.4 Challenges to Implementing One Health

There are a number of reasons why One Health successes have been relatively uncommon. These include professional segregation, difficulty engaging the human health sector, separation of regulatory functions, reliance on traditional animal disease models, funding priorities, and difficulty of incorporating environmental health aspects of an assessment or intervention.

Professional segregation is the phenomenon of human, animal, and environmental health professionals receiving training in separate institutions, with often little or no opportunity to interact with each other. Subsequently, in their professional lives, these individuals find no routine way to communicate or collaborate. The recent interest in One Health approaches may be breaking down such segregation among the upcoming generation of professionals.

Another challenge for One Health has been its differential visibility in veterinary medicine as compared to human medicine or environmental science. For example, the Journal of the American Veterinary Medical Association regularly features One Health, yet no medical or environmental health journals currently have an equivalent forum for One Health topics (https://www.avma.org/KB/Resources/

Reference/Pages/One-Health.aspx. Accessed 20 March 2014). Similarly, many veterinary medical schools have One Health courses or programs, but such curricula are currently absent from most medical schools and public health schools. Possible reasons for the difficulty in engaging human health in the One Health concept include lower levels of awareness and knowledge about zoonotic diseases among human health care providers and the historical professional segregation described above.

On the governmental level, there has been an historical separation of the agencies responsible for human, animal, and environmental health. Departments or ministries of agriculture traditionally deal with animal health and disease issues, while departments or ministries of public health oversee human health, while usually another separate department may handle environmental issues. In the US, there have been up until recently few examples of different agencies working together in a One Health model, although internationally, such models are being explored (Kenya One Health national Plan 2013). Again, with the growing discussion of One Health approaches at the international and national level, these traditional barriers between governmental agencies may be changing in the years to come.

12.5 Comparative Clinical Medicine and One Health

An important component of One Health is its comparative medicine approach to cross-species health and disease, noting the similarities and differences. Of course, human medicine has long relied on animal models of disease. However, most current comparative medicine research efforts involve laboratory animal models, and there has been little exploration of comparative clinical and observational approaches that could bring together human and animal health clinicians and epidemiologists. One example would be exploring the different expression of cancers in dog breeds to shed light on risk factors, both innate and environmental, for cancers in humans. Another example would be asthma in cats, a disease condition that closely resembles human asthma and could serve as a model for how the lung responds to the environment. The One Health related "Zoobiquity" initiative is an educational effort to bring human and animal clinicians together to identify clinical overlaps between human and animal medicine which remain largely unexplored but appear to hold great promise for providing new insights into disease treatment and prevention across species (Natterson-Horowitz and Bowers 2012). These efforts to move comparative medicine outside of the laboratory to consider naturally occurring diseases in multiple species could be termed "Comparative Clinical Medicine".

The lack of readily available funding or a dedicated funding agency for One Health projects is another serious barrier to implementation. The National Institutes of Health by law must prioritize human health above animal and environmental health, and has not established funding mechanisms encouraging the type of interdisciplinary approach provided by One Health. On the clinical level, medical insurance does not reimburse preventive visits to a veterinarian, even though such visits may help prevent zoonotic disease transmission from a pet to an owner.

Human and animal medicine share many of the same pharmaceutical types of interventions. One success has been the development of ivermectin, an antiparasitic agent used widely in veterinary medicine, and more recently the same medicine has been used in human medicine against lymphatic filariasis (Campbell 2012). Intervening on the environmental level requires a different set of skills and perspectives. For example, while both human and animal health professionals, including public health professionals, learn in their training how to apply the tools of disease epidemiology, they do not receive training in disease ecology and ecological principles. It has been proposed that to effectively work at the human-animal-ecosystem interface, there needs to be an appreciation of ecological concepts such as population dynamics, community structure, and ecosystem aspects including complexity, resilience, and geochemistry (Preston et al. 2013). Developing practical, evidence-based applications of such environmental and ecological principles will be a major challenge. Even when specific environmental interventions can be proposed, they can be costly in the short run. For example, a published analysis of a project controlling Tungiasis, a parasitic disease, acknowledged that better environmental control of water runoff could produce the greatest benefit on disease risk, but according to the authors would have the greatest short term costs compared to the medical expense of treating symptomatic patients (Pilger et al. 2008).

12.6 Future Needs

Despite the barriers listed above, there is a critical need for One Health approaches to a number of pressing problems. To develop such approaches, interdisciplinary cooperation along the lines described above will be essential.

As the world's human population increases, there has been a parallel rapid growth in the global population of food animals. There is also a trend toward intensification of animal production, with greater number of animals housed in production facilities. The intensification and crowding of food animals creates potential for amplification of disease outbreaks, evolution and emergence of zoonotic pathogens, occupational risks to workers, and risks for environmental contamination by animal waste and other facility emissions. There is a need to develop One Health models for food animal production that simultaneously maximizes the health of the involved humans (workers and consumers of food), the animals, and the local environment and minimizes the risks (Sarkar et al. 2012). Such models need to be applied both in small scale farming, where biosecurity is low, as well as large scale intensive animal agriculture, where biosecurity may be higher, but the scale of production amplifies some of the disease and environmental risks, as well as issues of animal welfare.

Another challenge is to define a One Health approach to the keeping of companion animals. There is growing evidence of the importance and possible health benefits of the human animal bond. In general, the positive aspects of pet ownership are believed to outweigh the risks of zoonotic diseases or other animal-associated health issues, even for immunocompromised individuals. However, increasingly urbanized lifestyles and the frequent presence in a home of an immunocompromised individual pose special challenges for pet ownership, as does the widespread ownership of exotic and wild pets. Keeping parks and beaches free of fecal contamination is an important environmental responsibility of pet ownership. Veterinarians and human health care providers, as well as environmental health specialists, need to collaborate to help define healthy models of living with companion animals.

Clinicians have an important role to play in One Health efforts. Sentinel cases of disease in humans or animals can indicate a risk in the environment that is shared and relevant for both humans and animals in the household. An example is when the diagnosis of a dog with the rickettsial disease Rocky Mountain spotted fever provides evidence of the presence of infected ticks in the environment and potential risk to humans. Clinical communication between veterinarians and human health care providers can also help with the early detection and prevention of zoonotic disease transmission. Lessons learned from diagnosing and treating one species, including the recognition of particular susceptibilities to environmental stressors, can be used to improve the health care of other species. Challenges to such communication include the professional segregation and financial obstacles mentioned above. There is therefore a need to define pathways and procedures for routine clinical collaboration between human health care providers and veterinarians.

There have been relatively few published examples of disease prevention and control interventions that effectively integrate human, animal, and environmental health. The examples cited in this chapter should serve as a basis for larger controlled interventions for a number of diseases shared between animals and humans. While One Health is a promising model, there is a need to develop the evidence base to support its integrated approach to disease prevention and control.

References

Bischoff K, Priest H, Mount-Long A (2010) Animals as sentinels for human lead exposure: a case report. J Med Toxicol 6(2):185–189

Campbell WC (2012) History of avermectin and ivermectin, with notes on the history of other macrocyclic lactone antiparasitic agents. Curr Pharm Biotechnol 13(6):853–865

Constantine DG (1962) Rabies transmission by nonbite route. Public Health Rep 77(4):287–289

Dennis DT, Staples JE (2009) Plague. Bacterial infections of humans. Springer, USA, pp 597–611

Dowsett R, Shannon M (1994) Childhood plumbism identified after lead poisoning in household pets. N Engl J Med 331(24):1661–1662

Favoretto SR, de Mattos CC, de Mattos CA, Campos AC, Sacramento DR, Durigon EL (2013) The emergence of wildlife species as a source of human rabies infection in Brazil. Epidemiol Infect 141(7):1552–1561. doi:10.1017/S0950268813000198, Epub 2013 Feb 22

Fitzpatrick MC, Hampson K, Cleaveland S, Meyers LA, Townsend JP, Galvani AP (2012) Potential for rabies control through dog vaccination in wildlife-abundant communities of Tanzania. PLoS Negl Trop Dis 6(8):e1796

Gray DJ, Williams GM, Li Y, Chen H, Forsyth SJ, Li RS et al (2009) A cluster-randomised intervention trial against Schistosoma japonicum in the Peoples' Republic of China: bovine and human transmission. PLoS One 4:e5900

Griggs A, Keel MK, Castle K, Wong D (2012) Enhanced surveillance for white-nose syndrome in bats. Emerg Infect Dis 18:530. doi:10.3201/eid1803.111751

UC Davis Children's Hospital patient becomes third person in US to survive rabies. *Health News*. Accessed 12 June 2011

Jackson AC (ed) (2013) Rabies, 3rd edn. Academic, San Diego, p 3

Kwan JL, Park BK, Carpenter TE, Ngo V, Civen R, Reisen WK (2012) Comparison of enzootic risk measures for predicting West Nile disease, Los Angeles, California, USA, 2004–2010. Emerg Infect Dis 18(8):1298–1306. doi:10.3201/eid1808.111558

Lembo T, Partners for Rabies Prevention (2012) The blueprint for rabies prevention and control: a novel operational toolkit for rabies elimination. PLoS Negl Trop Dis 6(2):e1388. doi:10.1371/journal.pntd.0001388, Epub 2012 Feb 28

Linthicum KJ, Anyamba A, Tucker CJ, Kelley PW, Meyers MF, Peters CJ (1999) Climate and satellite indicators to forecast rift valley fever epidemics in Kenya. Science 285(5426):397–400. doi:10.1126/science.285.5426.397

Mills JN, Ksiazek TG, Peters CJ, Childs JE (1999) Long-term studies of hantavirus reservoir populations in the southwestern United States: a synthesis. Emerg Infect Dis 5(1):135–142

Natterson-Horowitz B, Bowers K (2012) Zoobiquity. Vintage Books, New York

Pattnaik P (2006) Kyasanur forest disease: an epidemiological view in India. Rev Med Virol 16 (3):151–165. doi:10.1002/rmv.495

Pilger D, Schwalfenberg S, Heukelbach J, Witt L, Mencke N, Khakban A et al (2008) Controlling tungiasis in an impoverished community: an intervention study. PLoS Negl Trop Dis 2:e324

Kenya One Health national Plan (2013) http://zdukenya.org/strategic-plan/. Accessed 9 September 2013

Porras C, Barboza JJ, Fuenzalida E, Adaros HL, Oviedo AM, Furst J (1976) Ann Intern Med 85 (1):44–48

Preston ND, Daszak P, Colwell RR (2013) The human environment interface: applying ecosystem concepts to health. Curr Top Microbiol Immunol [Epub ahead of print]

Rabinowitz P (2009) Lisa Conti human-animal medicine, 1st edn. ISBN-9781416068372, Print book, Release Date: 2009

Randall DA, Williams SD, Kuzmin IV et al (2004) Rabies in endangered Ethiopian wolves. Emerg Infect Dis 10:2214

Rupprecht CE, Smith JE, Fekadu M, Childs JE (1995) The ascension of wildlife rabies: a cause for public health concern or intervention? Emerg Infect Dis 1(4):107–114

Sarkar A, Aronson KJ, Patil S, Hugar LB, vanLoon GW (2012) Emerging health risks associated with modern agriculture practices: a comprehensive study in India. Environ Res 115:37–50. doi:10.1016/j.envres.2012.03.005, Epub 2012 Apr 4

Velasco-Villa A, Reeder SA, Orciari LA et al (2008) Enzootic rabies elimination from dogs and reemergence in wild terrestrial carnivores, United States. Emerg Infect Dis 14:1849–1854

Wålinder R, Riihimäki M, Bohlin S, Hogstedt C, Nordquist T, Raine A et al (2011) Installation of mechanical ventilation in a horse stable: effects on air quality and human and equine airways. Environ Health Prev Med 16(4):264–272

Wang LD, Guo JG, Wu XH, Chen HG, Wang TP, Zhu SP, Zhang ZH, Steinmann P, Yang GJ, Wang SP, Wu ZD, Wang LY, Hao Y, Bergquist R, Utzinger J, Zhou XN (2009) China's new

strategy to block Schistosoma japonicum transmission: experiences and impact beyond schistosomiasis. Trop Med Int Health 14(12):1475–1483. doi:10.1111/j.1365-3156.2009.02403.x
Willoughby RE, Tieves KS, Hoffman GM, Ghanayem NS, Amlie-Lefond CM, Schwabe MJ, Chusid MJ, Rupprecht CE (2005) "Survival after treatment of rabies with induction of coma" (PDF). N Engl J Med 352(24):2508–2514. doi:10.1056/NEJMoa050382
Winkler WG, Fashinell TR, Leffingwell L, Howard P, Conomy JP (1973) Airborne rabies transmission in a laboratory worker. JAMA 226(10):1219–1221
Wohlsein P, Baumgartner W, Kreipe HH, Haverich A, Hori A, Stan AC (2011) Rabies transmission through organ transplantation. Pathologe 32(5):406–410

Index

A. Yamada et al. (eds.), *Confronting Emerging Zoonoses*,
DOI 10.1007/978-4-431-55120-1

MIX
Papier aus verantwortungsvollen Quellen
Paper from responsible sources
FSC® C105338

If you have any concerns about our products,
you can contact us on
ProductSafety@springernature.com

In case Publisher is established outside the EU,
the EU authorized representative is:
Springer Nature Customer Service Center GmbH
Europaplatz 3, 69115 Heidelberg, Germany

Printed by Libri Plureos GmbH
in Hamburg, Germany